the
UNIVERSITY
of
GREENWICH

Safety
of
Microbial
Insecticides

Editors

Marshall Laird
Honorary Research Fellow
Department of Zoology
University of Auckland
Auckland, New Zealand

Lawrence A. Lacey
Vector Biology and Control Project
Arlington, Virginia

Elizabeth W. Davidson
Department of Zoology
Arizona State University
Tempe, Arizona

CRC Press, Inc.
Boca Raton, Florida

Library of Congress Cataloging-in-Publication Data

Safety of microbial insecticides/edited by Marshall Laird, Lawrence
  A. Lacey, Elizabeth W. Davidson.
     p. cm.
   Bibliography: p.
   Includes index.
   ISBN 0-8493-4793-9
   1. Microbial insecticides—Safety measures. 2. Microbial
insecticides—Safety regulations. 3. Microbial insecticides—
Environmental aspects. I. Laird, Marshall. II. Lacey, Lawrence
A., 1946-   III. Davidson, Elizabeth W.
SB933.34.S24 1989
632'.951—dc20

                                       89-9921
                                          CIP

Direct all inquiries to CRC Press, Inc., 2000 Corporate Blvd., N.W., Boca Raton, Florida, 33431.

© 1990 by CRC Press, Inc.

International Standard Book Number 0-8493-4793-9
Library of Congress Card Number 89-9921
Printed in the United States

PREFACE

Assurances of complete safety in any connection are clearly matters of faith, and human error can always enter into the issue. Legend demonstrates this with Achilles, for he was invulnerable through his immortal mother's having immersed him in the River Styx (except for the heel that she grasped during the procedure). Ultimately, however, he found himself in the right place at the wrong time and struck in that heel by the right kind of arrow.

Nevertheless, while the question, "How safe *is* safe?", will remain endlessly debatable, a great deal has been learned about the health and environmental safety of microbial insecticides since the first widely commercialized products based on *Bacillus thuringiensis* Berliner were developed over 3 decades ago. Another sporeformer, *B. popilliae* Dutky, was already being used for field establishments in the U.S. since the early 1940s. However, until the international marketing of microbial insecticides based on *B. thuringiensis* focused attention on such new problems as possible adverse effects to, for example, the Japanese sericultural industry (Chapter 4), and highlighted the need for national registration requirements, the relative safety record was an assemblage of assurances that nothing particularly untoward had happened.

In the early years of the use of microbial insecticides, safety to humans and domestic animals was presumed based upon a lack of deleterious effects during laboratory and field testing. For example, the field introductions of the *Oryctes* baculovirus were begun in 1967 under the United Nations Development Programme/South Pacific Commission (UNDP/SPC) for the control of the rhinoceros beetle, while safety testing of this organism was not contemplated until the mid 1970s.[1] Trials of the fungus, *Coelomomyces stegomyiae*, in the Tokelau Islands involved treatment of the entire drinking water supply of one atoll for control of *Aedes polynesiensis* mosquito larvae, long before safety testing of this organism was undertaken.[2]

More than humans and livestock, though, must be considered here. For pertinent risks are somewhat different for at least a few of the invertebrates. Prominent among these are mankind's earliest domesticated insects, the honeybee (*Apis mellifera*) and the silkworm (*Bombyx mori*). It was well known "From very early times and before the Christian era",[3] that these are subject to fungal attack at economically devastating levels. Until the 1830s, the problem was ascribed to extraneous conditions such as humidity and temperature. For silkworms, these mycoses were shown to be due to *Beauveria bassiana* (Balsamo) Vuillemin. This deuteromycete fungus is now known to have more than 700 insect hosts.[4] In his 1835 work, "Del mal del segno, calcinaccio o moscardino, malattia che affligge i bachi da seta e sul modo di liberarne le bigattaie anche le piu infeste" (Parte I, Teoria. Orcesti, Lodi. 67, orig. Italian), Agostino Bassi demonstrated that the causal agent of silkworm "muscardine" was (later named for him) *B. bassiana*. Over the past 3 decades it has continued to be primarily these two anciently domesticated insects, the honeybee and the silkworm, which are center stage in terms of possible adverse effects of microbial insecticides on nontarget organisms (NTOs). Increasingly, though, the effects upon invertebrate predators and parasitoids have come under careful scrutiny. With regard to vertebrates, exposure of which "to these pathogens has been severe during epizootics of insect disease," relevant data and accumulating information on "the specificity of selected pathogens . . . comprise a vast body of circumstantial evidence that they are harmless."[5]

During 1971, a "Working Group on Safety of Microbial Control Agents" was founded during the Annual Meeting of the Society for Invertebrate Pathology (SIP) in Montpellier, France.[6] Reference to the index of that work will show the substantial advances in safety-related knowledge made over the decade, 1971 to 1981. The World Health Organization (WHO) convened a seminal Conference on the Safety of Biological Agents for Arthropod Control in Atlanta, Georgia, April 16 to 18, 1973.[7] This meeting led to a widening involvement of WHO in the support of studies, including safety-related ones, of potential microbial control agents of arthropod pests and vectors of public health importance. Meanwhile, Marshall Laird chaired the Society for Invertebrate Pathology (SIP) Safety Working Group, maintaining its documents repository at the Research Unit on Vector Pathology (RUVP), Memorial University of Newfoundland, St. John's, Newfoundland, Canada. Elizabeth Davidson succeeded him as chairperson, but the documents repository ceased to exist with the RUVP in 1983. With the passing of responsibility for the Working Group to Lawrence Lacey, attention turned to the preparation of an annotated bibliography on health and environmental aspects of the safety of microbial control agents, to be sponsored by SIP. However, this project was set aside when the present volume was conceived.

We thank our 27 co-authors, based in 12 countries, for furnishing authoritative and up-to-date statements on a dynamic topic. In acknowledging that dynamism we have chosen to allow more overlap between certain chapters than some might like, both to recognize differing viewpoints and to allow each individual contribution to stand alone. At the same time, we extend our apologies to those colleagues from whose chapters it has been necessary to delete material or make other changes so as to avoid excessive overlap. Our thanks are due to Dr. B. Dobrokhotov of WHO, Geneva, who was kind enough to establish the initial contact between us and Prof. N. V. Kandybin. Last but not least we are grateful to many colleagues unnamed here for stimulating suggestions as the book developed, and even more so, to our understanding spouses.

Marshall Laird
Lawrence A. Lacey
Elizabeth W. Davidson

REFERENCES

1. **Sturrock, J. M. and Ulbricht, T. L. V., Eds.,** The Rhinoceros Beetle Project: history and review of the research programme, in *Agriculture, Ecosystems and Environment,* Vol. 15, Elsevier, Amsterdam, 1986, 149.
2. **Laird, M.,** A coral island experiment: a new approach to mosquito control, *WHO Chron.,* 21, 18, 1967.
3. **Bell, J. V.,** Mycoses, in *Insect Diseases,* Vol. 1, Cantwell, G. E., Ed., Marcel Dekker, New York, 1974, 185.
4. **Li, Z.,** A list of the insect hosts of *Beauveria bassiana, 1st Natl. Symp. Entomogenous Fungi,* Gonzhuling, Jilin, People's Republic of China, 1987, 1.
5. **Burges, H. D.,** Safety, safety testing and quality control of microbial pesticides, in *Microbial Control of Pests and Plant Diseases, 1970-1980,* Burges, H. D., Ed., Academic Press, New York, 1981, 737.
6. **Laird, M.,** in *Microbial Control of Pests and Plant Diseases, 1970-1980,* Burges, H. D., Ed., Academic Press, New York, 1981, 913.
7. **Smith, R. F., Ed.,** WHO mimeographed document, Ser. WHO/VBC/73.1; and *Bull. WHO,* 48, 685, 1973.

THE EDITORS

Marshall Laird, Ph.D., D.Sc., retired from his Canadian Professorship in 1983, returning to his native country after 3 decades elsewhere to purchase a small farm 100 km north of Auckland from which he writes and consults on vector control. He is an Honorary Research Fellow in the Department of Zoology, University of Auckland, New Zealand. Dr. Laird attended a constituent college (now Victoria University of Wellington) of the then University of New Zealand, of which he is a graduate — M.Sc. (Hons.) (1947), Ph.D. (1949), D.Sc. (1954). His B.Sc. was completed before World War II service in the S.W. Pacific (Guadalcanal, Papua New Guinea) as Royal New Zealand Air Force Entomologist, a post that he continued to hold, while rising to the rank of Squadron Leader, until 1954. At that point, when he and his wife had been resident in Fiji while traveling widely in the Pacific on malariological research, he accepted an appointment to the faculty of the University of Malaya (now, National University of Singapore). In 1957 he moved on to the Institute of Parasitology, McGill University, Montreal, being promoted Associate Professor in 1958.

Dr. Laird, a member of the World Health Organization's Expert Advisory Panel on Insecticides since 1953 and Chairman of WHO's Expert Committee on Insecticides in 1960 (the session that laid the foundations of the prevailing aircraft disinsection recommendations in the International Health Regulations), transferred to administrative duties in Geneva, Switzerland in 1961, as Chief, Environmental Biology Unit, WHO. Having been increasingly interested in the biocontrol of vectors since his wartime research in New Britain, he was given the task of initiating a global program in this field. That accomplished, in 1967 he returned to academic life as Professor and Head, Department of Biology, The Memorial University of Newfoundland, St. John's, Canada. Other responsibilities at that time included Council Membership of the National Research Council of Canada (1968 to 1971) and Vice-Presidency of the World Federation of Parasitologists (1970 to 1974). In 1973 he was named Research Professor and Director, Research Unit on Vector Pathology, Memorial University. During the decade in which he held this post, RUVP was heavily involved in innovative approaches to vector control through integrated technologies stressing biocontrol, in West Africa, Central America, and latterly, Tuvalu, tropical Pacific.

Dr. Laird, who has served on the editorial boards of a number of scientific journals over the past 20 years, is a Fellow of the American Association for the Advancement of Science (1965), Honorary Member (Wellington Branch) of the Royal Society of New Zealand (1966), and Emeritus Member of the Society for Invertebrate Pathology (1984) and Pacific Science Association (1986). A Fellow of the Royal Society of Tropical Medicine and Hygiene and the American Mosquito Control Association, he has, to his credit, more than 240 scientific publications, including 12 edited and authored books, 5 published since "retirement".

Lawrence A. Lacey, Ph.D., is the Vector Biologist with the USAID Vector Biology and Control Project. Since obtaining his graduate degrees in 1978 from the University of California, Riverside, he has worked as an assistant professor at the Amazon Research Institute in Manaus, Brazil, consultant to the World Health Organization in the Onchocerciasis Control Program, Volta Basin, West Africa, and as a Research Entomologist with the U.S. Department of Agriculture at the Insects Affecting Man and Animals Research Lab in Gainesville, Florida before taking his current employment. His field experience thus far includes work in Central and South America, Asia, Oceania and West Africa.

Dr. Lacey is an active member of the Society for Invertebrate Pathology, American Society of Tropical Medicine and Hygiene, American Mosquito Control Association, and the Society of Vector Ecologists, and has served as either an officer, committee member, chairman of committees or on the editorial boards of these societies.

Dr. Lacey is the author of over 60 research papers and book chapters. His research interests are predominantly on the formulation and safety of microbial control agents, and the ecology and control of medically important insects using alternatives to insecticides.

Elizabeth W. Davidson, Ph.D., is Associate Research Professor in the Department of Zoology, Arizona State University, Tempe, Arizona. Dr. Davidson graduated from Mt. Union College, Alliance, Ohio in 1964, and received M.S. (1968) and Ph.D. (1971) degrees from Ohio State University. She held an NIH postdoctoral appointment in insect tissue culture at Ohio State University in 1971-72, and served as Instructor at The University of Rochester (New York) in 1972 to 1973. She joined Arizona State University as Research Associate in 1973, and was appointed Associate Research Professor in 1986. She has also served as Visiting Research Professor at Monash University, Melbourne, Australia (1980) and at The University of Zurich, Switzerland (1987).

Dr. Davidson is a member of the Society for Invertebrate Pathology, the Entomological Society of America, the American Mosquito Control Association, the Society of Vector Ecologists, the Tissue Culture Association, and Sigma Xi. She has served the Society for Invertebrate Pathology as Secretary (1984 to 1986), Vice President (1988 to 1990), and will serve as President (1990 to 1992).

Dr. Davidson is author of over 50 papers and has previously edited a book on pathogenesis of invertebrate microbial diseases. Her research interests are related to bacterial diseases of mosquitoes, microbial control of vector insects, and safety of microbial control agents.

CONTRIBUTORS

Keio Aizawa, Ph.D.
Institute of Biological Control
Faculty of Agriculture
Kyushu University
Fukuoka, Japan

Raymond J. Akhurst, Ph.D.
Division of Entomology
CSIRO
Canberra, Australia

Frederick S. Betz, Ph.D.
Hazard Evaluation Division
U.S. Environmental Protection Agency
Washington, D.C.

John A. Couch, Ph.D.
Senior Research Scientist
Environmental Research Laboratory
U.S. Environmental Protection Agency
Sabine Island
Gulf Breeze, Florida

Elizabeth M. Cozzi, Ph.D.
Regulatory Administrator
Department of Regulatory Affairs
Abbott Laboratories
North Chicago, Illinois

Claude Dejoux, D.Sc.
Director of Research
ORSTOM
La Teste, France

Beth-Jayne Ellis, Ph.D.
President
BioNet International Corp.
Manassas, Virginia

Jean-Marc Elouard, D.Sc.
Hydrobiological Laboratory
ORSTOM
Bamako, Mali

Sheila F. Forsyth, Ph.D.
Pesticides Directorate
Agriculture Canada
Ottawa, Ontario, Canada

Steven S. Foss
Biologist
Environmental Research Laboratory
U.S. Environmental Protection Agency
Sabine Island
Gulf Breeze, Florida

James R. Fuxa, Ph.D.
Department of Entomology
Louisiana State University Agricultural
  Center
Baton Rouge, Louisiana

Mark S. Goettel, Ph.D.
Research Scientist
Crop Sciences Section
Agriculture Canada Research Station
Lethbridge, Alberta, Canada

Albrecht Gröner, Ph.D.
Virology Research Department
Behringwerke AG
Marburg, West Germany

Nikolay V. Kandybin, Dr. Biol. Sci.
All-Union Institute for Agricultural
  Microbiology
Podbelskii Shausse 3
Leningrad, U.S.S.R.

Cynthia M. Lacey
Biologist
Falls Church, Virginia

Lawrence Lacey, Ph.D.
Vector Biologist
Vector Biology and Control Project
US AID/MSCI
Arlington, Virginia

Marshall Laird, Ph.D., D.Sc.
Honorary Research Fellow
Department of Zoology
University of Auckland
Auckland, New Zealand

Zengzhi Li, Ph.D.
Associate Professor
Department of Forestry
Agricultural College of Anhui
Hefei, Anhui, People's Republic of China

Brian E. Melin, Ph.D.
Eastern U.S. Manager
Forestry and Vector Products
Abbott Laboratories
North Chicago, Illinois

Mir S. Mulla, Ph.D.
Department of Entomology
University of California
Riverside, California

Frederick D. Obenchain, Ph.D.
Director, Research and Regulatory Affairs
Ringer Corporation
Minneapolis, Minnesota

Tadeusz J. Poprawski, Ph.D.
Insect Pathologist
European Parasite Laboratory
Agriculture Research Service
U.S. Department of Agriculture,
  International Activities
Behoust, Orgerus, France

Ray J. Quinlan, Ph.D.
Head of Technical Development
Verify Ltd.
London, England

Donald W. Roberts, Ph.D.
Insect Pathologist and Coordinator
Insect Pathology Resource Center
Boyce Thompson Institute
Ithaca, New York

Judith E. Saik, D.V.M.
Comparative Pathology Section
Veterinary Research Branch
National Institutes of Health
Bethesda, Maryland

John A. Shadduck, D.V.M., Ph.D.
Dean
College of Veterinary Medicine
Texas A&M University
College Station, Texas

Joel P. Siegel, Ph.D.
Department of Veterinary Pathobiology
College of Veterinary Medicine
University of Illinois at Urbana-Champaign
Urbana, Illinois

O. V. Smirnov, Candidate Biol. Sci.
All-Union Institute for Agricultural Microbiology
Podbelskii Shausse 3
Leningrad, U.S.S.R.

William E. Stewart, Ph.D.
Pesticides Directorate
Agriculture Canada
Ottawa, Ontario, Canada

John D. Vandenberg, Ph.D.
Research Leader
Bee Biology and Systematics Laboratory
USDA Agricultural Research Service
Utah State University
Logan, Utah

S. Bradleigh Vinson, Ph.D.
Department of Entomology
Texas A&M University
College Station, Texas

TABLE OF CONTENTS

IV. SAFETY OF MICROBIAL INSECTICIDES TO NONTARGET INVERTEBRATES

V. CONCLUSIONS

I. Registration Requirements and Safety Considerations for Microbial Pest Control Agents

Chapter 1

REGISTRATION REQUIREMENTS AND SAFETY CONSIDERATIONS FOR MICROBIAL PEST CONTROL AGENTS IN NORTH AMERICA

F. S. Betz, S. F. Forsyth, and W. E. Stewart

TABLE OF CONTENTS

I. INTRODUCTION

Microbial pest control agents have been in commercial use in North America for over 40 years. A variety of bacteria, fungi, protozoa, and viruses now comprise a small but steadily expanding group of products in the U.S. and Canada. Advantage has been taken of their host-specific mode of action to control mosquitoes (Culicidae), blackflies (Simuliidae), and numerous pests in agriculture and forestry. Regulatory oversight of microbial pesticides for many years was based on precedents and standards established for synthetic chemical pesticides. Now, however, both countries have registration requirements that address the distinctive characteristics of microorganisms. Recently, considerable attention has been paid to the evaluation of genetically engineered microbial pesticides and appropriate adjustments to the regulatory programs are underway.

Microbial pesticides have been viewed favorably from a safety perspective due to their usual narrow host range, lack of mammalian toxicity or pathogenicity, compatibility with beneficial organisms, and consequent value in integrated pest management schemes. However, commercial success of these products has been hindered by their narrow activity spectrum, short field life, and/or slow activity against the target pest. To address these problems, an intensive effort has been mounted to develop more environmentally stable formulations, to isolate superior strains, and to use both classical and modern genetic manipulation techniques, such as recombinant DNA, to produce more effective and commercially competitive products.

II. GENERAL SAFETY CONSIDERATIONS

Registration of a microbial pesticide product implies the possibility for widespread commercial use and environmental exposure. Therefore, regulatory authorities must consider potential impacts of pesticide exposure on public health and the environment. Human exposure may be direct for those mixing, loading, or applying the product, or indirect, as in the case of dietary exposure through consumption of treated food crops. Nontarget organisms (NTOs) already coexist in the environment with naturally occurring entomopathogens. However, large-scale applications of microorganisms may warrant evaluation because they present different numerical, spatial, and temporal relationships between microbial control agents and NTOs.

Risk assessment for microbial pesticides is the systematic scientific evaluation of hazard and exposure data, followed by the formulation of conclusions about the potential for human health or environmental risks as a result of using the pesticide (Hazard × Exposure = Risk). Hazard is expressed in terms of an adverse effect on an NTO, such as death resulting from a microbial toxin or other pathogenic action. Exposure of NTOs may be direct at the time of application or indirect as a result of microorganism reproduction and dispersal. Taken separately, neither hazard nor exposure poses a human health or environmental concern. Potential problems arise, however, when a susceptible NTO is exposed to a microbial pesticide capable of eliciting an adverse effect in it.

The above risk assessment model is the basis for tiered data requirement schemes set forth by Canada and the U.S. for the evaluation of naturally occurring microbial pesticides. In general, thorough product identification, plus human health and environmental effects data, are required in Tier I (Table 1). Usually, additional testing (e.g., Tier II) is required only when the results of Tier I tests identify a hazard to NTOs; i.e., the effects tests are positive. Tier II testing emphasizes development of exposure data, as well as further NTO effects testing. Thus far, none of the currently registered microbial pest control agents have required testing beyond Tier I (see also Chapters 8, 9, and 13).

Thorough identification of the microorganisms to be registered is the first step in eval-

TABLE 1

Data Required to Support Registration and Experimental Use of Microbial Pesticides in the U.S.[a]

Data requirement	When required	
	Registration	EUP
I. Product analysis		
Product identity and manufacturing process	R	R
Discussion of formation of unintentional ingredients	R	R
Analysis of samples	CR	CR
Certification of limits	R	R
Analytical methods	R	R
Physical and chemical properties	R	R
Submittal of samples	CR	CR
II. Residue analysis	CR	CR
III. Toxicology		
Tier I:		
Acute oral exposure	R	R
Acute dermal exposure	R	—
Acute pulmonary exposure	R	R
Acute intravenous exposure	R	R
Primary eye irritation	R	R
Hypersensitivity incidents	R	—
Tissue culture (viruses)	R	R
Tier II and III	Further tests required based on results obtained in Tier I	
IV. Nontarget organisms and environmental fate[b] Tier I—Ecological effects		
Avian oral	R[c]	R[d]
Avian inhalation	CR[d]	—
Wild mammal	CR[d]	—
Freshwater fish	R[c]	R[d]
Freshwater aquatic invertebrate	R[c]	R[d]
Estuarine and marine animal	CR[d]	—
Nontarget plants	R[d]	R[d]
Nontarget insects	R[c]	R[d]
Honey bee	R[c]	R[d]
Tier II—environmental fate	Data required based on results of Tier I tests	
Tier III--ecological effects	Data required based on results of Tier I tests	
Tier IV—simulated and actual terrestrial or aquatic field studies	Data required based on results of Tier I tests	

Note: R = required; CR = conditionally required; and — = generally not required.

[a] These are the normative data requirements. Fewer or additional data may be required in specific instances. In the case of EUPs, fewer data can often be justified based on the nature of the microorganism and limited use of the pesticide.

[b] Recommended protocols have been revised to bring them in line with recent research and experience. Statements on when data are required or conditionally required are to be updated to bring them in line with current knowledge and practice.

[c] Conditionally required for greenhouse and indoor uses.

[d] Not required for greenhouse and indoor uses.

uating human health and environmental effects. Identification also provides a basis for ensuring a uniform, uncontaminated product and test substance and provides the reference point for comparison with microorganisms for which there is a human health or environmental concern. In this way, many concerns can be put to rest before any hazard testing is conducted. For example, less rigorous testing may be required if the microbial pesticide is not a member of a species or group of microorganisms that are of clinical importance or that are known to cause disease in economically or ecologically important species.

Hazard evaluation requires submittal of information to demonstrate whether or not the microorganism has pathogenic, toxic, or other adverse effects on mammals or other representative nontarget species. As noted above, microorganism host range and relationship to known pathogens are important determinants to establish the extent of data needed and to identify which groups of NTOs require the most intensive evaluation. For example, NTO testing for a microbial insecticide probably would concentrate primarily on evaluation of beneficial insects related to the target pest or likely to be exposed in the treated areas.

Information needed for exposure assessment includes data on microorganism survival, growth, reproduction, and dispersal. Exposure assessment hinges on the ability to detect and quantify the microorganism in environmental media. Usefulness of exposure data in the risk assessment depends on how rigorously the lack of microorganism survival or growth can be established. Generally, it is very difficult to show that a microbial pest control agent will completely die out after application. Therefore, it is difficult to discount the possibility that the microorganism may survive at levels below detection, yet be able to grow and disperse when presented with favorable environmental conditions. These difficulties are one reason that the initial Tier I risk assessment emphasizes microorganism identification and hazard assessment. Then, only if significant hazards are foreseen from Tier I testing, is an in-depth exposure assessment undertaken in Tier II.

III. U.S. REGULATORY PROGRAM

National regulation of pesticides in the U.S. is administered by the Environmental Protection Agency (EPA) under the authority of the Federal Insecticide, Fungicide, and Rodenticide Act (FIFRA) of 1972, as amended October 24, 1988. Individual states may enact their own pesticide regulations to augment those set forth by EPA. FIFRA specifies that all pesticides (including microorganisms) must be registered before they may be sold for commercial use. The extent of EPA oversight depends on the degree of pesticide use and environmental exposure. Use and exposure generally correlate with the stage of product development. For example, small-scale field testing would be expected to result in minimal exposure and risk and, therefore, would warrant less EPA oversight. On the other hand, large-scale testing (and eventually, full commercial use) may result in more extensive exposure with a greater likelihood of risk. Therefore, EPA generally extends more comprehensive oversight in the latter situation. Large-scale preregistration testing of microorganisms to evaluate efficacy and gather other information to support registration usually requires an experimental use permit (EUP). After evaluating the submitted information and data, EPA may register those products that, when used as specified, will not pose unreasonable risks to human health or the environment.

Normative data requirements for experimental use and registration of microbial pesticides are set forth in a regulation entitled "Data Requirements for Pesticide Registration" (Part 158).[2] Registration requirements are organized in tabular form for each of the five major disciplinary areas where data are needed; i.e., product analysis, residue analysis, toxicology, ecological effects, and environmental fate (Table 1). Data requirements for experimental uses of pesticides are a subset of the data requirements for registration.

In many instances, the need to fulfill a particular data requirement depends on the pesticide use pattern. Therefore, data requirements in each area are specified as being required, conditionally required, or not required, for each of the major pesticide use patterns: terrestrial food and nonfood use, aquatic food and nonfood use, greenhouse food and nonfood use, forestry, and domestic outdoor and indoor use.

Provided along with each data requirement is a designation of the substance to be tested in developing the data and a cross reference to the Subdivision M Pesticide Assessment Guidelines.[3] The guidelines contain recommended test methods and protocols that can be used to develop the required data. The guidelines have been revised and updated to reflect recent improvements in test methods, experience gained in evaluating test results over the past 5 years, and to focus the document more directly on a microbiological perspective for the testing requirements. The most significant modifications include changes in recommended routes and levels of exposure of organisms to the pest control agent in order to provide test conditions that will more closely approximate anticipated exposure. The revised guidelines will be available in 1989. Procedural requirements for experimental use and registration of microbial pesticides are provided in the Code of Federal Regulations in Part 172[4] and Part 162,[5] respectively. These regulations specify the circumstances under which an EUP or registration must be obtained. They provide details concerning EUP and registration applications, product labeling, supporting data, and conduct of field trials for experimental programs. Major elements of the EUP application include identification of the product to be tested, quantity to be tested, test sites and acreage, proposed testing program, product label, and the supporting product analysis and safety data. Applications for product registration include similar, but expanded elements which are directed towards full commercial use of the pesticide product.

Products developed through the use of genetic engineering techniques, such as recombinant DNA, have advanced to the point where field testing is necessary in order to further evaluate their usefulness. In view of scientific questions and public concerns associated with some of these products, EPA determined that it would be prudent to evaluate genetically engineered microbial pesticides before application in the environment. This is in contrast to naturally occurring indigenous microorganisms and chemicals, which are typically not subject to review until used in large-scale experimental applications. EPA's policy on review of genetically engineered microbial pesticides was introduced in 1984[6] and reiterated in 1986[7] in a policy statement which also included statements from the other major federal agencies that have responsibilities in this area (i.e., Department of Agriculture, Food and Drug Administration, and National Institutes of Health).

Under the 1986 policy statement, EPA is to be notified at least 90 d before initiation of a small-scale field test. The applicant is requested to provide the information and data identified in the policy statement, including microorganism identity and how it has been manipulated; information on the parental and/or engineered strain(s), including growth, survival, and competitive characteristics; host range and potential effects on NTOs; and information on the purpose and design of the proposed field study.

Fewer or additional data may be needed depending on the specific microorganisms, nature of the genetic alteration, and inserted material and proposed use pattern. This case-by-case review allows EPA to conduct a substantive evaluation of the proposed test, determine whether it poses any significant human health or environmental risks, and determine whether an experimental use permit is required. At this initial stage, the Agency's scientific assessment emphasizes consideration of key risk factors that could lead to significant adverse effects from even a small-scale application of a microbial agent: ability to survive and become established in a new niche, ability to carry out a new function, transfer of inserted genetic material to other microorganisms, and competitiveness.

If the proposed field test poses no significant risks and no additional data are needed, then no EUP is needed before a field test. If risk questions arise or the need for additional data or monitoring is established, then the Agency may determine that an EUP is needed before the field test is initiated. In such a case, the applicable EUP requirements must be fulfilled.

Beyond the small-scale testing stage, the same basic EUP and registration requirements apply to both engineered and nonengineered microbial pesticides. Additional data may be required for engineered organisms but only to the extent warranted by any additional concerns that arise based on the nature of the genetic material that has been inserted or manner in which the pesticide agent has been altered or is to be used.

IV. CANADIAN REGULATORY PROGRAM

All pesticides, including microbial control agents, must undergo a scientific assessment of safety and efficacy data prior to registration, sale, and use in Canada. Registration of all pesticides comes under the mandate of the Pest Control Products (PCP) Act and Regulations and is the responsibility of the Pesticides Directorate in the Food Production and Inspection Branch of Agriculture Canada. In addition, a number of other federal acts impact on the use and regulation of pesticides. In particular, the Food and Drug Act sets maximum residue limits for pesticides for all food uses.

The primary purpose of the review process is to assure the safety, merit, and value of pesticide products. It stresses broad consultation and always strives to arrive at the best balanced decisions. Advisors from other federal agencies review parts of the data related to their expertise; e.g., Environment Canada comments on environmental fate and expression, and potential impact on terrestrial organisms; Fisheries and Oceans Canada reviews the potential impact on aquatic organisms; Health and Welfare Canada assesses occupational and bystander hazards and establishes maximum residue limits for foods; the Canadian Forestry Service comments on forestry and environmental concerns.

Registration guidelines, data requirements, and protocols for microbial control agents are presently under development. A tiered approach is being followed in order to accommodate and assess potential risk to human health and effects on the environment depending on the properties of the agent; e.g., naturally occurring vs. genetically engineered organisms, degree of genetic engineering, indigenous vs. nonindigenous, and intended use.

Interest and experimentation with microbial and biological control agents have steadily increased in Canada since the early 1960s. In 1962, two formulations of *Bacillus thuringiensis* were registered and represented the first microbial pesticides available for use in Canada. In 1982, formulations of *B. thuringiensis* ssp. *israelensis* (serotype H14) and two types of nuclear polyhedrosis virus (Douglas fir tussock moth (*Orgyia pseudotsugata*) virus and red-headed pine sawfly (*Neodiprion lecontei*) virus) were registered. New microbial control agents traditionally have been reviewed on a case-by-case basis; however, the early regulatory approach did not distinguish biological agents from chemical pesticides. Special draft guidelines for regulation of naturally occurring pesticides were published in 1986.

In response to the rapid growth of research and industrial developments in biotechnology, a concerted effort is underway to develop guidelines to address all types of microbial control agents. The challenge is to establish data requirements with a sound scientific basis which on the one hand will be reasonable for developers but which will adequately demonstrate and assure human health and safety of the environment.

At present, the registration process is a phased approach. It included a well-established permit system for field testing, the regulatory review, followed by full registration (1 to 5 years). In addition, plans are under consideration to develop a notification scheme and containment guidelines for the laboratory research stage.

A. RESEARCH PERMIT REQUIREMENTS

The current research permit requirements (Trade Memorandum, T-1-126[8]) do not differentiate between chemical pesticides and microbial agents and each research permit application is reviewed on a case-by-case basis. Guidelines for research permits for naturally occurring organisms have been developed and were distributed for comment, review, and use on a provisional basis for the 1988 season. The guidelines follow the same general requirements as for chemical products. A research permit application will be required for all microbial agents regardless of the size of the trial. Appropriate toxicology studies may be required before field testing along with an experimental label, characterization of the organism and a description of the target pest, an experimental protocol outlining details of the experiment, the amount of the product to be used, rates, size, and description of the test site(s). Procedures and data requirements for research permits for genetically engineered organisms are under development. In addition to the standard research permit requirements, the following data requirements are under consideration:

1. Details on the method of genetic engineering
2. New phenotypic and genotypic trait(s)
3. Comparison to parent organism(s) for biology, survival, and other environmental effects
4. Details of monitoring/detection procedures
5. Contingency plans in the event of an accident

B. REGISTRATION REQUIREMENTS

Interim requirements for the registration of naturally occurring microbial pesticides are outlined in R-memorandum, R-1-229, "Guidelines for the Registration of Microbial/Biological Pesticides".[9] Basic data and information requirements include

1. Product identification and characterization, including specification of the technical active ingredient (most concentrated form), the manufacturing, sampling, and quality control methodology
2. (Tiered) toxicology requirements
3. Residue studies
4. Environmental expression/fate tests
5. Environmental toxicology studies
6. Efficacy/performance data

These guidelines are currently being revised to be more microbiologically appropriate and to provide a more organism-specific selection of species to test for nontarget effects. Genetically engineered microbial pest control agents will be evaluated on a case-by-case basis using the guidelines for naturally occurring pest control agents as a start until background research and studies for development of guidelines for registration of genetically engineered agents are completed. An announcement of intent to regulate has been published in R-memorandum R-1-231.[10] Draft guidelines are targeted for release for comments in 1988.

REFERENCES

1. Federal Insecticide, Fungicide, Rodenticide Act of 1972, 7 U.S.C. 136 et. seq., as amended October 24, 1988.
2. Data Requirements for Pesticide Registration, Title 40, Code of Federal Regulations, Part 158, 1988.

3. Pesticide Assessment Guidelines—Subdivision M, National Technical Information Service, Springfield, Virginia, #PB83-153965 (revised guidelines to be available in 1989).
4. Experimental Use Permits, Title 40, Code of Federal Regulations, Part 172, 1988.
5. Regulations for the Enforcement of the Federal Insecticide, Fungicide, and Rodenticide Act, Title 40, Code of Federal Regulations, Part 162, 1988.
6. Microbial Pesticides; Interim Policy on Small-Scale Field Testing, FR 49 (202), 40659, 1984.
7. Coordinated Framework for Regulation of Biotechnology; Announcement of Policy and Notice for Public Comment, FR 51 (123), 23302, 1986.
8. Control Product Research Programs, Trade Memorandum T-1-126, Pesticides Directorate, Agriculture Canada.
9. Guidelines for the Registration of Microbial/Biological Pesticides, Memorandum to Registrant R-1-229, Pesticides Directorate, Agriculture Canada.
10. Regulation of Pesticides Produced by Biotechnology, Memorandum to Registrants R-1-231, Pesticides Directorate, Agriculture Canada.

Chapter 2

REGISTRATION REQUIREMENTS AND SAFETY CONSIDERATIONS FOR MICROBIAL PEST CONTROL AGENTS IN THE EUROPEAN ECONOMIC COMMUNITY

R. J. Quinlan

TABLE OF CONTENTS

I. THE EUROPEAN ECONOMIC COMMUNITY (EEC)

The EEC, otherwise known as the "Common Market", is an association of European countries with no tariffs or trade controls within it, and with a common external tariff for imported materials. There is free movement of capital and labor within the market, and a fair degree of fiscal harmonization. At the formation of the EEC in 1958, six countries were members — Belgium, France, Holland, Italy, Luxembourg, and West Germany. By 1987, the community had, over the intervening years, expanded to include Britain, Ireland, Denmark, Greece, Portugal, and Spain.

While some degree of legal and bureaucratic coordination has been introduced, and much attempted, many aspects of EEC activity remain discrete and idiosyncratic. This is typified by the community's approach to the registration or approval of pesticides. Despite attempts to produce an EEC-wide equivalent of the U.S. Environmental Protection Agency (EPA), pesticide registration remains the right of individual country authorities with no agreed principles or real collaborative activity.

II. EEC GUIDELINES FOR MICROBIAL CONTROL AGENTS

Despite a lack of intergovernmental coordination in this area, the Council of Europe has produced a document entitled "Guidance for Registration of Biological Agents used as Pesticides". The document was produced to give broad guidance to manufacturers on what data were likely to be required by registration authorities for approval of naturally occurring bacteria, protozoa, fungi, or viruses, or their mutants, or other biological agents for the control of crop pests. The guidelines specifically omitted microbes designed for use against mammalian pests or those that had been genetically manipulated.

The general principle of the guidelines was stated as follows: " . . . that a notifier should demonstrate that the use of the agent will not present an unacceptable hazard to users, consumers of treated products or wildlife and the environment." This is a principle which is applied not only to biological agents, but to chemical agents as well. It is also one which is universally applied throughout the countries under review for pest control agents generally.

The document includes a request for data normally considered to be outside conventional toxicology, but which nevertheless relate specifically to the safety of any microbial agent. Such information is required by all the registration authorities in the EEC. These data include:

1.   Details of the procedures and criteria used for identification of the organism under review.
2.   Data on the formulation and likely chemical impurities.
3.   Data on microbiological purity including history, methods of analysis, and a demonstration that "any contaminants can be controlled to an acceptable level."
4.   "Methods to show freedom from any human and mammalian pathogen" — usually a one-dose batch toxicology test, e.g., high dose of pathogen given to mice by intraperitoneal injection (see below).
5.   Specificity data, including a discussion of taxonomic relationship to vertebrate pathogens, and mutation rate.
6.   Data on residues.

The toxicology data required are not discussed in detail. However, mammalian studies on infectivity, multiplication *in vivo*, and allergenic potential (especially for airborne fungal spores) are considered essential. The document is less stringent on acute studies: "*some* assessment of acute or subacute toxicity will *normally* be required." Carcinogenicity and teratogenicity studies are considered to be appropriate only if toxic metabolites are know to be produced.

No protocols are provided in the EEC document, and no country applies the recommendations directly. However, the document must be considered, in conjunction with those from the World Health Organization, the International Group of National Associations of Agrochemical Manufacturers, the International Organization of Biological Control, and also the EPA — all of which have served as essential discussion material for the development of procedures within individual countries.

III. UNITED KINGDOM (U.K.) REGULATIONS

The U.K. took the European lead in developing guidelines specifically for microbial control agents, releasing them in 1979. At that time British pesticide registration was voluntary and free. Although virtually all companies complied, the concept of a voluntary scheme eventually ceased to satisfy critics. This situation was rectified with the Food and Environment Protection Act of 1985, which empowered the government to set legally enforceable guidelines and to charge fees. Thus, microbial agents now have to be legally registered. As the U.K. microbial pesticide guidelines were among the first and remain the best developed in Europe, they will be discussed in somewhat more detail than most of the others.

A. DEFINITIONS
The U.K. defines a pesticide as "any substance, preparation, or organism which fits it for use as a pesticide." Thus, all organisms which can replace a conventional pesticidal chemical theoretically require registration whatever their mode of action. However, if the same species is used for another purpose, e.g., as a fertilizer, it does not.

In practice, the guidelines apply only to bacteria, viruses, fungi, and protozoa. Nematodes slip through as they are "multicellular" and are lumped with other biological agents such as predators and parasites. The guidelines specifically exclude genetically engineered organisms or non-native organisms. These are considered by a different authority on a case by case basis.

B. THE REQUIREMENTS
Information about the submitted microbial agent is presented in the "data sheet" as answers to 13 categories of questions. Registration criteria are provided, but the extent of the information (and hence the length of the submitted data sheet) is up to the submitter. Most of the information required for the data sheet will already be known to the applicant, e.g., methods of production, biological properties, target species, dose rates, etc.

C. SAFETY DATA
The toxicological data required and the protocols for the experiments are presented in outline. The U.K. authorities encourage the applicant to discuss the protocols with them, and are hence able to review the type and nature of the data required case by case. The guidelines state that: "Each application will be considered on its merits and expert advice will be sought both on the requirements of testing and the methods to be employed." The guidelines also suggest that tests on viral products will differ from those for other microbial agents.

1. Acute Toxicity and Infectivity
This group of tests on both formulated product and active ingredient includes (a) acute oral dose, (b) acute inhalation, (c) acute subcutaneous injection, (d) an eye irritation test, and (e) a skin irritation test.

The acceptable protocols for these tests are similar in many respects to those recommended by the EPA. However, the following points of difference are noteworthy:

1. The U.K. requests a "maximum practicable dose" instead of setting a figure of, e.g., 5 g/kg. In practice, the 5-g rate is acceptable to the U.K. authorities.
2. The U.K. requests subcutaneous administration tests not usually required elsewhere.
3. The U.K. requests that the acute tests last for 28 d, as compared with a 14-d test for most other countries. It is worth noting that some authorities do not accept a 28-d, single sacrifice test as it allows too lengthy a period for an animal to recover. In practice, 14-d tests are acceptable in the U.K.
4. The U.K. requests histopathology on major organs and an attempt to reisolate the organism from lungs and conjunctival fluid.

Infectivity under immunosuppression is also requested but is rarely undertaken. No protocols are given in the guidelines, and the value of extra data is usually considered to be marginal if there are no indications of infectivity from any other source.

2. Subacute Toxicity

Unlike the conditions prevailing in the U.S., a 13-week subacute feeding test (on mice) and a 20-d inhalation test (on guinea pigs) are required. The study need only be on unformulated material, if contraindications in the acute studies are not found. The feeding study requires necropsy for clinical observations, cause of death (if any), gross pathology, and full histopathology. The inhalation study requires "a suitable range of doses using (a) particle size within the range that will reach the lungs. Exposure for one hour in the morning and one hour in the afternoon 5 days per week for not less than 20 days."

There are no formal mechanisms described in the guidelines for waiving the subacute tests. However, if the microorganism is shown to be innocuous in the acute tests, and if there are no other contraindications in tests or in the literature, it may not be necessary to perform subacute tests. This has certainly been the case for *Verticillium lecanii* and the various subspecies of *Bacillus thuringiensis*. If tests are performed with viruses, viable active virus must be assessed by insect bioassay or cell culture. Nucleic acid probe or monoclonal antibody techniques may be used following discussion.

3. Allergenicity

The guidelines state that "there are at present no tests that can be recommended for general application." Nevertheless, applicants are asked to look for evidence of sensitization in workers (sought by questionnaire) and in potentially relevant animals. The applicant should be prepared to carry out a Magnusson/Kligman test for delayed contact hypersensitivity. This test should probably be done in any case, as it is inexpensive and can be used in a variety of countries, especially West Germany, to underpin any claims about nonallergenicity.

4. Known Toxins

If there are known toxins in or produced by the microbe, the U.K. requires data on tumorgenicity based on *in vitro* cell transformation, teratogenicity, and chemical structure of the toxin and its stability. A mutagenicity test using a prokaryotic or eukaryotic microorganism is also requested if a toxin is known to be produced. In practice, none of these tests have been required for *B. thuringiensis* products. The same group of tests are said to be required for viruses "if so indicated by the 90-day animal studies." Precisely what the trigger for these additional studies is is not indicated.

IV. REGULATIONS IN OTHER COUNTRIES

The development of guidelines for registration of microbial agents outside the U.K. has on the whole been slow. Most EEC countries have no formally written procedures for these organisms. However, in some countries (e.g., the Netherlands and Italy) a draft application form has been prepared. In others, only memoranda or a record of conversations with the relevant staff are available to guide the applicant.

Despite the lack of guidelines, *Bacillus thuringiensis* products for caterpillar control (e.g., Dipel®, Thuricide, and Bactucide) are now cleared for commercial use in most countries in Europe. Many other microbials are either on sale in individual countries or under development. Thus, in spite of a lack of specific registration criteria, products are being approved for use. Systems must therefore, be operational for clearance to be obtained. In most of these countries microbial pest control agents have been cleared for use by reference to the system operated for the registration of chemical pesticides.

The lack of specific national registration guidelines for microbial pesticides does not necessarily indicate a lack of interest by the countries concerned in the registration of such products. The situation probably more correctly reflects those microbials which have been submitted for commercial clearance — usually only one or two, most notably *B. thuringiensis* ssp. *kurstaki*. Familiarity with this organism and its extensive use within the U.S. has eased clearance in most European countries.

In only one country, Denmark, is it possible to sell microbial pesticides without reference to the pesticide registration authorities at all. The regulations in Denmark state that any pest control agent that kills by infection, rather than by a toxin, is a biological control agent and hence does not require registration. While this definition clearly encompasses fungi and viruses, the inclusion of *B. thuringiensis* must be questionable. However, at least one product was being sold for use on vegetables in 1986. This situation may change in the near future, as the authorities are indicating that they intend to introduce a registration scheme similar to that being used in the U.K.

A. The Operational Procedures

In practice, countries lacking specific guidelines will normally register a microbial product which has a data package similar to that required for clearance with the EPA. In some countries, specific concerns on allergenicity have been voiced (see below). In most, a certificate of registration in the country of origin is essential; although this by itself will not automatically guarantee approval. Evidence that the organism is naturally occurring within the country is also normally an essential prerequisite.

The safety data required by these countries always includes a request for mammalian toxicology studies. In most cases, dose rates are not stated. Usually, registrants will follow the recommendations of the EPA, i.e., a one-dose experiment using 5 g of test material per kg body weight.

The range of tests normally sufficient to obtain commercial clearance in these countries (for example in Italy) is (1) acute oral toxicity, (2) acute subcutaneous toxicity, (3) acute intraperitoneal toxicity, (4) subacute (90-d) feeding studies (rarely actually undertaken), (5) skin and eye irritation studies, (6) inhalation toxicity, and (7) allergic sensitization.

However, the data requested by most countries do not acknowledge a number of points, notably

1. That many of the questions raised are inappropriate for microorganisms, e.g., it is quite often requested that data be supplied on vapor pressure, solubility in solvents, or boiling point.
2. That other, specific problems may arise which are related to the use of microorganisms, e.g., allergic responses or contamination of the product by human or animal pathogens.
3. That some data usually regarded as essential for assessing chemical pesticides, e.g., chronic feeding studies, may be unnecessary or inappropriate.

For example, submissions for clearance of microbial pesticides in Spain must include data on metabolism in animals. No such data for any microbial product are known to exist in a form suitable for registration.

The approaches of the individual countries will now be discussed in outline.

1. Italy

Italy has produced an application form for microbial pesticides. In most respects, the points raised are similar to those of other countries. The approach appears to be founded on that devised by the U.K. authorities. However, the following points differ:

1.   Information on the organoleptic (taste) qualities of treated edible products is requested.
2.   A subcutaneous injection study is required.
3.   A 90-d subacute feeding study is required.

In practice, very few microbials have been registered in Italy. The time taken for clearance can run to 5 to 7 years. This delay appears to relate more to the bureaucracy involved, rather than to any particular scrutiny on products.

2. The Netherlands

The Dutch authorities have produced an application form similar to the Italian in approach and layout. However, the information it requests on a number of aspects is similar to that required by the U.K. and U.S. authorities. The Dutch registration system allocates responsibility to three separate governmental branches. The country's Department of Health is responsible for all safety aspects.

3. France

The French have no specific microbial control guidelines. They have, however, registered several microbials (see below). The French authorities have (compared to others) been particularly aware of the potential of fungal mycotoxin production. Evidence for the absence of such toxins is required.

4. Federal Republic of Germany

Germany produced guidelines on the registration of entomopathogenic viruses in 1983. However, none have yet been registered. Other microbials are dealt with "by negotiation". Although no specific requirements have been stated, the West German authorities have paid attention to potential allergenicity problems. These authorities have also considered many aspects of environmental safety, and the tests requested are often much more stringent than those for other countries.

5. Greece, Spain, and Portugal

Registration in these southern European countries has been undertaken following procedures for chemical pesticides. There is no evidence that any special consideration is given to microbial pesticides, although *B. thuringiensis* ssp. *kurstaki* is registered in all three countries (presumably on less data than would be otherwise required for a chemical). Registration of the mosquitocidal *B. thuringiensis* ssp. *israelensis* has, however, taken much longer. Times taken to register products in Greece have been particularly lengthy, extending to periods beyond 5 years. Again, this does not appear to relate to discussions on the relative safety of microbial pesticides, but rather to bureaucratic delay.

V. QUALITY CONTROL

In most EEC countries, additional data in support of the standard toxicology package are required to demonstrate the safety and purity of individual production batches. The precise nature of these quality control tests is not detailed by any authority in the EEC and is thus left to the individual applicants.

One scheme devised for this purpose by Tate & Lyle PLC in the early 1980s has been approved by the International Union of Physics and Chemistry (IUPAC) for bacterial insecticides. The scheme was based on the IUPAC microbiological contamination standards for animal feeds and consists of a series of microbiological purity tests and a single batch toxicology test. The standards for the microbiology test are

1. Viable mesophiles $<1 \times 10^5/g$
2. Viable yeasts and molds $<100/g$
3. Coliforms $<10/g$
4. *Staphylococcus aureus* $<1/g$
5. *Salmonella* <1 in 10 g
6. Lancefield Group D Streptococci $<1 \times 10^4/g$

The batch toxicology test consists of administering 5×10^6 viable spores or colony forming units of the active ingredient into each of five mice subcutaneously. After a week, animals are killed and subjected to a postmortem. A batch is rejected if (1) there are any mortalities, (2) there is evidence of organ damage, and (3) there are reversible pharmacological effects that may be considered detrimental.

This batch test serves both as one of quality control (it is considered to be the final arbiter on the occurrence of any mammalian pathogens not detected in the microbiological test) and as a test for mutations or for cultural conditions which may have resulted in a production of a hitherto unknown toxic substance by the microorganism being used.

VI. REGISTRATION STATUS

The availability of many microbial pesticides has varied considerably over the last 10 years as a result of product failures, company insolvency, etc. The list in Table 1 includes products that were available at the time of this writing or which are likely to be available in the near future. Organisms with experimental clearance only are not included. It is assumed, though, that if they were, the number of organisms in the list would increase dramatically.

VII. CONCLUSION

The development in Europe of registration and safety data guidelines specifically for microbial pesticides has been slow. In addition, there is no EEC-wide scheme for registration. Only the U.K. has published procedures directly applicable to the present topic although several other countries have produced suitable application forms. Most countries rely on the application of guidelines originally developed for chemical pesticides. It is likely that more countries will develop procedures and recommendations specifically for microbial pesticides when the number of products increases and the pressure on the authorities is greater.

In practice, most of the countries of the EEC, with respect to most of the organisms under development for use as microbial pesticides, will accept safety data produced to meet the guidelines and specifications of the EPA. Because of this, commercial companies will always tend to develop their data package in response to the EPA rather than any other registration authority.

TABLE 1
Microbial Pesticides Cleared for Use in the EEC

Austria (not EEC)
Bacillus thuringiensis ssp. *kurstaki* Bactospeine (Solvay)   Bacterial insecticide
Cyprus (not EEC)
B. thuringiensis ssp. *kurstaki* Bactospeine (Solvay)   Bacterial insecticide
France
Agrobacterium radiobacter Galeine A (Corvagi)   Bacterial antagonist
Arthrobotrys irregularis Royal Champignon 350 (Royal Champignon)   Fungal nematicide
B. thuringiensis ssp. *kurstaki* e.g., Dipel (Abbott), Thuricide HP and SC (Sandoz),   Bacterial insecticide
  Bactospeine (Solvay)
B. thuringiensis ssp. *israelensis* e.g., Vectobac (Abbott), Teknar (Sandoz), Bactimos   Bacterial insecticide
  (Solvay)
B. thuringiensis ssp. *aizawai* B-401 (Sandoz)   Bacterial insecticide
Trichoderma viride Phior P (Orsan)   Fungal fungicide
Greece
B. thuringiensis ssp. *kurstaki* e.g., Dipel (Abbott), Thuricide HP (Sandoz) Bactospeine   Bacterial insecticide
  (Solvay)
B. thuringiensis ssp. *israelensis* Teknar (Sandoz)   Bacterial insecticide
Italy
B. thuringiensis ssp. *kurstaki* e.g., Bactucide (CRC), Dipel (Abbott), Bactospeine   Bacterial insecticide
  (Solvay)
B. thuringiensis ssp. *israelensis* e.g., Bactis (CRC), Vectobac (Abbott), Teknar (San-   Bacterial insecticide
  doz), Bactimos (Solvay)
B. thuringiensis ssp. *aizawai* B-401 (Sandoz)   Bacterial insecticide
The Netherlands, Belgium, Luxembourg
B. thuringiensis ssp. *aizawai* B-401 (Sandoz)   Bacterial insecticide
B. thuringiensis ssp. *kurstaki* e.g. Dipel (Abbott), Thuricide HP (Sandoz), Bactospeine   Bacterial insecticide
  (Solvay)
B. thuringiensis ssp. *israelensis* Bactimos (Solvay)   Bacterial insecticide
Spain and Portugal
B. thuringiensis ssp. *kurstaki* e.g., Dipel (Abbott), Thuricide HP (Sandoz), Bactospeine Bacterial insecticide
  (Solvay)
B. thuringiensis ssp. *aizawai* B-401 (Sandoz)   Bacterial insecticide
B. thuringiensis ssp. *israelensis* Bactimos (Solvay)   Bacterial insecticide
U.K.
B. thuringiensis ssp. *aizawai* B-401 (Sandoz)   Bacterial insecticide
B. thuringiensis ssp. *kurstaki* e.g., Dipel (Abbott), Thuricide HP (Sandoz), Bactospeine Bacterial insecticide
  (Solvay)
B. thuringiensis ssp. *israelensis* e.g., Teknar (Sandoz), Bactimos (Solvay)   Bacterial insecticide
Neodiprion sertifer NPV Virox (Oxford Virology)   Viral insecticide
Trichoderma viride Binab T (Bio-Innovation)   Fungal fungicide
Verticillium lecanii e.g., Mycotal and Vertalec (no current producer)   Fungal insecticide
Federal Republic of Germany
B. thuringiensis ssp. *kurstaki* e.g., Dipel (Abbott), Thuricide HP (Sandoz)   Bacterial insecticide
B. thuringiensis ssp. *kurstaki* Pyrethrum Celamerck "Natural Insecticide"   Bacterial insecticide
B. thuringiensis ssp. *israelensis* Teknar (Sandoz)   Bacterial insecticide

ACKNOWLEDGMENTS

The author would like to thank Dr. Denis Burges, Denis Goffinet (Abbott, Belgium), Dr. Willem Ravensberg (Koppert BV), and Jan Wolffhechel (Plantekemi, Denmark) for their help in obtaining information on the products currently registered in Europe.

Chapter 3

REGISTRATION REQUIREMENTS AND SAFETY CONSIDERATIONS FOR MICROBIAL PEST CONTROL AGENTS IN THE U.S.S.R. AND ADJACENT EASTERN EUROPEAN COUNTRIES

N. V. Kandybin and O. V. Smirnov

TABLE OF CONTENTS

I. INTRODUCTION

The matter under discussion has two aspects. On the one hand, scientific assurances are required of the health and environmental safety of microbial control agents, and of their efficacy, as compared with chemical pesticides, against target pests. On the other, judicial questions must be considered: the regulation of the use of these agents, ways of controlling their application, the criteria for their formulation and for granting use permits, and provision for the revision of their official registration as necessary.

At a time when the impact of chemical pesticides on man and nature has become a matter of widespread concern, the present topic is clearly timely. For rapid developments in biotechnology are offering us not only a diversity of microbiological means of plant protection, but also relevant biologically active substances including bioregulators, hormones, and antibiotics.[1-3] Growing evidence from the intensive investigations of dedicated scientists has already made it clear that in general, entomopathogenic microorganisms are far safer for man and nontarget organisms (NTOs) than are conventional chemical control agents. These, following their massive production and application, have sometimes proved hazardous to the health of humans and animals;[4-11] circumstances that have not arisen with respect to microbial control agents, used for plant protection. It is thus being argued that every effort should be made to substitute conventional chemical control agents with microbial ones.

The chief reason for considering the latter less dangerous than the former is their frequent close host specificity, a consequence of evolutionary processes that have tailored entomopathogens to particular hosts. Furthermore, the fermentative characteristics and optimal temperature requirements of entomopathogens sharply differentiate them from the microorganisms of homeothermic animals, in the bodies of which microbial pathogens of poikilotherms can hardly be expected to develop.

Finally, as microorganisms in general and entomopathogens in the present context are among the myriad components of natural ecosystems within which biological buffering prevents their oversaturation of the environment, and as their mass production only becomes feasible following isolation, it seems unlikely indeed that heavy field applications will result in major adverse ecological effects of the kind that have followed the massive spraying of crops and forests with persistent synthetic chemical pesticides.

II. SCIENTIFIC PREREQUISITES FOR SAFETY OF MICROBIAL INSECTICIDES FOR CONTROLLING INSECT PESTS

A. HEALTH REQUIREMENTS DURING THE PRODUCTION OF MICROBIAL INSECTICIDES

The main argument in favor of applying microbial control agents against insect pests is their above-mentioned host specificity and consequent target selectivity. Their use is thus like that of a rifle, to pick off a single victim without harming its close associates as a shotgun blast would do.[12]

Such a concept cannot but emphasize the overriding importance of correlating safety and efficacy with respect to any candidate biocontrol product. Interestingly too, those in daily contact with entomopathogens by virtue of their involvement in the manufacture of such products, are necessarily far more exposed to these microbials than is the population at large or indeed those employed in the practical field application of innovative control agents of this kind.

Health protection of U.S.S.R. citizens, including all those personally involved in the control of pest insects of agricultural importance, is the paramount consideration as regards the development and practical use of relevant microbial control agents in our country.

Therefore, searching preliminary analyses are in progress with respect to every group and individual strain of entomopathogens serving as the base of candidate products. Such analyses include every aspect of possible consequence to humans, in accordance with all demands of contemporary medicine.[13,14]

It is therefore a pleasure to report that the results of many experiments, conducted throughout the world, have shown to date that pertinent viral and bacterial preparations pose no threat to people.[12,15-21] Moreover, while certain fungal products have sometimes initiated allergic responses, entomopathogenic fungi, in general, seem harmless, too.[12,20,28,29]

The recognized leader among the entomopathogens on which microbial control products have so far been based is obviously *Bacillus thuringiensis*. The industrial production of preparations based on this microorganism has been in progress for almost 3 decades, without any harm to humans having been identified (nevertheless, the possibility of mutational changes to the bacterium taking place in an undesirable direction, remains under close scrutiny).

Environmental pollution due to field applications of *B. thuringiensis*-based products has yet to be demonstrated, and indeed, America's Environmental Protection Agency (EPA) has cancelled the earlier need for specifying maximal quantities of the organism in agricultural products.[12] Actual plant residues following *B. thuringiensis* application, are approximately 2 mg/kg of the plant mass, and within a week this figure is reduced by some two- to threefold. Such concentrations are far below the levels recognized by the World Health Organization (WHO) as of health concern (and these figures have a 50-fold inbuilt safety factor). There is thus no need for the establishment of specific environmental safety standards.[19,20] The same applies to products founded upon the viral entomopathogens investigated to date.[20]

Whether or not the harmlessness of such viral and bacterial preparations is held to be absolute, searching investigation of every newly isolated strain under consideration as the basis of a biocontrol product of course remains necessary. In this context, the problem of the practical application of the β-exotoxin of *B. thuringiensis* must now be summarized. It was in the late 1970s that following intensive research, field use of Bitoxybacillin (which contains 0.6 to 0.8% of β-exotoxin) was authorized in the U.S.S.R.[30,31] Further preparations were then developed consisting of β-exotoxin only; 1% in the case of Turingen-1, and 10% in that of Turingen-2. Such products have long been denied registration in the U.S. (see Section III. A, 2nd para., Chapter 8) and other countries, but by 1983 American investigations were suggesting their acceptability, e.g., that of ABG-6146 which contains 55% β-exotoxin[32] (see Chapter 11 for a full appraisal of relevant safety aspects). Again, as a result of lengthy investigations by Carlberg et al.,[33] which also demonstrated the field safety of β-exotoxin, Muskabac has been produced in Finland, and preparations containing the β-exotoxin have become available in Romania (Turintox) and Italy (Exobac).

B. EFFECTS OF MICROBIAL INSECT CONTROL PRODUCTS ON PLANTS

Most microbial control products are applied against insect pests of plants by spraying the latter, which thus are the first NTOs to receive these biocontrol agents. Therefore, the phytotoxicity (ability to cause "burns", etc.) of all new formulations of microbial insecticides must be carefully assessed. Any harmful effects revealed will necessitate appropriate modification of candidate products, or even their abandonment.

C. EFFECTS OF MICROBIAL INSECT CONTROL PRODUCTS ON ENTOMOPHAGA AND HONEYBEES

The problem here concerns possible hazards to beneficial insects closely related to the target hosts of particular entomopathogens. The safety of the latter to NTOs reflects, for example, the absence of enzymatic abilities to transform *B. thuringiensis* protoxin into toxin, or the capacity to dissolve polyhedra (with consequent virion release) in the gut juices.

Additionally, the manner of feeding of NTOs, the nature of their food, phenological circumstances, and a wide variety of biotic and abiotic factors, all influence the selectivity of entomopathogens and the microbial insecticides based on them.

The shotgun vs. rifle allegory of Section II.A is again relevant here. While the application of chemical pesticides harms "both saints and sinners",[12] highly specific biocontrol agents such as Bitoxybacillin, when used, for example, on potatoes against the Colorado potato beetle (*Leptinotarsa decemlineata*) or on cotton against the corn earworm (*Heliothis zea*), present no hazard to the entomophaga already present on these crops. No *B. thuringiensis* preparations are known to adversely affect insect parasitoids, predators, and pollinators. Most of this entomopathogen's subspecies destroying only a relatively small number of insect pest taxa, an acceptable level of selectivity can surely be claimed here. The same can be said for virus preparations. It must be emphasized that these remarks on safety aspects of microbial insecticides are based not only upon experience following the application of recommended doses against the caterpillars of plant pests, but also on experimental results from the exposure of entomophaga and bees to massive doses.[34,35]

While the general harmlessness of such microbials to the NTOs under discussion is thus established, special problems will always call for vigilance; for example, where pest control is required on sericultural plants, for viruses and *B. thuringiensis* preparations in particular can endanger the silkworm (*Bombyx mori*).

It must also be emphasized that a sharp fall in the incidence of insect pests of plants, due to successful microbial control, will in itself be likely to disrupt the populations of entomophaga and parasites already limiting pest numbers. The planned destruction of phytophagous pests must inevitably diminish host availability and food supply to beneficial insects. The resultant chain reaction decline in their numbers must not be mistaken for a direct lethal effect of the application of microbial control materials.

D. EFFECTS OF MICROBIAL INSECT CONTROL PRODUCTS ON FISH AND THEIR FOOD ORGANISMS

Such an effect can result from either of two distinct causes. First, drift from aerial spraying of microbials against crop pests can contaminate nearby water. Second, biocontrol agents (at this time notably *B. thuringiensis* ssp. *israelensis*, = serotype H-14) may be directly applied to aquatic sites against mosquito and blackfly pests and vectors.

In Eastern Europe and the U.S.S.R., comprehensive testing of entomopathogens representing various groups of microorganisms has not demonstrated survival in the tissues of fish. However, the development and introduction into practical use of such products as Bactoculicid (U.S.S.R.) and Moscitur (Czechoslovakia) provided the occasion for in-depth studies of *B. thuringiensis* ssp. *israelensis* in aquatic environments, but again without revealing any hazard to fish or their food organisms, aside from larval Culicidae and Simuliidae. Fluctuations in NTO populations in waters to which these products were applied proved to reflect migration and seasonal phenomena.

To conclude our Section II, we must point out that depending upon their particular qualities microbial control agents can be applied either inundatively (as in the case of *B. thuringiensis* ssp. *israelensis*) or inoculatively (as in the case of microorganisms applied live, and able to become established in the target host population). In the latter event the use of entomopathogens can be contemplated for creating artificial epizootics. Also, the consequences of treatment of an earlier generation of the target species with microbial insecticides can be exploited by further selective applications. Such matters are further developed in the following section.

III. REGISTRATION OF MICROBIAL INSECT CONTROL AGENTS IN EASTERN EUROPE AND THE U.S.S.R.

The socialist countries of Eastern Europe and the U.S.S.R. are much concerned with environmental quality issues, and each of their governments has appropriate legislation,[36-48] the details of which differ from one nation to another. Only in the German Democratic Republic (GDR), though, is there a single comprehensive legal enactment (which has been in force since 1971). In other countries of the region, environmental quality matters are the responsibility of various branches of state establishments; the control of harmful insects, including microbial measures, being implemented under agricultural boards and offices, with constant liaison with medical authorities.[49,50]

The Council of Mutual Economic Aid (CMEA) unites the relevant individual efforts of the Eastern European member states and the U.S.S.R., regional interaction being coordinated through the Council of Representatives of all the countries involved, towards "discovering new types of pesticides, developing biological and other means of plant protection, and promoting comprehensive studies of ways of improving environmental protection." Poznan, Poland is the site of the administrative center of the latter council, the chief scientific research institute of which is in Bratislava, Czechoslovakia.

The two Councils collaborate in problem solving concerning health and environmental safety criteria for bioinsecticides, their work envisaging successive stages of attention to the possible impact of microbial control practice on people, including:

1. Investigations of the primary infectivity and toxicity of entomopathogens and the toxicity of formulations based upon them; also studies relating to dosage, timing of applications and efficacy
2. Establishment of health standards relating to the environmental persistence of such preparations
3. The improvement of health requirements concerning the industrial manufacture of relevant products based upon entomopathogenic viruses and microbial toxins.

This program's aim is not only to elucidate the direct hazards of infection and toxicity to humans posed by all such biopreparations, but also to identify the range of possible risks arising from mutagenicity, carcinogenicity, teratogenicity, etc. *Inter alia*, it has furnished the opportunity for scientists of the GDR and Poland to elaborate (bearing in mind the opinions of scientists in other countries) "health requirements for the registration of new pesticides in CMEA countries and Yugoslavia."[51-60]

For the better understanding of the procedure for registering microbial insecticides in our region, it will be useful to describe the manner of fulfilling relevant requirements in the U.S.S.R., which uses the greatest variety and broadest nomenclature of bioinsecticides of all the socialist countries.

Questions relating to environmental quality and care are under the immediate responsibility and control of the Soviet government. Their chief legal foundation is Article 18 of the U.S.S.R. Constitution, which established a solid base for nature protection by enunciating clear principles for the management and scientific development of natural resources. All of our country's individual Soviet Republics have adopted laws providing for nature protection. The responsible government bodies have arranged for the preparation and publication of documents which are fundamental to the proper regulation of the application of control measures against harmful insects. A number of these documents envisage the protection of environmental quality by the provision of alternatives to chemical pesticidal pollution via microbial control products through appropriate biotechnology.

In this context, the U.S.S.R. maintains worldwide communications at both country and international levels. For example, Soviet scientists play active roles towards attainment of the goals of the international program, "Man and Biosphere", questions relating to the latter being dealt with by a special Scientific Council of the Supreme Council of the Academy of Science of the U.S.S.R. Our country has also played an active part in the United Nations Environment Programme (UNEP) and the International Organization on Biological Control (IOBC). The U.S.S.R. has been initiatory as well in developing and implementing a "Convention for forbidding military or any other harmful influence upon nature" which was adopted by the Decree of the Supreme Council of the U.S.S.R. on 16 May, 1978. Furthermore, the State Agroproductive Committee (Gosagroprom) is required to supervise the orderly conduct and correct application of chemical and biological control procedures in agriculture, while actual permission to use microbial insecticides calls for mutual decision making and necessary action on the part of a number of different offices and organizations.

Initially, applications for the production of entomopathogenic preparations are considered by the Commission on Microbiological Means for Plant Protection. This is established within the All-Union Agricultural Academy, which bears the name of V. I. Lenin and is part of Gosagroprom (see above). All applications presented to this Commission are considered under its "Rules for examination of new preparations and strains." The Commission evaluates both the merits of submitted products with respect to their efficacy against harmful insects, and the appropriateness of data submitted concerning their safety to humans and other NTOs.

Once experts appointed by the Commission on Microbiological Means for Plant Protection have reached their conclusions, the microorganism in question and/or a sample of a microbial control preparation based upon it is/are passed over to specialized research centers (accompanied of course by the pertinent documentation) for more detailed investigation of all aspects of its/their possible health and environmental impacts. These centers include:

1. The All-Union Institute of Hygienic and Toxic Aspects of Pesticides and Polymers, Ministry of Health of the U.S.S.R. (located in Kiev). This Institute is the chief one undertaking studies towards establishing the safety of candidate microbial control agents to the health of humans and other homeotherms, both at the stage of manufacture and in practical use. Experts there assign each test sample to its proper toxicity class as adopted in the U.S.S.R. State Standards for "Biological Security", and designate its maximum allowable concentration in the atmosphere of an industrial manufacturing plant. Their conclusions follow acute and chronic experiments on laboratory animals to determine the possible adverse effects of entomopathogens and/or microbial control products derived from them, under the various circumstances of exposure; i.e., via enteric, intranasal, aerial, and other routes (including skin lesions and the eyes). The searching investigations of these scientists also include relevant assessments of the likelihood of allergenicity, embryotoxicity, sensitizing, etc.

   The Institute of Epidemiology, Microbiology, Pathology, and Infectious Diseases (bearing the name of L. V. Gromashevsky, and located at Kiev), is specifically responsible for evaluations of entomopathogenic viruses in the above contexts.

2. The All-Union Scientific Research Institute of Biological Methods of Plant Protection (located in Kishinev) and the Belorussian Scientific Research Institute of Plant Protection (located in Minsk) are jointly responsible for efficacy tests of candidate entomopathogens and microbial control preparations against insects that are harmful as economic pests or disease vectors.

3. The Scientific Research Institute of Fish Culture (located in Azov) tests candidate entomopathogens and microbial control preparations against fish.

4. The Institute of Apiculture (located in Ryazan), Moscow's Veterinary Academy, and the Research Institute of Sericulture all participate in investigations of the possible

harm of candidate entomopathogens and microbial control preparations to bees and silkworms.

At the broader level, the findings of all these groups spread through several Republics of the U.S.S.R. are being channeled into our country's industry. The State Standards of the U.S.S.R. require the implementation of safety requirements in this context:

1. An admissible range of toxicity levels, in accordance with the ''Biological Security'' requirement of the above State Standards
2. An acceptable level of maximum concentration of any microbial insecticide, for field use
3. Full protection of all staff concerned
4. Exemption of personnel whose personal medical history has suggested the undesirability of their working closely with the microbial insecticides under consideration.

For any microbial control agent in course of assay under the above requirements, there must clearly be adequate supportive documentation assuring that the health of all concerned in both production and practical use of any microbial control agent will not be prejudiced. The listing by priority of all responsible bodies is assured under the appropriate State Standards of the U.S.S.R. This guarantees that not one aspect of either product development or subsequent practical application will endanger any of the workers involved in microbial control research and practice. The responsible bodies referred to include:

1. Those utilizing any particular biopreparation, any of the Gosagroprom bodies involved in field application
2. The supervisory authorities, those subordinate groups of the Board of Health Security of the particular U.S.S.R. Republic where the biopreparation in question is produced
3. All concerned with health and environmental safety aspects of transportation from a production site to the field area where a specific microbial control agent is to be applied.

Gosagroprom and the Board of Medical and Microbiological Industry are responsible for giving final approval to all resultant permits for the manufacture and use of microbial preparations. All such permits are time limited, their period of validity usually being 5 years. Moreover, they are subject to review and amendment as necessary in the best interests of health and environmental safety. As in the case of any other manufactured goods, the production of biopreparations must be precisely in accordance with the terms of the permit (as originally issued or officially amended), and this document must be valid. Any manufacturing enterprise infringing on these requirements would be subject to penalty under the law.

A State Commission now considers all data from the proceedings outlined above. Having ratified the earlier conclusions as to a product's health and environmental safety from the initial laboratory studies of the entomopathogen in question through the manufacturing stage to field testing, this body duly arranges for state testing via its branches and laboratories located in various regions of the U.S.S.R.[61]

During this state testing phase, experiments are conducted with each approved microbial insecticide to determine the optimal dosage and manner of application against each selected insect pest. One of several expert working groups (which consist of leading plant protection specialists on the staff of the State Commission) is permanently assigned to such evaluation of biopreparations.

Both of the Commissions mentioned above constitute something of a ''two-chambered legislature'', the appropriately specialized members of which have the duty of supervising every aspect of the development, manufacture, and practical application of microbial insecticides — from the first isolation of each promising strain through to regulatory measures ensuring the resultant biopreparations' safe and effective use against insect pests.

Results from the state testing program are analyzed by the State Commission's biopreparations working group, the reports from the meetings of which approve those microbial insecticides to be added to the "list of chemical and biological agents, and growth regulators, adopted for application against pest insects and plant diseases in agricultural and forestry economy." The "list" for 1986 to 1990 includes: insecticides, acaricides, and molluscicides (collectively totalling 101 products); fungicides (67); nematicides (12); rodenticides (chemical, 9 and biological, 2); Biopreparations (19 — these are discussed in detail later); pheromones (2); herbicides (184); defoliants (11); and plant growth regulators (23). The "list" also includes products adopted for use in greenhouses, on crops of pharmacological importance, and on stored agricultural products.[62,63] In addition, it includes instructions for optimal timing, number of applications, and maximum/minimum dosages, on a country-wide basis. These instructions must be obeyed by all regional scientific research institutes seeking to develop application methodologies specifically suited to the needs of their own areas and problems within the broad national requirements.

The Plenum of the State Commission confirms the above "list" in conjunction with the Ministry of Health of the U.S.S.R. and the State Committee of Hydrometeorology and Environmental Quality. This document thus gains full official status as the register of (among the many other products referred to above) all approved microbial insecticides, with the overall regulations for their safe and effective use. It is published in the journal, *Plant Protection*, and annually reviewed over the 5-year period of its validity. A comprehensive revision towards the end of its currency ensures that its successor will be fully up-to-date with respect to the results of state testing conducted since it was issued.

As already indicated, the "list" for 1986 to 1990 includes 19 biopreparations. Two further such products appearing therein, being within the category of rodenticides, fall outside our present mandate. So do 4 of the 19 biopreparations referred to, these having been developed against phytopathological problems. The "list" for 1986 to 1990 therefore includes 15 microbial control agents developed specifically against pest insects:

1.   Bacterial Preparations
      Bitoxybacillin
      Dendrobacillin (two variants)
      Lepidocid
      Entobacterin
      Gomelin
      BIP (two variants)
      Bactospein
      Dipel®
2.   Viral preparations (six products)
3.   Fungal preparations (Boverin)

The importation of foreign microbial insecticides and authorization of their application in the U.S.S.R., are preceded by expert evaluation and State testing. The authors of this chapter participated in the organization of such testing and the analysis of its results with respect to Bactospein and Dipel, after which both products were duly cleared for use in our country.

Similar procedures are followed in the U.S.S.R. with respect to microbial insecticides for use against insects of medical importance, as have now been detailed for agricultural and forestry entomology. The responsible body is a Commission of Regulation dealing with vector control matters within the Ministry of Health of the U.S.S.R. This Commission meets four times annually to consider applications from interested organizations desiring to develop appropriate biopreparations for practical use. On receiving expert assurance that a particular microbial insecticide is indeed safe as well as effective,[64] it grants permission for the development of a model for the necessary testing methodology (the participating organizations and research workers being approved by the Commission). After the testing has been conducted and its results evaluated, a further application is made to the Commission, this time for the manufacture of biopreparation batches in commercial quantity, and for permission to apply these in the field (the question of whether this will be done from the ground or by aerial spraying is settled separately). Experts from the Commission examine the proposed test procedures, participate in the preparation of documentation defining standards and methodology, and pay particular attention to matters of safety to humans and other homeotherms.[64]

The next step is for the Commission of Regulation to grant the required permission if their deliberations so justify, another office of the Ministry of Health of the U.S.S.R. now preparing the appropriate permit as an Order of the Board of Public Health. Each such Order includes specific requirements concerning application methodology, and following its promulgation the biopreparation in question is officially adopted for use and added to the list of insecticides (including microbial ones) approved for application in medical entomology. This nomenclatural list is a basic official document[65] published under the authority of the Chief Chemist's Office, Board of Public Health of the U.S.S.R., of equivalent standing to other such lists of approved preparations in the fields of plant protection and medical disinsection, and to the document defining the "period of secure care of plants after application of pesticides." A special passage in the list of insecticides approved for application in the field of medical entomology regulates all necessary safety measures for workers returning to cultivated fields following the use of entomopathogens and products derived from them.

Finally, it should be stated that a number of documents embody regulations on safety precautions to be followed in the handling and acceptable entry into the environment of insecticides, including microbial ones.[66-73] *

IV. CONCLUSIONS

While the procedure for registration of microbial insecticides outlined herein may not be altogether faultless, it has proved effective to date and seems likely to do so into the future because of its combination of dialectical flexibility and strictness. In the first place, its mobility and dynamism allow for the rapid correction of shortcomings in the approved methodologies and early incorporation of recent advances in the documentation supporting permits for individual biopreparations. In the second, safety is assured by precisely defined limits of tolerance across the spectrum from product development and manufacture to all aspects of field application.

It is nevertheless clear that we must continue to be vigilant in the present context, as in all other spheres of life, for the consequences of rapidly developing biotechnology have highlighted the second half of this century and in the future they will assuredly spread to the mainstream of industry. Microorganisms, which for centuries has been considered as rather distant from "real life" in terms of their positive aspects, have thus received an unexpected "historical chance". This has stimulated fears based on theory, that the artificial replication and/or reconstruction of microorganisms could lead to unforseeable adverse con-

---

* In Section II of this chapter it is mentioned that in the U.S. requirements for analysis of residues of microbial products based on *Bacillus thuringiensis* and its various subspecies have been abolished, a fact that bears witness to the accumulating evidence of their health and environmental safety.[12]

sequences to human and environmental health. In particular, genetically engineered organisms (beyond our mandate in this chapter) are viewed by some as potentially more dangerous than synthetic chemical pesticides; for the reasons that such entities, having harmful qualities, could multiply and spread in nature. This argument must be kept under consideration as pest control continues to develop.

However, scientists who are specializing in microbial pest control and who have satisfied themselves, on existing evidence, that this approach as currently practiced is altogether safe, clearly must not relax their guard on safety issues. The field of possible although quite unforeseen adverse impact of microbial metabolites on human and environmental health, must never be neglected. Objective consideration of this issue, which could become paramount in the working out of mankind's destiny, must remain central to the continuing development of safe and effective microbial measures for the suppression of our insect enemies.

Modern man being resident in environments oversaturated by pesticides, sometimes up to the danger point, has a wonderful opportunity to utilize microorganisms in an ecological perspective, for the better control of insect pests and vectors of disease. We must continue to further develop our field in expectation of its announcing itself aloud on the threshold of the third millenium of our era.

REFERENCES

1. **Bykov, B. A.,** To work in a new way, creatively, *Biotechnology,* 3 (9) 1, 1986 (Russian).
2. **Ogarkov, V. I.,** The basic biotechnological problems of the microbial industry, *Biotechnology,* 4 (10) 1, 1986 (Russian).
3. **Kadar, B.,** Achievements of biotechnology for the benefit of man, *Int. Agric. J.,* 4, 3, 1986 (Russian).
4. **Weiser, J.,** Microbial insecticides: present status and perspectives, *Inf. Bull. EPS, IOBS,* 6, 17, 1983.
5. **Kandybin, N. V. and Shekhurina, T. A.,** The technology of the use of microbial means of plant protection and measures for its improvement, *Inf. Bull. EPS, IOBS,* 6, 27, 1983.
6. **Kravtsov, A. A.,** *Preparations for Plant Protection at the Personal Level,* Rosselkhosisdat, Moscow, 1986 (Russian).
7. **Larsky, P. P., Dremova, V. P., and Brinkman, L. I.,** Medical disinsection, *Medicine,* 1985 (Russian).
8. **Weiser, J., Videnova, E., Kandybin, N. V., and Smirnov, O. V.,** The technical characteristics and standardization of entomopathogenic microbial preparations, *Inf. Bull. EPS IOBS,* 16, 44, 1986.
9. **Weiser, J., Kandybin, N. V., Barbashova, N. M., and Smirnov, O. V.,** Catalogue of entomopathogenic microorganisms deposited at the Warsaw establishment of the Eastern Palearctic section of IOBC, Prague, 1980.
10. **Anon.,** *Chemical and Biological Means of Plant Protection,* Reference book of Soviet normative-technical documents, international and foreign standards and bibliographical information, Moscow, 1985 (Russian).
11. **Saikina, N. N.,** The perspective for microbial preparations, *Plant Prot.,* 4, 32, 1987 (Russian).
12. **Heimpel, A. M.,** Safety of insect pathogens for man and vertebrates, in *Microbial Control of Insects and Mites,* Burges, H. D. and Hussey, N. W., Eds., Academic Press, New York, 1971, 469.
13. **Anon.,** Mammalian safety of microbial agents for vector control: a WHO memorandum, *Bull. W. H. O.,* 59, 857, 1981.
14. **Burges, H. D.,** Safety, safety testing and quality control of microbial pesticides, in *Microbial Control of Pests and Plant Diseases 1970—1980,* Burges, H. D., Ed., Academic Press, New York, 1981, 737.
15. **Mursa, V. I.,** The sanitary-hygienic characteristics of the bacterial preparations: Entobacterin, Dendrobacillin and Insectin, in *Arthropod Pathology and Biological Means of Control of Harmful Organisms,* Kiev, 1974, 110 (Russian).
16. **Mursa, V. I.,** Information substantiating the maximum permissible concentration of bacterial insecticides in the industrial atmosphere. Current health problems in the use of pesticides, in *Proc. 5th All-Union Sci. Conf.,* Kiev, 1975 (Russian).
17. **Melnikova, E. A.,** Health aspects of the microbial means of plant protection, in *8th Int. Congr. Plant Protection,* Moscow, 1975 (Russian).
18. **Melnikova, E. A. and Mursa, V. I.,** Studies of the safety of industrial strains of microorganisms and microbial insecticide preparations, *J. Hyg. Epidemiol. Microbiol. Immunol.,* 24, 425, 1980 (Russian).

19. **Melnikova, E. A. and Mursa, V. I.**, Toxicological and health aspects of microbial means of plant protection, in *The Supervision of Forest Pests and Diseases, and Improvement of the Measures for their Control, Proc. All-Union Sci.-Tech. Conf.,* Pushkino, Moscow, 1981, 136 (Russian).

20. **Guli, V. V., Lescova, A. J., Mursa, V. I., Sternshis, M. V., and Ivanov, G. M.**, The problems of safety of microbiological preparations with respect to the health of man and his environment, Melnikova, E. A., Ed., *Inf. Bull. EPS, IOBS,* 17, 19, 1986.

21. **Israjlet, L. I. and Kogaj, R. E.**, The health tolerances of Dendrobacillin in the atmosphere of industrial buildings and the ambient air, *Hyg. Sanit. (USSR),* 7, 32, 1978 (Russian).

22. **Karpov, E. G. and Chusnutdinova, F. I.**, On the survival ability of *Bacillus thuringiensis* in warm-blooded animals, in *Problems of Veterinary Arachno-Entomology,* Tümen, 1973, 13 (Russian).

23. **Baturin, V. V.**, The toxic effect on warm-blooded organisms is not registered, *Plant Prot.,* 10, 31, 1973 (Russian).

24. **Butko, M. P., Frolov, B. A., and Kasymjanova, G. S.**, The veterinary/sanitary evaluation of the meat of fowls fed preparations based on *Bacillus thuringiensis* in *Pap. of All-Union Res. Inst. Vet. San.,* 60, 137, 1978 (Russian).

25. **Baranovsky, B. I.**, Experimental interactions of the entomopathogenic baculoviruses with warm-blooded animals, in *Proc. 1st All-Union Conf., Microbial Methods of Plant Protection* Kishinev, 1976, 168 (Russian).

26. **Vasiljeva, V. L. and Gural, A. L.**, Present and possible future problems of the use of entomopathogenic viruses for plant protection, *Mol. Biol. (USSR),* 22, 7, 1979 (Russian).

27. **Vasiljeva, V. L., Gural, A. L., and Trusov, V. I.**, The sanitary-hygienic evaluation of the viral insecticide, Virin-ENS, *Pap. Latvian Agric. Acad. Jelgava,* 181, 9, 1980 (Russian).

28. **Israjlet, L. I., Drosdova, L. V., and Eglite, M. E.**, The health evaluation of Boverin and some perspectives of the use of microbial measures for plant protection, *Hyg. San.,* 11, 91, 1975 (Russian).

29. **Nesterenko, L. P. and Rudnichenko, V. F.**, Data on the fungal flora of the air at production sites for Boverin, *Hyg. Toxicol.* (Kiev), p. 23, 1967, (Russian).

30. **Talanov, G. A., Tonkonogenko, A. P., and Karavaeva, T. M.**, Investigation of the effect of the thermostable exotoxin of entomopathogenic bacteria on warm-blooded animals, *Prob. Vet. Health (Moscow),* 43, 170, 1972 (Russian).

31. **Baranova, V. P. and Kandybin, N. V.**, Studies of the toxic effect of Bitoxybacillin and exotoxin on rabbits, *Bull. All-Union Res. Inst. Agric. Microbiol.,* 17, 1, 1974 (Russian).

32. **Cantwell, G. E., Dougherty, E., and Cantelo, W. W.**, Activity of the β exotoxin of *Bacillus thuringiensis* var. *thuringiensis* against the Colorado potato beetle (Coleoptera: Chrysomelidae) and bacterial mutagenic response as determined by Ames test, *Environ. Entomol.,* 12(5), 1424, 1983.

33. **Carlberg, G., Hautala, J., Huhtinen, O., and Lehti, P.**, Bacterial fly control, in *British Crop Protection Conf., Pests and Diseases,* Vol. 1, The Lavenham Press Limited, Lavenham, Suffolk, 1984, 363.

34. **Guli, V. V., Talpalacky, P. L., and Rybina, S. U.**, The influence of the standard microbial insecticidal preparation on honey bees, *Bull. Siber. Res. Inst. Agric. Microbiol.,* 17, 1, 1974 (Russian).

35. **Klimpinja, A. E. and Kasanskaja, V. A.**, Surveillance of the viral preparation Virin-JM on bees, *Pap. of Latvian Agric. Acad., Jelgava,* 181, 43, 1980 (Russian).

36. **Anon.,** *The International Reference System on Sources of Information Concerning the Environment,* Moscow, 1982 (Russian).

37. **Anon.,** *Concerning the Protection of the Environment,* Political Book Publishing House, Moscow, 1979 (Russian).

38. **Anon.,** *Nature and Law,* Kajnar, Alma-Ata, 1984 (Russian).

39. **Sharicov, L. P.** (compiler), *The Protection of the Environment,* Shipbuilding, Leningrad, 1978 (Russian).

40. **Mitrushkin, K. P., Ed.,** *Reference Book on Nature Protection,* Forest Industry, Moscow, 1980 (Russian).

41. **Shahaev, V. G. and Sherbitsky, B. V.,** *Reference Book on Environmental Protection,* Budivelnik, Kiev, 1986 (Russian).

42. **Anon.,** *Socialism and Nature, the Scientific Foundations of the Socialist Approach to Utilization of Nature,* Misl, Moscow, 1982 (Russian).

43. **Madar, S.,** *CSSR Law and Management in the Field of Environmental Protection,* Progress, Moscow, 1981 (Russian).

44. **Bgesinsky, B.,** *The Legal Foundations of Environmental Protection,* Progress, 1979 (Russian).

45. **Stajnov, P.,** *The Legal Problems of Nature Protection,* Progress, 1974 (Russian).

46. **Anon.,** *Hungarian Legislation and Management in the Field of Environmental Protection,* Progress, Moscow, 1983 (Russian).

47. **Leontieva, A. V.,** *The Legal Foundations of the Cooperation of Socialist Countries in the Field of Environmental Protection,* Progress, Moscow, 1982 (Russian).

48. **Mikolash, J. and Pitterman, L.,** *The Management of Environmental Protection,* Progress, Moscow, 1983 (Russian).

49. **Churaev, I. A. and Kulakov, E. P.,** *The Organization of Plant Protection Services in Some Socialist Countries,* Ministry for Agriculture, Moscow, 1975 (Russian).

50. **Weiser, J.,** Registration of bioinsecticides in CSSR, manuscript, 1987.
51. **Aleksandrin, N.,** Scientific-technical cooperation of the member-countries of the CEA in agriculture and forestry, *Int. Agric. J.,* 6, 7, 1986 (Russian).
52. **Bogan, J. and Bena, M.,** Cooperation of member-countries of the CEA as regards information in the field of agriculture, *Int. Agric. J.,* 6, 65, 1986 (Russian).
53. **Rosanovskaja, N.,** International cooperation on health aspects of pesticide use, *Int. Agric. J.,* 1, 64, 1983 (Russian).
54. **Rostovoj, L.,** Scientific-technical cooperation in the field of health and toxicological aspects of pesticides, *Int. Agric. J.,* 1, 43, 1982 (Russian).
55. **Sidorenko, G. I., Litvinov, N. N., Rachmanina, N. A., and Vachkova, V. V.,** The cooperation of the member-countries of the CEA in the field of environmental health over the period 1981—1985, *Hyg. Sanit.,* 9, 36, 1986 (Russian).
56. **Sidorenko, G. I.,** International health cooperation, *Hyg. Sanit.,* 5, 4, 1982 (Russian).
57. **Nikonov, V.,** The cooperation of the USSR with other member-countries of the CEA in the agricultural-industrial area, *Int. Agric. J.,* 5, 2, 1986 (Russian).
58. **Litvinov, N. N., Rachmanina, N. A., Vershkovich, N. A., and Vashkova, V. V.,** The cooperation of the member-countries of the CEA in the problem, "Environmental Health": achievements and perspectives, *Int. Centre Sci.-Tech. Information, Natural Resources and Environment,* 11, 9, 1984 (Russian).
59. **Krygianovskaja, M.,** Current health problems concerning the use of plant protection measures, *Int. Agric. J.,* 4, 40, 1985 (Russian).
60. **Reichert, G.,** Environmental protection and the increase of economic effectiveness, *Int. Agric. J.,* 2, 3, 1987 (Russian).
61. **Anon.,** On the new staff of the State Commission (editorial note), *Plant Prot.,* 4, 39, 1987 (Russian).
62. **Anon.,** Measures for plant protection—defoliants, dessicators and growth regulators, *Plant Prot. Overseas Rev.,* 1, 30, 1987 (Russian).
63. **Anon.,** (Insecticidal) preparations adopted for use, *Plant Prot. Overseas Rev.,* 2, 50, 1987 (Russian).
64. **Kandybin, N. V. and Smirnov, O. V.,** The toxicological evaluation of Bactoculicid, in *Proc. 1st Repub. Sci. Conf., Insect Pathology and Biological Means of Controlling Harmful Organisms,* Kanev, 97, 1982 (Russian).
65. **Anon.,** The time-intervals for the safe treatment of plants after pesticide application, *Plant Prot. Overseas Rev.,* 4, 54, 1987 (Russian).
66. **Porchachevskaja, O. A.,** The means of protection of the individual in agriculture, *Plant Prot. Overseas Rev.,* 12, 60, 1987 (Russian).
67. **Bychovec, A. I. and Scurjat, A. F.,** Safety precautions in working with pesticides, in *Reference Book on the Protection of Agricultural Plants Against Pests, Diseases and Weeds,* Antonov, G. B. and Korol, I. T., Eds., Minsk, 1983 (Russian).
68. **Anon.,** Methodology for the identification of Gomelin in objects in the environment, using microbiological tests, in *Methods of Identification of Microoamounts of Pesticides in Food, Feed and Environment,* Klisenko, M. A., Ed., Kolos, Moscow, 1983, 251 (Russian).
69. **Anon.,** Methodology for the identification of β-exotoxin in the yield of agricultural crops, in *Methods of Identification of Microoamounts of Pesticides in Food, Feed and Enviroment,* Klisenko, M. A., Ed., Kolos, Moscow, 1983, 253 (Russian).
70. **Geniatulin, K. V.,** Problems of the biological contamination of the environment, and relevant control, *Hyg. Sanit.,* 1, 58, 1982 (Russian).
71. **Anon.,** *Instructions for Safety Measures During the Storage, Transportation and Application of Pesticides in Agriculture,* Agropromisdat, Moscow, 1985 (Russian).
72. **Vasiljeva, V. L. and Melnicova, E. A.,** Analytical methods to detect microbiological agents used in plant protection, in *Methods for the Identification of Microoamounts of Pesticides,* Klisenko, M. A., Ed., Medicine, Moscow, 1984 (Russian).
73. **Antonov, G. B. and Korol, I. T., Eds.,** *Reference Book on the Protection of Agricultural Plants Against Pests, Diseases and Weeds,* Minsk, 1983 (Russian).

Chapter 4

REGISTRATION REQUIREMENTS AND SAFETY CONSIDERATIONS FOR MICROBIAL PEST CONTROL AGENTS IN JAPAN

K. Aizawa

TABLE OF CONTENTS

I. INTRODUCTION

Three microbial insecticides have been registered in Japan; first, a *Bacillus moritai* preparation for the control of the housefly, *Musca domestica* (as a feed additive, by the Ministry of Agriculture and Forestry which became the Ministry of Agriculture, Forestry and Fisheries in 1969) and later as a medical insecticide (by the Ministry of Health and Welfare in 1973). Second, a cytoplasmic polyhedrosis virus preparation for the control of the pine caterpillar, *Dendrolimus spectabilis*, was registered by the Ministry of Agriculture and Forestry in 1974. And third, a native spore-killed *Bacillus thuringiensis* preparation was registered by the Ministry of Agriculture, Forestry and Fisheries in 1981; native and foreign living spore-crystal mixture preparations were registered by this Ministry in 1982. This chapter will review safety aspects of these microbial insecticides with respect to their registration in Japan.

II. AGRICULTURAL CHEMICALS REGULATION LAW IN JAPAN*

The purpose of this law, its definitions of agricultural chemicals and "natural enemies" (including parasites, predators, and entomopathogenic microorganisms) and the registration of agricultural chemicals by manufacturer and importer, are described. Quotations of certain items from this law of particular relevance to this review follow.

A. PURPOSE
Article 1 — The purpose of this law is to assist stabilization of agricultural production and protection of human health, and to contribute towards conservation of the human environment, by normalizing the quality of agricultural chemicals and ensuring their safe and due use by establishing a registration system and exercising control over their sale and application.

B. DEFINITION
Article 1-2 — (1) The term "agricultural chemicals" as employed in this law shall mean fungicides, insecticides, and other chemicals (including those prescribed by the Cabinet Order among the products using agricultural chemicals as raw material or material for said control) used for control of fungi, nematodes, mites, insects, rodents, or other animals and plants, or viruses which are injurious to crops (hereinafter referred to as "diseases and insect pests"); also growth accelerators, germination depressors, and other chemicals used for the promotion or deceleration of physiological functions of the treated crops (including trees and agricultural and forestry products, hereinafter collectively referred to as "crops"). (2) In the application of this law, natural enemies used for control as prescribed in the preceding paragraph shall be deemed agricultural chemicals.

C. REGISTRATION BY MANUFACTURER AND IMPORTER OF AGRICULTURAL CHEMICALS
Article 2 — Application for registration as defined above shall be made by submitting an application in which the following items are presented; a document stating the test results concerning effectiveness, phytotoxicity, toxicity, and persistence of each agricultural chemical, accompanied by a sample of that agricultural chemical.

The Plant Protection Division, Ministry of Agriculture, Forestry and Fisheries issued

---

* Translated by the Society of Agricultural Chemical Industry (Tokyo), and improved by the editors of this book, but being no more than a compilation of English translations, this does not necessarily represent an exact and officially authorized English version.

"Guidelines on toxicology study data for the application of agricultural chemical registration" on January 28, 1985. Since April 1, 1985, it has been mandatory to submit toxicological study data on agricultural chemicals in accordance with the "Requirements for safety evaluation of agricultural chemicals." These basic data for safety evaluation are as follows: (1) acute oral toxicity; (2) acute dermal toxicity; (3) acute inhalation toxicity; (4) primary eye irritation; (5) primary dermal irritation; (6) dermal sensitization; (7) acute delayed neurotoxicity; (8) subchronic oral toxicity; (9) subchronic dermal toxicity; (10) subchronic inhalation toxicity; (11) subchronic neurotoxicity; (12) teratogenicity; (13) mutagenicity; and (14) pharmacological study data. Toxicological study data required for safety evaluation of persisting residues, are as follows: (1) acute oral toxicity, (2) acute dermal toxicity, (3) subchronic toxicity, (4) chronic toxicity, (5) oncogenicity, (6) reproduction, (7) teratogenicity, (8) mutagenicity, (9) metabolism, and (10) pharmacological study data.

Registration of the three microbial insecticides mentioned in the Introduction were applied for prior to 1973. At that time, requirements for safety data were less strict than they are today; and the consideration of applications (in which emphasis was placed on the fact that microbial control agents are naturally occurring) was on a case-by-case basis.

III. *BACILLUS MORITAI* PREPARATION FOR THE CONTROL OF *MUSCA DOMESTICA*

Bacillus moritai Aizawa and Fujiyoshi which is effective for the control of the housefly was isolated in 1962 from soil. *B. moritai* shows pathogenicity against the glossy black muscoid fly, *Ophyra leucostoma*, and a greenbottle fly, *Lucilia* sp. as well. However, it is not pathogenic to the silkworm, *Bombyx mori* or to the honeybee, *Apis mellifera*.[1,2]

For the registration of *B. moritai* preparation for the control of the housefly as a biocide against insects of public health importance, safety data (particularly subchronic toxicity and dispersal of bacterial cells into organs) were requested by the Ministry of Health and Welfare. The following safety tests were duly undertaken.

A. TESTS WITH MICE[2]

Acute oral toxicity — *B. moritai* spore suspension (1.7×10^5, 1.7×10^6, and 1.7×10^7 spores per mouse, five mice per each concentration) was admininstered by stomach probe. No death, no adverse effects, and no abnormal body weight gain were observed. *B. moritai* spores were administered with diet and spore suspension for 7 d and mice were reared with normal diet for 1 month further. Three levels of total spores administered per mouse for 7 d were as follows: 1.8×10^6 spores by diet + 3.5×10^7 spores by water, 1.3×10^7 spores by diet + 2.0×10^8 spores by water, 1.5×10^8 spores by diet + 2.6×10^9 spores by water (five mice were used for each level). Neither death nor any abnormal body weight gain took place.

Application of spores on wounded skin — Three hundred milligrams of paste containing 1.0×10^8 spores were applied on wounded skin of mice. The skin injury healed completely.

Intracutaneous injection — Spore suspension (1.1×10^7 spores per mouse) was injected intracutaneously to five mice, however, no sign of adverse effects was observed.

Intraperitoneal (i.p.) injection — Spore suspension (1.9×10^6, 2.0×10^7, 2.4×10^7, 6.0×10^7, and 1.8×10^8 spores per mouse) was inoculated i.p. Mortality after 7 d was 0/8, 2/8, 4/5, 5/5, and 11/11, respectively. In another experiment, acute toxicity by oral administration of spores was shown ($LD_{50} > 4.3 \times 10^{10}$ spores per kg body weight) but LD_{50} value by intraperitoneal injection was 1.2×10^9 spores per kg body weight, which is comparable to that of *B. thuringiensis*.[3]

Inhalation — Mice were exposed to spore powder (1.8×10^{11} spores per g; 242 to

263 g/m³) for 2 h or 1 h/d for 7 d (ten mice were used for each test). No death and no adverse effects were observed. *B. moritai* cells in liver, kidney, spleen, and heart disappeared 1 week after the termination of inhalation, but bacterial cells in lung decreased gradually.

Subchronic toxicity test —

Pathogenicity — B. moritai spores (5.0×10^8, 1.0×10^9, and 5.0×10^9 spores per g diet) were administered for 7 months. Twenty mice were used for each spore concentration and data on body weight, body temperature, and diet consumed were recorded every week. Five mice after 3 months' feeding, and ten mice after 7 months' feeding, were sacrificed to ascertain the weight of their organs (liver, spleen, kidney, heart, and lung), to make bacterial cell counts and for histopathological observations. The remaining mice were sacrificed for the same tests as mentioned above after 7 months' feeding of spores with additional 2 weeks' rearing with normal diet. There was no difference between the control and the treatments. However, the number of mouse deaths over 3 to 7 months was one in the test with 5.0×10^9 spores per g diet, two in control 1 (no treatment), and one in control 2 (diet added with culture medium), respectively. The death of mice was not considered to be caused by the feeding of *B. moritai* spores. Similar results were obtained in another experiment.[4]

Disappearance of bacterial cells from mouse organs and blood — In the bioassay of *B. moritai* spores at concentrations of 5.0×10^8, 1.0×10^9, and 5.0×10^9 spores per g diet, numbers of bacterial cells per organ in liver, spleen, kidney, heart, and lung of mice at 7 months' feeding were lower (0 to 250) than those at 3 months' feeding (0 to 3400). However, bacterial cells in organs disappeared 2 weeks after the termination at 7 months' feeding except for one case (19 bacterial cells in liver at a concentration of 5.0×10^9 spores per g diet). Colony numbers per milliliter in blood of mice fed for 3 months, 7 months, and 2 weeks with normal diet after 7 months of experimentation were 0 to 1.5. Similar results were obtained from another experiment.[4]

Maebashi[3] reported the disappearance of *B. moritai* cells in blood after i.p. and i.v. injections of spores and after three serial i.v. injections.

Disappearance of *B. moritai* cells in organs and in blood was further tested every month for 12 months' feeding of spores to mice (5.0×10^9 spores per g diet). During the first month, dispersal of bacterial cells into organs was observed but cell number in organs began to decrease gradually after 2 to 3 months and the decrease was clearly observed after 7 to 8 months, except for lung. Disappearance of bacterial cells was observed after 12 months.

B. TESTS WITH RABBITS[3]

Acute oral toxicity — Rabbits were administered 2.0×10^8 or 1.5×10^9 spores per g diet daily for 10 d. Five rabbits were used for each dietary level and for the control. No adverse effects were observed in body temperature, hematologic examinations, and body weight gains. No bacteria were found in the blood.

Subchronic test — Rabbits were administered *B. moritai* spores for 100 d under the same experimental condition as the acute oral toxicity test. During the test, one rabbit fed 1.5×10^9 spores per g diet showed weight loss, hyperthermia, lower values in the hematologic examinations, and alveolar septitis in lung at the end of the test. This rabbit was excluded from the data in some of the final results because the sickness was not considered to be a result of the feeding of spores.

No significant difference in feed consumption or body weight gains was observed in any group including the control. Rectal temperatures recorded weekly were normal. In the early stage of the test, some values in hematologic examinations at a level of 2.0×10^8 spores per g diet and in the control were in the lower ranges. Later, however, normal values were obtained in gravity of whole blood, gravity of plasma, hemoglobin concentration, hematocrit, red blood cell count, and white blood cell count.

No significant difference was observed in the organ-body ratios in liver, kidney, spleen, heart, and testes between the tests and the control. Bacteria were detected in the blood 1 month before the termination of the test and at the end of the test, however, none of the few colonies formed could be identified as *B. moritai*.

C. ALLERGY AND ANAPHYLAXIS TESTS WITH GUINEA PIGS[2]

Guinea pigs were sensitized by ten intracutaneous injections of spore suspension but the induction of allergy was not remarkable and the same degree of response was shown by *Bacillus subtilis* (= *Bacillus natto*) which is used for "natto" fermentation (soybean fermented with *B. subtilis*). Allergy was not induced in guinea pigs which were fed with a diet containing 1.0×10^9 spores per g for 1 week, or those exposed to spore powder by inhalation (2.3×10^{11} spores per g, 15 g/8 1, 15 min/d for 7 d). Anaphylaxis was not induced in either the test using *B. moritai* spores or the control using *B. subtilis* spores.

D. SAFETY TEST WITH LAYING HENS[2]

Spore suspension (1.0×10^9 spores per ml) was administered to laying hens (4 ml/d) for 1 month. No adverse effects were observed and organs were normal. Bacterial infection of eggs was not observed.

Laying hens (11 months old) were fed with spores (1.1×10^7 and 2.2×10^7 spores per g feed, 40 laying hens per spore concentration) for 4 months. No difference was observed in feed consumed, egg production, and egg weight between the tests and the control. Hens (4 weeks old) were fed with spores (5.0×10^7 and 2.5×10^8 spores per g feed, 20 hens per spore concentration) for 6 months. No difference was observed in growth, feed consumed, and egg productivity between the tests and the control.

E. SAFETY TEST WITH PIGS[2]

Three pigs were fed with spores (2.4×10^7 spores per g feed) for 140 d. No adverse effects were observed and average body weight gain and feed consumed were normal.

F. SAFETY TEST WITH CATTLE[2]

One Holstein cow was fed with spores (2.0×10^8 spores per g feed) for 3 months. Excretion of bacterial cells into feces and urine, transfer of bacterial cells into blood, and quality test of milk were examined. The number of bacterial cells excreted into feces was 1.7×10^7 spores per g feces during the feeding of spores but decreased to 0.01% 3 d after the termination of the experiment. Bacterial cell number in the blood was 1.5, 8.8, and 1.5 cells per ml after 1, 2, and 3 months, respectively. The blood picture was normal after 3 months' feeding.

G. SAFETY TEST WITH FISH[2]

Ten carp were reared in water containing spores (5.0×10^6 spores per ml) for 48 h but no abnormal signs were observed after 72 h.

H. SAFETY TEST WITH BIRDS[2]

Four canaries were fed with spores (4.0×10^7 to 2.5×10^8 spores per canary, four feedings at each concentration, one canary per concentration) and observed for 2 months. No adverse effects were observed.

I. HUMAN FEEDING TEST[5]

This test was conducted by four male volunteers, aged 31 to 48. Three were placed in the experimental group and one in the control. One gram of *B. moritai* spores (1.7×10^9 spores) in a gelatin capsule was given to 3 volunteers before every breakfast for 5 d. The same amount of autoclaved spores was ingested by the control.

Physical examination — There was no sign of any adverse effects due to ingestion of *B. moritai* spores.

Laboratory examinations — Blood analysis (erythrocytes, hemoglobin, color index, leukocytes, hematocrit, and plasma protein), urinalysis (protein, glucose, urobilinogen, and urobilin), liver function (icterus index, bilirubin, zinc sulfate turbidity, thymol turbidity, total cholesterol, GOT, GPT, LDH, and ALP), and renal function (BUN, Na, Cl, K, uric acid, and PSP) tests were all normal and showed no sign of bacterial infection.

Spores excreted in the feces — Excretion of spores began the day after ingestion of spores. The following 4 d the spore numbers were 1.6 to 1.7×10^9/d. After the termination of ingestion there was a rapid reduction in the excretion of spores and no *B. moritai* cells were detected on the seventh day. The total recovery of ingested spores in the feces for 7 d after the final ingestion was 96%, suggesting the complete excretion of spores.

IV. CYTOPLASMIC POLYHEDROSIS VIRUS PREPARATION FOR THE CONTROL OF THE PINE CATERPILLAR, *DENDROLIMUS SPECTABILIS*

A cytoplasmic polyhedrosis virus of the pine caterpillar, *Dendrolimus spectabilis,* was found to be effective for the control of this pest in forests.[6] A commercial preparation was registered in Japan in 1974. The following safety tests were conducted.

A. INJECTION OF CYTOPLASMIC POLYHEDRA OR VIRUS PARTICLES INTO TEST ANIMALS

Cytoplasmic polyhedra or virus particles were injected into mice, rats, hamsters, and rabbits by i.v., i.p., and intracerebral injections, but no pathogenicity was observed.[7] According to Katagiri,[6] no pathogenicity resulted from serial passages of the virus through mice and chick embryos.

B. ACUTE AND SUBACUTE TOXICITIES

Rats and mice were fed with more than 3.3×10^{10} polyhedra per kg body weight for 7 d. No clinical signs and pathological changes were observed. Rats and mice were inoculated with cytoplasmic polyhedra by stomach probe for 90 d at concentrations of 1.8×10^{10}, 9.0×10^9, 4.5×10^9, and 2.3×10^9 kg per body weight. Growth was normal except for slight suppression at a concentration beginning with 4.5×10^9 polyhedra per kg. There was no abnormal weight increase or pathological abnormalities of internal organs, except for spleen and ovary of female rats which were slightly enlarged. However, there was some blood congestion in the intestines and kidney in treated mice and a small amount of leukocyte infiltration into the stomach of treated rats.[6]

C. TOXICITY TO FISH

No carp or killifish died after being held in water to which polyhedra had been added (10^3, 10^4, 10^5, and 10^6 polyhedra per ml for 72 h).[6]

V. *BACILLUS THURINGIENSIS* PREPARATIONS

A bacterium was isolated by Berliner in 1911 from diseased larvae of the Mediterranean flour moth, *Ephestia kühniella,* and it was named *Bacillus thuringiensis* in 1915.[8,9] However, *B. thuringiensis* had already been discovered by Ishiwata in 1901 in diseased larvae (severe flacherie) of the silkworm, *Bombyx mori,* in Japan. Ishiwata[10] named this sporeformer "sotto bacillus" (sotto, meaning fainting or sudden collapse in Japanese). It is now known as *B. thuringiensis* subsp. *sotto* (serotype 4a:4b).

When *B. thuringiensis* preparations were first developed in the U.S. and France, they were imported into Japan. However, owing to their toxicity to the silkworm, the importation of *B. thuringiensis* strains isolated outside Japan and preparations derived from them have been forbidden by plant quarantine, except for experimental uses approved by the Ministry of Agriculture and Forestry (since renamed Ministry of Agriculture, Forestry and Fisheries).

Due to fears at that time that *B. thuringiensis* might cause outbreaks of diseases prejudical to silkworm farming, it became prerequisite to the practical use of these microbial insecticides in Japan that rigorous investigations be undertaken to elucidate their possible adverse effects on *Bombyx mori* and sericulture generally. In this context, the following subjects were investigated:

1. Susceptibility of silkworm races against *Bacillus thuringiensis* strains and preparations.
2. Rearing of larval *Bombyx mori* on mulberry leaves sprayed with sublethal doses of *B. thuringiensis*.
3. Isolation of this microorganism from sericultural farm litter.
4. Relationship between the incidence of silkworm diseases and the existence of spores of *B. thuringiensis* on sericultural farms.
5. Distribution of serotypes of this microorganism.
6. Multiplication of the latter in dead *Bombyx mori* larvae.
7. Multiplication of *B. thuringiensis* in soils.
8. Treatment (with formalin and other disinfectants) of equipment and rearing facilities used in sericulture.
9. Elucidation of dangers posed by drift of relevant sprays to mulberry plantations.
10. Selection of a *B. thuringiensis* strain of low toxicity to the silkworm for utilization against pest insects in Japan.

Investigations as outlined above duly proceeded for 10 years. Their results permitted the 1971 exemption from Japanese plant quarantine, of foreign *B. thuringiensis* strains and preparations. Due provision for registration of the latter in Japan had now to be made. Accordingly, in 1972 the Japan Plant Protection Association established the Study Committee on *Bacillus thuringiensis* Preparations, its five sections each dealing with a topic from the following: relevant fundamental research, the effectiveness and safety of these microbial insecticides, and sericultural and apicultural aspects of their practical employment.

The committee investigated the efficacy and safety of *B. thuringiensis* preparations and their effects on environments, particularly sericultural ones. If carefully applied, *B. thuringiensis* preparations can be disseminated in sericultural environments without harming silkworms. The incidence of silkworm diseases is possible to be determined whether or not this had been caused by the dissemination of *B. thuringiensis* preparations. The committee also explored the means of evaluating the potency of *B. thuringiensis* preparations, and it adopted the following procedures for the regulation of the potency of the latter.[11,12]

Larvae of the silkworm (*Bombyx mori,* race Shunrei × race Shogetsu) are reared on an artificial diet at 24 to 25°C. Dilution of *B. thuringiensis* preparations is usually from $^4/_3$ to twofold, 0.5 ml of the diluted sample and 10 g of artificial diet being mixed in a Petri dish. Thirty larvae (ten larvae × three Petri dishes) in the second day of the third instar, are used for each dilution of the sample. Larvae are reared for 3 d on a diet incorporated with the sample, and for an additional 2 d on a normal diet. Mortalities are recorded. If more than 2 out of 30 control larvae die, the results are discarded.

A "self-standard" is prepared by each formulator for deposition with the Agricultural Chemicals Inspection Station, Ministry of Agriculture, Forestry and Fisheries. The potency of each self-standard has 1000 BmU/mg (*Bombyx mori* units), irrespective of toxicity to the silkworm. This potency is calculated by the following formula:

$$\text{Potency of sample} = \frac{\text{LC}_{50} \ (\mu\text{g/ml}) \ \text{self-standard}}{\text{LC}_{50} \ (\mu\text{g/ml}) \ \text{sample}} \times 1000 \ \text{BmU/mg}$$

The potency of commercial preparations is required to remain between 85 and twice the 100% listing of the indicated BmU, when the samples are bioassayed by the Agricultural Chemicals Inspection Station. Each self-standard must be expressed as IU (international units) per milligram; if necessary, using E-61 (international standard) or U.S. standard (HD-1-S-1971).

This procedure is used only for the regulation of the indicated potency of preparations. The practical amount of the preparation for the control of the target insects is to be determined by separate tests.

When application for the registration of native and foreign *B. thuringiensis* preparations was made to the Ministry of Agriculture and Forestry before 1973, published data and reliable information from outside Japan were submitted for consideration. Data on acute oral toxicity, acute i.p. toxicity, acute dermal toxicity, acute inhalation toxicity, and primary eye irritation toxicity were all reviewed.

Most of the pertinent data by Japanese manufacturers have yet to be published. It is, nevertheless, possible to estimate, from such information as is available, the average numbers of *B. thuringiensis* cells in organs of mice after 3 and 7 months' feeding of the microorganism's spores at 1.0×10^9 spores per g diet; and at 2 weeks after the termination of 7 months' such feeding.

This information reveals that the relevant cell numbers per organ after a 3-month feeding program were liver (670), spleen (360), kidney (1100), and heart (75). A 7-month feeding program produced sequential figures of 16, 0, 17, and 0. Moreover, *B. thuringiensis* cell numbers per ml of blood were 0 after 3 months' feeding, 1.1 after 7 months' feeding, and 1.3 2 weeks after the completion of 7 months' feeding.[2] Similar results pointing to the nonpathogenicity to mice of *B. thuringiensis* after 7 months' feeding have been reported from another experiment.[4]

Environmental impact of *B. thuringiensis* has been reviewed in the following connections: (1) consequences to NTO plants and animals, (2) microbial ecology of the agent in soils and water, (3) aerial drift experiments with *B. thuringiensis* preparations, and (4) the application of the latter to sericultural environments.

Since some *B. thuringiensis* strains cause fluid accumulation in rabbit ileal loops,[13] enterotoxin production in those strains actually used in the development of commercial preparations was examined. Therefore, it took close to a decade for the registration of *B. thuringiensis*-based products in Japan. This was, of course, a special case, weighted by Japan's economic interest in sericulture.

VI. SUGGESTIONS FOR THE FUTURE

While the following suggestions concern Japanese internal issues, they are presented nonetheless because of our conviction that the dissemination and further development of biocontrol and microbial insecticides depend at least as much upon international cooperation as upon national circumstances.

The registration of microbial insecticides for use against insect pests of agricultural and forestry importance in Japan has been the subject of close scrutiny by the Ministry of Agriculture, Forestry and Fisheries (Agricultural Chemicals Inspection Station and Plant Protection Division, Ministry of Agriculture, Forestry and Fisheries) — in close cooperation with the Ministry of Health and Welfare and the Environmental Agency.

Specifically, the registration of a microbial insecticide for the control of medically important insects — *Bacillus thuringiensis* ssp. *israelensis* — has been under the responsibility of our Ministry of Health and Welfare.

Should applications for registration of microbial insecticides increase, it seems inevitable that a special administrative organization for the registration of *all* biocontrol preparations (clearly including the microbial insecticides) will have to be set up. Since the mode of action of all biocontrols is entirely different from that of chemical pesticides, their registration (and that of related innovative agents such as the insect growth regulators [IGRs]) must obviously merit attention from an appropriately conceived new organization charged with relevant responsibility.

Finally, with respect to the field application of genetically engineered microorganisms, this has been discussed officially in Japan and, of course, this matter should be investigated too, by a new organization as is ventured in the preceding paragraph.

REFERENCES

1. **Aizawa, K. and Fujiyoshi, N.,** Selection and breeding of bacteria for control of insect pests in the sericultural countries, in *Proc. Joint U.S.-Japan Seminar on Microbial Contr. Insect Pests* , U.S.-Japan Committee Scientific Cooperation, 79, 1968.
2. **Fujiyoshi, N.,** Studies on the utilization of sporeforming bacteria for the control of houseflies and mosquitoes, Research Report Seibu Chemical Industry Co., Special Issue, 1, 1, 1973.
3. **Maebashi, H.,** On the pathogenicity to mammals of the new bacterial insecticide — *Bacillus moritai, Jpn. J. Hyg.,* 27, 267, 1972.
4. **Maebashi, H. and Fujiyoshi, N.,** Transient bacillemia by bioinsecticides in chronic feeding tests on mice *Bacillus moritai* Series No. 2.) *Jpn. J. Hyg.,* 27, 342, 1972.
5. **Aizawa, K., Kawamura, A., Fujiyoshi, N., and Maebashi, H.,** Human feeding test using the bacterial insecticide — *Bacillus moritai (Bacillus moritai* Series No. 3), *Jpn. J. Hyg.,* 29, 275, 1974.
6. **Katagiri, K.,** Pest control by cytoplasmic polyhedrosis viruses, in *Microbial Control of Pests and Plant Diseases,* Burges, H. D., Ed., Academic Press, New York, 1981, 433.
7. **Tanaka, N., Naiki, M., and Mitsui, T.,** Toxicity of a cytoplasmic polyhedrosis virus of the pine caterpillar, *Dendrolimus spectabilis* against warm-blooded animals. I. Pathogenicity of purified virus, *Abstr. Annu. Meet. Jpn. Soc. Appl. Entomol. Zool.,* 36, 1967.
8. **Berliner, E.,** Über die Schlaffsucht der Mehlmottenraupe, *Z. Gesamte Getreidewes.,* 3, 63, 1911.
9. **Berliner, E.,** Über die Schlaffsucht der Mehlmottenraupe *Ephestia kühniella* Zell.) und ihren Erreger *Bacillus thuringiensis* n.sp., *Z. Angew. Entomol.,* 2, 29, 1915.
10. **Ishiwata, S.,** On a kind of severe flacherie (sotto disease). I, *Dianihon Sanshi Kaiho,* 114, 1, 1901.
11. **Aizawa, K.,** Recent development in the production and utilization of microbial insecticides in Japan, in *Proc. 1st Int. Colloq. Invertebr. Pathol.; 9th Annu. Meet. Soc. Invertebrate Pathology,* Queen's University, Kingston, 1976, 59.
12. **Aizawa, K.,** Microbial control of insect pests, in *Advances in Agricultural Microbiology,* Subba Rao, N.S., Ed., Oxford and IBH Publishing, New Delhi, 1982, 394.
13. **Spira, W. H. and Goepfert, J. M.,** *Bacillus cereus*-induced fluid accumulation in rabbit ileal loops, *Appl. Microbiol.,* 24, 341, 1972.

II. Ecological Overview: Potential Impact of Microbial Insecticides

Chapter 5

POTENTIAL IMPACT OF MICROBIAL INSECTICIDES ON BENEFICIAL ARTHROPODS IN THE TERRESTRIAL ENVIRONMENT

S. B. Vinson

TABLE OF CONTENTS

I. INTRODUCTION

Biological control of insect pests has had a long history that includes the use of pathogens. With the advent of synthetic insecticides the use of pathogens and other biocontrol agents failed to keep pace with the effective and inexpensive chemical approaches available. The development of resistance by important pest species, increasing cost of new compounds, and increased concern for the environmental consequences of chemical pesticide use has encouraged research and investment in biological control. Although the use of pathogens for pest control still forms a small part of the pest control industry,[1] their use is increasing and many opportunities for pathogen improvement exist.[2] However, there are misgivings about the safety of microbial control agents, particularly for the nontarget invertebrates (NTOs).[3] Of particular concern is the potential adverse impact of microbial use on the beneficial arthropod community.

The possible effects that insect pathogen use might have on the beneficial insect community are complex and have received only moderate attention. Pathogens not only result in a reduction of host or prey resources, but may also directly infect beneficials resulting in death. Also, the disease may reduce the fitness of the host or prey as a nutritional resource. Yet, the complexity goes beyond a reduction in the fitness of the beneficial insect population. Some parasitoids and predators can mechanically disseminate pathogens among their host or prey population and some parasitoids may even transovarially transmit certain pathogens. Thus, beneficials may play an important role in the dissemination of pathogens to distant populations in both space and time. However, the dissemination of pathogens often has a profound negative effect on populations of beneficials involved.

Some of the early work indicating that pathogens might disadvantage beneficial insects was the result of efforts to rear beneficials for use in biocontrol programs. Paillot[4] reported that microorganisms, which presumably gained entry into the host via parasitoid ovipositional wounds, resulted in parasitoid losses in his laboratory colony. King and Atkinson[5] reported that (*Gonia*) = *Salmacia* sp., a tachinid parasitoid that attacks the larvae of the red-backed cutworm, *Euxoa ochrogaster,* reduced the host's vitality leading to septicemia which led to the parasitoid's death. The stress of parasitism presumably increases the susceptibility of the host to infection in a similar way to that reported for physical stress.[6] The spread of pathogens within parasitoid cultures led a number of researchers[7-10] to the realization that parasitoids and pathogens may not be compatible.

Much of the concern and research has focused on the effects of insect pathogens on the parasitic Hymenoptera. Whether or not this is due to the particular worries of early researchers, the fact that the parasitic Hymenoptera are most often laboratory reared (their susceptibility to diseases thus being more often encountered), or to these entomophaga having an unusually high disease risk is not clear. Other NTOs may also be at risk during the use of disease organisms for control. These NTOs include scavengers, species important as food, and presumably a large group of rarer insects. With the exception of the honeybee, pollinators have also been neglected. As regards predators, only a select few have been studied.

I shall sketch the complex interaction between beneficial insects (i.e., parasitoids, predators, and pollinators) and the major pathogens of insects (bacteria, protozoa, fungi, nematodes, and viruses). There are a number of approaches to this literature, but the interrelationships between all types of pathogens, parasitoids, and predators will first be considered, followed by an examination of the effects of pathogens from the standpoint of beneficial insects including pollinators.

II. DISEASE-CAUSING AGENTS AND THEIR INTERACTIONS WITH PREDATORS AND PARASITOIDS

A. BACTERIA

Two of the earliest microbial agents to be commercially produced for insect control were *Bacillus popilliae* Dutky and *Bacillus thuringiensis* Berliner which have been used for over 40 and just over 30 years, respectively. Much of the research regarding beneficials and bacteria has involved *B. thuringiensis* in one particular commercial product, Dipel® (Abbott Laboratories, Chicago, IL), Hamel[11] examined the effects of a Dipel WP formulation on both the western spruce budworm, *Choristoneura occidentalis*, the spruce coneworm, *Dioryctria reniculelloides*, and their parasitoids. After an application of Dipel, parasitism of first instar budworms by *Cotesia fumiferanae* and *Glypta fumiferanae* increased while parasitism of late instars and pupae by species such as *Phaeogenes harrolus*, *Ceromasia auricaudata*, and *Madremyia saundersii* decreased.[11] This decrease in parasitism of late instar larvae and pupae was presumed to be due to the pathogen killing the host which reduced host availability, or by selectively killing the parasitized host.

Hamel[11] speculated that the increased parasitism by (*Apanteles*) = *Cotesia fumiferanae* and *Glypta fumiferanae* following Dipel application was due to the parasitoid's early development in emergent second instar hosts and the altered behavior of parasitized insects that reduced Dipel exposure. Similarly, gypsy moth, *Lymantria dispar* larvae are also more heavily parasitized by (*Apanteles*) = *Cotesia melanoscelus* after exposure to Dipel.[12,13] Such results were supported by Wollam and Yendol[14] who indicated that Dipel and *C. melanoscelus* synergistically reduced a gypsy moth larval infestation. This synergism was due to an increase in the absolute number of second generation parasitoids[14] in contrast to the effect of *B. thuringiensis* on *C. fumiferanae*[11] in which only an increase in the percent parasitism of the western spruce budworm was observed. An explanation for the synergism[14] was provided by Weseloh and Andreadis[15] who found that while *C. melanoscelus* females did not distinguish between *B. thuringiensis*-exposed or healthy hosts 3 d after exposure, they showed preference for 10-d-old survivors. The latter were delayed in development and were available for parasitism[16] for a longer period. Since sublethal doses of *B. thuringiensis* were found to extend the gypsy moth developmental time, thus increasing parasitoid success, the effects of *B. thuringiensis* on the establishment of similar exotic species of gypsy moth parasitoids was undertaken. *Rogas lymantriae* was chosen because *R. lymantriae*, like *C. melanoscelus*, prefers first through third instars.[17] Although the number of adult parasitoids increased when reared from *B. thuringiensis*-treated caterpillars, the smaller host size skewed the sex ratio towards a male bias.[17]

Although some benefits accrued, exposure of *C. melanoscelus*-parasitized gypsy moth larvae to *B. thuringiensis* caused increased host mortality, and thus parasitoid mortality. *B. thuringiensis* exposure also caused a 3-d delay in the development time of the surviving *C. melanoscelus*.[18] Further, decreased emergence and reproductive potential of parasitoids that egress from *B. thuringiensis*-exposed hosts appears common.[19,20] Parasitized hosts also appear to be particularly vulnerable to *B. thuringiensis*.[21,22] Although Dipel use is not compatible with parasitic Hymenoptera, the preparation does not directly infect larvae of the latter but kills the host, resulting in death of the developing parasitoid.[23] While mortality of adult parasitoids fed Dipel has been reported,[24-26] others found that low doses of Dipel fed to adults has no effect.[27]

Reduced percent parasitism of *B. thuringiensis*-exposed hosts has been reported.[20] It is unclear whether the reduced parasitism is due to the increased vulnerability of parasitized hosts to *B. thuringiensis*,[22] or whether parasitoids can distinguish between similar-sized healthy and diseased hosts. The latter alternative would appear to be of adaptive advantage to the parasitoids.

The bacterium *Serratia marcescens* Bizio also has an impact on beneficials. The mechanical transmission of *S. marcescens* into *Galleria mellonella* pupae and *Heliothis zea* larvae by the ovipositor of the ichneumonid *Itoplectis conquisitor* and of the braconid *Microplitis croceipes*, respectively,[29] has been reported.[28] While adult *M. croceipes* fed the bacterium were susceptible,[29] the host was not, unless stressed. Only when *S. marcescens*-infected *H. zea* larvae were parasitized, did mortality of the host (and of the developing parasitoid too) occur. Another species, *Serratia liquefaciens*, resulted in the loss of an *Ernestia consobrina* culture which was being reared on *Mamestia*.[30] Unlike parasitoids developing from Dipel-exposed hosts, parasitoids developing from *S. marcescens*-infected hosts may also be infected,[29-31] and while some may egress from the host they often fail to pupate or emerge as adults.

Similarly, survival of dipterous parasitoids is depressed in *S. marcescens*-infected hosts.[32] *Blepharipa protensis* exposed to *Lymantria dispar* infested with a *S. marcescens* strain that was nonpathogenic for the host did infect the parasitoid, resulting in 12% mortality.[33] The parasitoid also influenced the susceptibility of *L. dispar* to *S. marcescens*. The eggs of *B. protensis* hatch in the gut of *L. dispar* and then penetrate the gut. Godwin and Shields[33] suggested that the injury caused by *B. protensis* as it penetrates the gut of its host was the reason for the 30% increase in the infectivity of *S. marcescens* in parasitized hosts.

Studies of the effects of bacterial pathogens on predators have primarily involved *Bacillus thuringiensis*. Both *Chrysopa carnea* and *Coccinella undecimpunctata* consumed less food, and the duration of their immature stages increased when fed *B. thuringiensis*-infected *Spodoptera littoralis* larvae.[34] However, other predators such as *Paederus alferii* or *Geocoris punctipes* were not affected.[20,35] While field tests using *B. thuringiensis* to control *S. littoralis* resulted in a reduction in the predator population, the effect was attributed to reduced host availability rather than to any direct pathogenic effect on the predators.[36] In an attempt to determine if *B. thuringiensis*-infected prey were detrimental to feeding predators, infected cabbage looper *Trichoplusia ni* larvae were fed to young Chinese praying mantids for 5 d with no ill effect.[37] Wicht and Rodriguez[38] also reported that *B. thuringiensis* had no serious effect on two predatory mites, *Fuscuropoda vegetans* or *Macrocheles muscaedomesticae*, which are predators of housefly eggs and first instar larvae. Hamed[25] also reported that *B. thuringiensis ssp. kurstaki* had no effect on the predatory bug, *Picromerus bidens*. Even the oil carriers used for the emulsifiable Dipel suspension did not alter the safety of *B. thuringiensis* for *Chrysopa carnea* and *Hippodamia convergens*.[39]

Although bacteria such as *B. thuringiensis* may not be detrimental to predators, they may survive in them and occur in the fecal discharge.[40,41] For example, Capinera and Barbosa[42] reported that the gypsy moth predator, *Calosoma sycophanta*, fed diseased larvae contained cocci typical of *Streptococcus faecalis*, a disease of *L. dispar*.[43] From the available data, predators appear to be resistant to infection of their bacterial-infected prey, but may play a role in disease dissemination.

B. PROTOZOA

Brooks[44] reviewed the literature concerning interactions between parasitoids, protozoa and the parasitoids' hosts. He noted that Paillot[9] was the first to suggest that parasitoids may serve as vectors of protozoan diseases. Paillot[9] had observed that all *Agrotis pronubana* infected with the flagellate, *Leptomonas chattoni* were also parasitized by the ichneumonid *Amblyteles armatorius*. Paillot[45,46] also reported that protozoan-infected *Pieris brassicae* (the European large white butterfly) were usually parasitized by *Cotesia* (= *Apanteles*) *glomerata*. Paillot's ideas were later confirmed by Issi and Maslennikova,[47] who demonstrated that *C. glomeratus* was responsible for the distribution and transmission of *Nosema mesnili* (= *N. polyvora*) within a cabbage butterfly population. Following the work of Paillot, many authors suggested that parasitoids were capable of transmitting protozoan diseases[7,41,48-57] but they supplied little data. Since the early report of Issi and Maslennikova,[47] several authors have demonstrated that parasitoids can mechanically transmit protozoa from infected to noninfected hosts.[58-62]

However, apparently not all parasitoids are capable of protozoan transmission. McLaughlin and Adams[63] were not able to demonstrate transmission of the neogregarine, *Mattesia grandis,* by *Bracon mellitor* from infected to healthy cotton boll weevil (*Anthonomus grandis*) larvae. The inability of the parasitoid to transmit protozoa from infected to healthy hosts has also been reported by several other authors.[64-66] One explanation has been provided by Masera[67] who reported that *P. brassicae* infected with *Nosema bombycis* were not parasitized by *C. glomerata,* while all parasitized larvae were *N. bombycis*-free, suggesting that *C. glomeratus* could select healthy host larvae.

Although McLaughlin and Adams[63] were not able to demonstrate the transmission of protozoa by parasitoids from infected to healthy hosts, i.e., mechanical transmission, they did provide evidence that *M. grandis*-diseased *B. mellitor* adults could transovarially transmit the disease. Lipa[52] earlier had claimed that *Apanteles* could transovarially transmit *N. apariae* to its host and suggested spores from the host entered the developing parasitoid and migrated to the ovaries where they were incorporated into the egg.[53] The lack of infection in parasitoid tissue led Brooks[44] to question the suggestion of Lipa[53] as to how the spores entered parasitoid eggs. Blunck[49] had also observed spores of microsporidia in the eggs of infected parasitoids. Since these reports, a number of authors[56,59,60,68] reported that parasitoids can transovarially transmit protozoa to healthy hosts. One of the more detailed studies was undertaken by Brooks and Cranford[59] who reported that infected *Campoletis sonorensis* could transovarially transmit *N. heliothidis* to healthy *Heliothis zea* larvae.

While parasitoids may disseminate protozoa during oviposition to healthy hosts by both mechanical and transovarian transmission, the development of both organisms within a host is not compatible. The impact of protozoa on parasitoids and their hosts was first indicated by Allen and Brunson[69] who reported that a microsporidian infection of a potato tuber moth (*Phthorimaea operculella*) culture resulted in malformed and short-lived parasitoids of reduced fecundity. The pathogen that infected the potato tuber moth was later identified as *N. destructor*[70,71] which was shown to result in unhealthy parasitoids.[72] As noted by Brooks,[44] it was Grosch[73] who was apparently the first to suggest that a parasitoid might be adversely affected when developing in protozoa-infected hosts. Working with a neogregarine, *Mattesia dispora,*[44] Grosch[73] stated that mature *Habrobracon* larvae were unable to metamorphose due to spores filling their midguts. Later, Thompson[74] reported that two parasitoids of the eastern spruce budworm, *Choristoneura fumiferana,* failed to pupate when developing in *Glugea fumiferanae*-infected hosts, attributing the failure to a nutritional imbalance caused by the accumulation of protozoan spores in the gut. The accumulation of spores is a serious problem for developing parasitoid Hymenoptera because the fore- and midgut form a blind sac that is not connected to the hindgut until pupation. Since these early reports there have been a number of studies demonstrating that protozoa can have a serious impact on parasitoids developing within diseased hosts.[58-62]

Cotesia marginiventris produced in microsporidian-infected *Spodoptera mauritia* ssp. *acronyctoides* usually died as larvae or pupae, or produced small and short-lived adult parasitoids.[58] Microsporidian infection of *Pieris rapae* (the now widely distributed European small white butterfly), also led to the death of a developing parasitoid, *Pteromalus puparum.*[64] On the other hand, the microsporidian, *Vairimorpha* sp., had only a minor effect on *Microplitis croceipes* and *C. marginiventris* developing in infected *Heliothis zea.*[60] While some authors[58,60,64] had not observed direct infection of the parasitoid, Brooks and Cranford[59] showed that *Nosema heliothidis* would infect the ichneumonid, *Campoletis sonorensis,* when it developed in parasitized *H. zea* larvae. They also reported that infection of *H. zea* reduced *C. sonorensis* emergence.[59] Direct systemic infection of *Macrocentrus grandii* by *N. pyr-*

austae-infected European corn borer (*Ostrinia nubilalis*) larvae adversely affected parasitoid pupal development and adult longevity.[65]

The hymenopterous parasitoids are susceptible to protozoan pathogens,[75-77] but whether these infections occur orally or by some other route is unknown. Larsson[78] reported that *Cotesia glomerata* were infected with *N. mesnili* only in hosts parasitized by this microsporidian and that the infection was always more advanced in the host than in the parasitoid. Moreover, the parasitoid infection did not begin in the gut. He suggested that protozoa gained entrance to the parasitoid by way of the anal vesicle. But as noted by Maddox,[79] there are few detailed studies of protozoan infections in parasitoids; and when a parasitoid becomes infected with a pathogen of its host, the epizootiology of the disease in the former is tied to that of the disease in the latter. When the pathogen is exclusively a disease of the parasitoid, horizontal transmission is difficult to evaluate.[79] In some cases the development of the protozoan is similar in both hosts. For example, the development of the pathogen *N. pyraustae* in both the host, *O. nubilalis,* and the parasitoid, *M. grandii* appeared identical.[65] However, the problem becomes much more complex when the developmental morphology of the disease-causing organism differs in the two hosts. For example, the developmental morphology of *N. heliothidis* differed in the parasitoid, *Campoletis sonorensis,* by comparison with development in the host *H. zea.*[59] Although both *C. sonorensis* and the parasitoid *Cardiochiles nigriceps* are subject to infection by *N. heliothidis,* which reduces their survival, each species also harbors its own species of *Nosema.*[59] While *N. heliothidis* infected both *C. sonorensis* and *H. zea,* these parasitoid-specific *Nosema* species do not appear to infect their hosts (*H. zea* or *H. virescens*). *Nosema campoletidis* did not show any detrimental effect in its host, *C. sonorensis.* In contrast, *N. cardiochiles* caused larval death and abnormal adult ovipositional activities, i.e., superparasitism, by *C. nigriceps.*[59]

McNeil and Brooks[62] examined the interaction between both *N. heliothidis* and *N. campoletidis* and two hyperparasitoids of *C. sonorensis, Catolacus aeneoviridis* and *Spilochalcis side.* Neither hyperparasitoid was infected with *N. heliothidis* when parasitizing infected *Campoletis sonorensis,* but *Catolacus aenoviridis* became infected with *N. campoletidis* during its development within infected *Campoletis sonorensis* and transovarially transmitted *N. campoletidis.*[62] In contrast, *S. side* also became infected with *N. campoletidis* when developing within infected *C. sonorensis.* However, development of microsporidians was arrested at the sporoblast stage.[62] Several hyperparasitoids of *P. brassicae* and *Aparia crataegi* were heavily infected with microsporidia of their host, and the presence of spores within eggs suggests that like the primary parasitoids, hyperparasitoids are also capable of transovarian transmission.[48-50]

Although most studies have involved larval hosts, Huger[80] reported that the egg parasitoid *Trichogramma evanescens* would readily oviposit in *N. pyrausta*-infected *O. nubilalis* egg masses. While parasitoid development within infected egg masses was not affected, the F_1 parasitoids were infected and their fecundity was half that of healthy females.[80]

The available information indicates that protozoan diseases reduce either the survival of parasitoids or their longevity and fecundity. *Pediobius foveolatus* introduced against the Mexican bean beetle was found highly susceptible to two *Nosema* diseases.[77] Andreadis[81] reported that infection of the introduced parasitoid *Macrocentus grandii* by *N. pyrausta* exceeded 45% and suggested that *Nosema* infection may explain the reduction and loss of effectiveness of *M. grandii* and other exotic parasitoids for *O. nubilalis* control. However, interactions between protozoan diseases and parasitoids are not always detrimental to the latter. *Cotesia fumiferanae* was not affected when its host was infected with *Nosema,* because *C. fumiferanae* emerge from early host stages occurs before the disease kills the host.[66]

Much less information is available for predators. As noted by Maddox,[79] microsporidia have been described from many insect predators.[82-86] However, no evidence was provided to indicate that the infections were attributable to the prey,[87] though such routes of infection have been suggested.[84] The available evidence suggests that predators may be resistant to infection by the protozoan pathogens of their prey.[88,89] However, when predators were fed on hosts infected with *Vairimorpha* sp. the predators were not infected but the accumulation of digestion resistant spores in their digestive system did cause starvation.[88,89] If these predators were fed healthy hosts after ingesting infected ones, they recovered.[88]

C. FUNGI

Some species of insect fungal pathogens such as *Aspergillus* attack a wide variety of insects, but as noted by Chapman et al.[3] such genera should be avoided as candidate biocontrol agents because of potential mycotoxin production. However, many of the entomopathogenic fungi generally have a narrow host range,[2] which reduces the risk of infection of beneficial species in different taxonomic classes from the target. Thus, the direct exposure of adult pollinators, predators, and parasitoids to fungal agents is not expected to cause serious problems. However, problems may occur when the predators and parasitoids interact with infected hosts.

Fungal pathogen interference with parasitoids, as with the other diseases, has been reported by several authors.[90-96] This incompatability is due to fungal invasion and death of the host[90,94-96] rather than fungal invasion of the developing parasitoid within the host.[90,92] Ullyett and Schonken,[97] similarly reported that tissues of the parasitoid, *Angitia* sp., were not invaded by the fungus *Entomophthora sphaerosperma* (= *Zoophthora radicans*) attacking the diamond-back moth, *Plutella maculipennis*. Keller[98] also concluded that fungi did not generally invade the tissues of endoparasitoids within infected hosts although he did report 2 cases out of 26 where a fungus had invaded the parasitoid.

There are reports that hosts harboring parasitoids exhibit enhanced susceptibility to fungal infection. Führer et al.[99] stated that entomophagous parasitism of *Pieris brassicae* (European large white butterfly) predisposed the invaded larvae to infection by *Beauveria bassiana*. Similarly, the infection of *H. zea* larvae by *Nomuraea rileyi* was increased if the larvae were either previously or simultaneously parasitized by *Microplitis croceipes*.[93] Powell et al.[90] also reported that 2-d-old parasitized hosts were predisposed to fungal infection. El-Sufty and Führer[100] suggested that the fungal pathogens may be better able to infect parasitized hosts than healthy ones. They later reported that the cuticle of unparasitized *Cydia pomonella* larvae responded to the penetration of *B. bassiana* by melanization but that this defensive reaction was suppressed by the parasitoid *Ascogaster quadridentatus*.[101] However, this increased host susceptibility to fungi may only occur for a short period following oviposition.

When 4-d-old *Encarsia formosa*-parasitized greenhouse whiteflies (*Trialeurodes vaporariorum*) were exposed to *Aschersonia aleyrodis*, the fungus was less successful in infecting the hosts.[95,96] A similar situation was reported involving both the rose-grain aphid, *Metopolophium dirhodum*, and its parasitoid, *Aphidius rhopalosiphi*.[90] Fungal development was impaired[90] when aphids parasitized for at least 4 d were exposed to the fungus *Erynia neoaphidis*.

Decreased susceptibility of parasitized hosts to fungal infection may be due to an antifungal agent released during parasitoid development. Thus Führer and El-Sufty[102] discovered that a fungistatic substance was present in the hemolymph of *Pieris brassicae* after hatching of larval *Cotesia glomerata*. It proved that this substance was released by the teratocytes (cells of the embryonic membrane of the parasitoid, which are liberated with the larva at hatching and then proceed to grow independently in the host's hemolymph). A similar antibiotic substance has also been found in the hemolymph of *P. brassicae* pupae attacked by *Pimpla turionellae,* an ichneumonid that does not release teratocytes.[103] In this case the active substance enters the host's hemolymph as an anal secretion of the parasitoid larvae,[104] its activity being attributed to its inhibitory effect on DOPA-tyrosinase.[105]

Although parasitoids do not appear to detect recently infected hosts, they may be able to single out those at more advanced stages of infection. After *Trialeurodes vaporariorum* had been infected for 7 d, *Encarsia formosa* females responded by inserting their ovipositor in infected hosts but did not deposit an egg.[96] Thus, female *E. formosa* are able to selectively refrain from ovipositing in hosts infected for more than 7 d, but contact with infected hosts opens up the possibility of transmitting the infection. Voukassovitch[106] speculated that parasitoids transmitted the fungus *Spicaria farinosa* to pupae of *Polychrosis botrana* during oviposition, but no evidence was provided. While it is generally accepted that parasitoids can function as mechanical vectors of some pathogens, evidence for the transmission of fungi is lacking. Fransen[96] and van Lenteren were unable to demonstrate that *E. formosa* transmitted the fungus from infected to healthy hosts even though the parasitoids regularly probed fungal-infected hosts.

The effects of fungal-infected prey on predators has received less attention. Although *Aschersonia aleyrodis*-parasitized whiteflies were consumed by the predatory mite, *Phytoseiulus persimilis*, no detrimental effects were observed on the mite.[95] The transmission of fungal vegetative stages by probing parasitoid Hymenoptera or feeding predators is not likely to occur, but the dissemination of spores, like pollen by insects, would be expected if predators or parasitoids contact hosts in the advanced stage of the disease. It would be surprising if some of the spores of some of the pathogenic fungi were not attractive to certain predators or parasitoids which could mechanically carry the spores to other potential host populations.

Of related interest is the observation that benomyl, a carbamate benzenidazole fungicide, reduced the emergence of the parasitoid *Cotesia marginiventris* from its host, *H. zea*.[105] Similarly, Sewall and Croft[108] reported that benomyl was not toxic to the third instar orange tortrix, *Argyrotaenia citrana*, at 300 ppm but resulted in the death of the solitary endoparasitoid, *Apanteles aristoteliae*, within its host.

D. NEMATODES

Ishibashi et al.[109] treated *Malacosoma neustria* with a DD-136 strain of *Steinernema feltiae** (= *Neoplectana carpocapsae*). They were then exposed to a predator. Twenty percent of the predators were killed after consuming 24-h-infected prey but did not feed on inactive prey that were present in later stages of infection. Likewise, when *Pieris rapae* ssp. *crucivora* were exposed to DD-136 for 4 h they also become inactive and were not attacked by the parasitoid *Trichomalus apanteloctenus*.[93] However, nematodes do have an effect on parasitoids. When nematode-infected larvae were exposed before they became inactive they were parasitized and, although the reason was not clear, parasitoid emergence was reduced.[109] Kaya,[110] using the DD-136 strain, reported that *Glyptapanteles* (= *Apanteles) militaris* died due to death of the host before the parasitoid could egress. However, if the parasitoid was near egression and could spin a cocoon, it appeared to be safe from infection.[111] In later studies[111,112] cocoons of several parasitoid Hymenoptera were reported resistant to infection because the intact cocoon consists of a pore-free layer of silk that is impenetrable to the infective nematode stages. Kaya and Hotchkin[112] extended earlier studies and found that when *Hyposoter exiguae* were exposed 8 d after parasitism within the host to nematodes, between 14 and 34% of the parasitoids were infected at 500 or 1000 nematodes per ml, respectively. Triggiani reported that *A. ultor* was unable to complete its development in *G. mellonella* infected by either *S. feltiae* or *Heterorhabditis heliothidis*.[113]

Parasitoids appear to have no effect on the development of nematodes.[110] Nematodes have been found developing in living hosts from which parasitoids have emerged,[114] as occurs after parasitism with certain braconids such as the *Apanteles* group.

Kaya[115] examined the effects of *S. feltiae* on the tachinid *Compsilura concinnata*. He

---

*   According to Akhurst (Chapter 16), now properly classified as *Steinernema carpocapsae*.

reported that parasitized hosts exposed to the nematode 1, 2, or 3 d after parasitism resulted in the death of host and parasitoid. However, if the parasitoid pupated it was not affected. Mrácek and Spitzer[116] found that if the tachinid, *Myxexoristops* sp., was exposed to the nematode, *Steinernema kraussei,* early in its development within its host, it was invaded by the nematode and died. However, if the tachinids were exposed late in development to the nematode the tachinids were resistant to infection and were able to complete development.[116]

H. heliothidis and the A-11 strain of *S. feltiae* were not found harmful to carabid, staphylinid, or coccinellid beetles, or to labidural earwigs, whether directly or via their being fed with infected hosts.[117] The predatory larvae of *Thereva handlirschi* (Diptera: Therevidae) and *Rhagio* spp., which attack larval sawflies, are invaded and killed by *Steinernema kraussei* when exposed to this nematode in soil, but these predators do not serve as a host.[116] The available data suggest that if predators are exposed to infective stages of the nematode, some infection may occur, but infection through feeding on infected prey appears uncommon.

E. VIRUSES

Viruses appear to be the pathogens most intimately involved with both parasitoids and predators and their hosts and prey. There are many types of viruses in insects which vary in their pathological effect from detrimental to beneficial. Most of the available data concerning insect viruses, beneficials, and pests involve baculoviruses.

The baculoviruses include a group of symbiotic viruses that occur in certain parasitoid Hymenoptera and in certain tissues of their insect hosts.[118] These are known as the polydnaviruses,[119] a family of viruses having polydisperse DNA genomes.[120-123] These polydnaviruses replicate in parasitoid calyx epithelial cells and are released into the lumina of the reproductive system from where they are injected into hosts along with eggs of the parasitoids.[118] Once in the host the polydnaviruses are expressed in host tissue.[124] These polydnaviruses appear to play an important role in the parasitoid-host relationship[125-127] and do not appear to cause any harm to the parasitoids harboring them; and while they affect the host of the parasitoid, they do not replicate in host tissues.[128] It is not established that the polydnaviruses have any biological control potential. However, there are some other viruses that can be found in the reproductive system of parasitoids that do replicate in the host of the parasitoid and may have control potential.

A nonoccluded filamentous baculovirus-like particle (Cm FV)[129,130] has been found in female parasitoid calyx cells and replicates in the hypodermal and tracheal cells of several noctuid larvae parasitized by *Cotesia marginiventris*. No specific effects of Cm FV could be distinguished in the host as these were also infected with a polydnavirus. Another virus has been reported that consists of a large cylindrical nucleocapsid surrounded by two unit membranes, similar to the ichneumonid polydnaviruses,[118] and that replicates in host hemocytes and parasitoid tissue.[131] This virus, Cm V2 from *C. melanoscelus*,[131] consists of a single 125-kbase single double-stranded DNA molecule that is maternally transmitted vertically.[132] No specific pathogenic effects were observed and the role of this virus in the host is unknown.

There are a number of pathogenetic baculoviruses which are being considered for use in biological control[2] and researchers have suggested that these viruses can be vectored by parasitoids.[133-136] However, not all hosts infected by virus are accepted as oviposition sites. Kelsey[137] reported that *C. glomerata* would not oviposit in virus-infected laboratory *Pieris rapae* larvae, and Versoi and Yendol[138,139] reported *C. melanoscelus* was less likely to attack virus-infected *Lymantria dispar*. Thus the source of viral infection of adult parasitoids may be limited to parasitoids that attack hosts in the early stages of infection or develop within diseased hosts. However, the progeny of parasitoids which oviposit in hosts in the early stages of viral infection are not likely to survive. The development of the ichneumonid *Campoletis sonorensis* rarely proceeded beyond the 1st instar when its host, *H. virescens* was infected with nuclear polyhedrosis virus (NPV).[133] In most cases parasitoid mortality is simply due to the premature death of the viral-infected host. For example, *C. sonorensis*,[133] *Hyposoter exiguae*,[140] and *Cotesia marginiventris*[141] died at the death of their host if the latter was infected with NPV at or near the time of parasitoid oviposition. The egg-larval parasitoid, *Chelonus insularis*, also dies in NPV infected hosts even though the host may become infected some days after parasitism.[140-142] When parasitism preceeded NPV exposure by 48 h many parasitoids appeared to survive unaffected,[132,135] although polyhedra accumulated in the midgut of larvae during their development. These polyhedra are voided with the meconium during pupation[132] which occurs inside cocoons of the parasitoids. The presence of polyhedra in the meconium may be responsible for the ability of females to sometimes transmit NPV, although the female parasitoids do not appear to be infected and have not contacted a diseased host. It is possible that as the female escapes from the cocoon, her ovipositor can become contaminated by virus from the meconium.

The recently discovered enveloped, double-stranded DNA virus, *Ascovirus*,[143] also appears to be mechanically transmitted by parasitoids such as *Cotesia marginiventris* from infected *Spodoptera* spp. to healthy larvae.[136] Like the NPV virus, when hosts were infected with *Ascovirus* at the same time or before parasitism, parasitoids fail to develop. When hosts were infected with *Ascovirus* 4 to 5 d after parasitism some parasitoids emerged. However, mortality of parasitoids failing to emerge did not appear to be simply due to the virus killing the host before the parasitoids could complete their development. The virus may compete for nutrients, as virus-infected host larvae lived longer than nonvirus-infected controls yet parasitoid larvae that emerged from virus infected hosts were smaller.[136] In some cases parasitoids develop faster in viral exposed hosts,[140] perhaps because the virus weakens the hosts' defense or makes nutrients more available. Hotchkin and Kaya[142] also found that *Hyposoter exiguae* had a shorter development time in viral exposed hosts. In contrast, development time took longer in virus-infected *Campoletis sonorensis* and *Cotesia marginiventris*.[142] These results also suggest that the virus may compete for essential nutrients.

However, competition for the limited host nutrients or infection does not always explain the effects of virus infected hosts on developing parasitoids. *Glyptapanteles* (= *Apanteles*) *militaris* dies in viral infected hosts before the host.[142] Whether these parasitoids die because of an infection is not clear. Just because the developing parasitoid dies within and before the host, does not imply the virus directly killed the parasitoid. For example, Kaya[144] reported that the hemolymph of the armyworm, Mythimna (= *Pseudaletia*) *unipuncta* infected with a Hawaiian strain of granulosis virus (HGV), was toxic to *G. militaris*. The effect of the viral induced toxic factor was found to extended to *Phanerotoma flavitestacea* but another parasitoid, *Venturia canescens*, was unaffected.[145] Kaya and Tanada[146] found that the hypertrophy strain of a nuclear polyhedrosis virus (HNPV) also produced a toxic factor in the hemolymph of its host. Hosts with the viral-influenced toxic hemolymph caused *G. militaris* to become encapsulated, the parasitoid dying before the host.[141]

Parasitism may also influence the host's susceptibility to viral infection. Tower[147] reported that larvae of the armyworm, *Mythimna unipuncta*, parasitized by *G. militaris*, consumed only half the food of nonparasitized larvae while larvae of the imported *Pieris rapae* parasitized by *C. glomerata* consumed more food than nonparasitoid larvae.[148] Such effects of parasitism on host feeding are common[149] and this differential food consumption may play a role in the exposure and susceptibility of the host to viral diseases. Yet this differential host exposure to viral inoculum caused by parasitism has rarely been examined. Nonparasitized larvae were reported to be twice as susceptible to NPV as those parasitized by *Hyposoter exiguae*,[140] which may be due to the parasitized host feeding less and receiving less viral inoculum. In contrast, a positive correlation between NPV and a hymenopteran parasitoid, *C. melanoscelus*, has been reported.[150] Gypsy moth larvae also appear more susceptible to NPV if parasitized,[151] but the effect cannot be attributed to increased feeding because the parasitized hosts feed less. A positive interaction between a dipteran, *Parasetigena silvestris,* and the development of NPV symptoms in its hosts has also been reported.[150] Similarly, *Porthetria dispar* parasitized by the dipteran *Blepharipa pratensis* (Meigen) were more susceptible to NPV.[33] It is possible that penetration of the gut of the host by parasitoid larvae, aids the NPV infection as suggested for bacterial infection.[33]

In studies of dipterous parasitoids, *Compsilura concinnata* emerged in lesser numbers from virus-infected hosts than from controls.[142] This was attributed to the reduced oviposition of *C. concinnata* into the less active virus-infected hosts.[152] Such hosts also yielded healthy *Voria ruralis.*[153] When *V. ruralis* females were allowed to oviposit in early instars of recently infected hosts or hosts infected soon after oviposition, the developing parasitoid pupariated earlier than expected, just before or soon after the host died.[153] The parasitic Diptera appear to be better adapted to respond to the premature death of their host.[154,155]

NPV-infected prey do not appear to have a serious effect on predators. However, several authors have suggested that predators may serve as mechanical vectors of baculoviruses.[40,41,156-160] Burgess[161] noted that *Calosoma sycophanta* were apparently unaffected by feeding on diseased gypsy moth larvae, but he did not consider the possible dissemination of virus. This question was considered by Capinera and Barbosa,[42] who reported that *C. sycophanta* to which diseased gypsy moth larvae were fed, defecated polyhedra in sufficient quantity to infect third instar *Porthetria dispar* larvae feeding on the contaminated material.

Beekman[162] examined the ability of a sucking predator, *Nabis tasmanicus,* to consume and excrete NPV. He and others found that NPV had little effect on the predator and that NPV was excreted.[162-164] Young and Yearian[163] reported that *N. roseipennis* could transmit the disease via the fecal material to the prey population. Biever et al.[165] used *Podisus maculiventris* contaminated with cabbage looper NPV to initiate an epizootic in the *Trichoplusia ni* population. Two species of scavenger beetles, *Ateuchus histeroides* and *Trox suberosus,* fed NPV-infected *Spodoptera* also showed no effects,[166] but whether these scavengers could also disseminate the virus was not studied.

III. POTENTIAL INTERACTION BETWEEN BENEFICIALS AND INSECT PATHOGENS

There are many nontarget organisms (NTOs) that will be exposed to insect pathogens when and in whatever manner they are used. However, many of these pathogens have a degree of specificity that should reduce their potential impact on the many distantly related NTOs. The greatest concerns are taxonomically closely related nonpest species and pathogens with a very broad host range. Secondly, the use pattern of the pathogen impacts the species of NTOs which are exposed. The pathogen may be used at a time or in a location that reduces its potential impact on susceptible NTOs. Thirdly, most pathogens are rather specific with regard to the stages they infect thus further reducing the possibility that the NTOs will be present and exposed at the susceptible stage in their life cycle.

However, many pathogens are likely to come into conflict with the predators and parasitoids that utilize the same host species.[167] This conflict can be detrimental or beneficial. Tamashiro[168] divided the pathogen/parasite (parasitoid)-host interaction into four effects: direct deleterious; indirect deleterious; beneficial; and none. However, the effects depend greatly on the situation and perspective. From a broad perspective the effects may range from no interaction, throughout a direct or indirect beneficial, to a compatible or neutral interaction; or to a direct or indirect deleterious interaction.

From the pathogens' perspective, the interaction tends to be beneficial. It not only leads to an increased host-range, but can increase host susceptibility, pathogen dissemination, spread, and success. From the host perspective, the interaction is generally negative. The host dies either from the disease or in consequence of the parasitoid or predator. The effect of the interaction from the perspective of the predators and parasitoids is the most complex one.

The other group of beneficials, the pollinators, may also be affected by the use of pathogens for biological control of pest species. At issue is whether the pollinators are likely to contact the pathogen, whether the pathogen can directly infect the pollinator, and in the case of social species, whether the pathogen is carried back to the colony to infect the immature stages.

A. POLLINATORS

Many species of insects act as pollinators. They include moths, dipterans, and beetles as well as hymenopterans.[169] However, there is very little information on any except the honeybee, *Apis mellifera*. This lack of information makes it difficult to predict the impact of the use of pathogens on such organisms. Pollinators are generally adult while many of the pathogens are more infective to the immature stages,[170] which along with the specificity of many pathogens probably reduces the risk of direct effects. These risks are further reduced if the pathogen is not used on flowering plants. However, pollinators may be at risk if they contact contaminated plants.[171]

The honeybee, has been one of the few nontarget insects and the only pollinator that has been consistently examined for effects from pathogens.[172] For example, Dutky and Hugh[173] reported that adult bees were not susceptible to the nematode, *Steinernema feltiae*, but that larvae directly exposed to the nematode in the laboratory were susceptible. When *S. feltiae* were sprayed directly onto brood frames, the brood was not seriously affected and mortality of adult bees only occurred during the first 3 d.[174] After feeding nematodes in a sugar solution to field honeybees, workers in the hive became infected although the brood did not.[175] The brood may not have become infected due to the higher temperatures in the brood-rearing area of the hive.[174]

A number of granulosis and nuclear polyhedrosis viruses have been fed to honeybees in a sucrose solution[176,177] with no effects observed in the bees or the brood in the colony. The bacterial pathogens, primarily *B. thuringiensis,* have been extensively studied. Of the tests of many subspecies, strains, and combinations of various *B. thuringiensis* preparations, only those with the β-exotoxin showed any activity against honeybees in laboratory tests.[176,178-183] As discussed by Cantwell et al.[172] in a review of the effects of biological insecticides on the honeybee, no investigator has reported adverse effects of *B. thuringiensis* used as a foliage spray. Cantwell et al.[172] calculated that a standard application of at least 22 lb/acre would be needed to expect an effect. *B. thuringiensis* has even been used to control *G. mellonella* in beehives.[2,184]

The protozoan diseases have received the least attention, although *Nosema apis* causes a serious bee disease.[185] Cantwell et al.[172] reported testing diseases due to *Nosema,* including *N. apis,* and also one caused by a species of *Thelohania* that attacks mosquitoes, which they tested against honeybees. With the exception of *N. apis,* no serious pathogens developed.[172] Cantwell et al.[172] concluded that of the various biological insecticides tested none showed any serious potential danger to *Apis mellifera*. While the honeybee appears resistant to many of the better-studied insect pathogens, little is known concerning the many other species of pollinators. Before widespread use of insect pathogens features in pest control, the other species of pollinators likely to be exposed should be examined.

B. PREDATORS

Predators, whether sucking or chewing, generally appear resistant to infection. While Salama et al.[34] reported that *C. undecimpunctata* ingested less when feeding on hosts exposed to *B. thuringiensis,* the reductions of predators in the field have been attributed to a reduction in prey[36] rather than any direct effect of pathogens. Nematode-infected prey do not appear to invade feeding predators,[116,117] although some predators exposed in moist soil to infective stages of nematodes may become infected.[116] Such exposure would generally be expected to be minimal. Predators also do not appear susceptible to virus-infected hosts.[42,161-164] Part of this resistance may be due to the more acid digestive systems of predators[186] to which the occluded viruses are resistant, thus the virions are not freed.[187]

Little is known about the fungi, but it is also unlikely that predators would be susceptible to the vegetative stages developing in diseased prey. Similarly, the effect that the interaction of protozoa and prey have on predators has lacked serious study. Microsporidia have been described from many predatory insects,[82-86,188] but whether these diseases have a relationship with their prey has not been addressed. It may be significant that it is the viruses and bacteria that have been found intact in the fecal material of predators, viruses being resistant to digestion and release of the virions,[189] and the bacteria being poor invaders.[190]

C. PARASITOIDS

1. Dipterous Parasitoids

A majority of the parasitic Diptera either oviposit a large number of eggs, some of which are consumed by the host; oviposit a few eggs on or near the host, these hatching soon after laying; or larviposit.[191] In each of these situations it is the larval stage that penetrates the integument or gut of the host in order to feed. Many of the developing parasitoid dipteran larvae must also obtain an air source. Their tapping into the host's tracheal system, or penetration of its integument, may offer additional routes of pathogen entry, although this is speculative. Especially in the later stages of their development,[192] such Diptera tend to be saprophytic; some of them may even eventually emerge from hosts that had died prematurely.

Nevertheless, should a host in the advanced stage of disease be encountered, some adult females of dipterous parasitoids avoid oviposition.[142] It is known that such adults will feed on dead or dying hosts; and by so doing, acquire infections.[158-160] These infected adults may now spread pathogens via their contaminated mouthparts and tarsi, for example, by contaminating foliage.[155,193]

2. Hymenopterous Parasitoids

Starr[194] discussed the role of "stingers" in the evolution of sociality among the Hymenoptera. It should not be overlooked, though, that the evolution of a "stinging" mechanism probably also led to a radiation of the parasitic habit in Hymenoptera and to a vectorial capacity. The latter holds good for entomopathogenic viruses,[133,135,136,140] bacteria[28,29,195] and protozoans,[44,58,60,62,65,196] although neither fungi nor nematodes appear to be readily disseminated or vectored by parasitoids.

Importantly, though, the transmission of viruses (except for the symbiotic polydnaviruses[197]) and other causal agents of insect diseases is generally detrimental to hymenopteran parasitoids.[198] Competition for the nutritional resources that a host represents may also influence the effects that pathogens may have when occurring in the same host as the parasitoid; and some parasitoids attack hosts that represent a suitable, but finite, quantity of nutritional resource.[199]

Furthermore, bacterial and protozoan pathogens affect host sex ratios, sometimes even converting one sex to another. For example, a bacteria-like microorganism converts males to females in one strain of isopod,[200] and two microsporidian taxa are known to convert males of an amphipod to functional females, able to disseminate these disease agents further.[201,202] In other cases the sex ratios are altered through differential sex mortality,[203] with cessation of male production. Huger et al.[204] declared that a systemic and chronic bacterial infection in adult *Nasonia* (= *Marmoniella) vitripennis* is associated with a son-killer trait. Over 80% of the male eggs of infected females fail to hatch, and the agent is maternally — and contagiously — transmitted.[202] And again, some entomopathogens slow down the growth of the host,[15] the smaller host-size shewing the sex ratio toward a male bias.[205-207]

IV. SUMMARY

When considering the honeybee, most entomopathogens appear to be safe if used carefully, for the adult bees and brood do not appear susceptible to most microorganisms tested. However, the effects of the various entomopathogens on solitary bees and other pollinators is unknown and clearly merits consideration. Restricted use of microbial insecticides on flowering plants commonly pollinated by species other than *Apis*, unless information regarding the pathogen and pollinator is available, will minimize problems.

Predators also appear generally resistant to contact with diseased hosts and to most entomopathogens. Protozoans may be of some concern, though, since various species have been reported to be pathogenic to predators. Although the protozoan diseases of predators do not appear to be due to horizontal transmission from prey to predator, their causal agents are difficult to identify. Predators in most situations may be susceptible to infection by generalist nematodes, but except under heavy exposure most predators are not infected by feeding on infected hosts. While the use of entomopathogens can reduce the prey resources for predators, their use is decidedly less hazardous than that of chemical pesticides.

The parasitoids are undeniably disadvantaged by microbial insecticides. Whether dipteran or hymenopteran, they are vulnerable to the premature death of their host due to entomopathogens. The parasitoid may become infected by some entomopathogens, primarily the viruses and protozoa, so even if the host fails to die before the parasitoid emerges, the adult parasitoid is infected and generally less fit. If the parasitoid is not infected, its competition with the entomopathogen for the limited host resource during development, often leads to less fit adults. Adult Hymenoptera, infected or not, may also mechanically transmit bacterial, protozoan, and viral pathogens and transovarially transmit many viral, protozoan, and (possibly some) bacterial diseases to healthy hosts as they parasitize. This leads to the premature death of their progeny and presents a problem in the laboratory propagation of many parasitoids. The transmission of diseases to the host during oviposition probably occurs in the field and the use of many diseases may have a serious impact on the parasitic Hymenoptera. Only with the polydnaviruses is an infection an advantage to the parasitoid.

Parasitoids also stress their hosts, in some cases causing the latter to succumb to lower doses of an entomopathogen, again to the detriment of the parasitoid. The opposite effect, the ability of the parasitoid to slow or reduce the effects of the pathogen, occurs primarily with the fungal diseases. Teratocytes, cells released upon egg hatch along with parasitoid larval secretions, appear to stop fungal and possibly bacterial infections, but these effects are limited. However, additional antipathogen secretions may await discovery.

The parasitic Diptera appear to be slightly more resistant to pathogens because they can sometimes complete their development despite the host's death. While some Diptera are capable of disseminating certain entomopathogens, they are much less effective in doing so, than are the parasitic Hymenoptera. The development of the parasitic Diptera may also promote pathogen infection. The parasitic Diptera generally penetrate the host as larvae, causing wounds that provide routes of infection for bacteria, protozoa and viruses.

Although parasitoids are at risk with the use of pathogens, there are exceptions.[11-15,95,96,109,150] Even in these cases, the parasitic Hymenoptera are at risk but the pathogen can be carefully used to minimize the effects. For example, Fransen and von Lenteren[95] reported that the parasitoid *Encarsia formosa* and the fungus *Aschersonia aleyrodis* could be used together for the control of the greenhouse whitefly (*Trialeurodes vaporariorum*) if the parasitoids were only released 4 or more days after the use of the fungus. The use of Dipel to control the gypsy moth also resulted in an increase in the production of *Cotesia melanoscelus*,[16] which is, however, also killed by Dipel.[18] Unfortunately the parasitic Hymenoptera are not generally compatible with the use of pathogens. Studies to minimize the impact of pathogens on these beneficial insects should therefore be encouraged.

ACKNOWLEDGMENT

The author expresses appreciation to Eric Baehrecke and Donald A. Nordlund for reading and making suggestions in the manuscript. Approved as TA 23739 by the Director of the Texas Agricultural Experiment Station.

REFERENCES

1. **Jutsum, A. R.**, Commercial application of biological control: status and prospects, *Philos. Trans. R. Soc. London Ser., B*, 318, 247, 1988.
2. **Payne, C. C.**, Pathogens for the control of insects; where next? *Philos. Trans. R. Soc. London Ser., B*, 318, 225, 1988.
3. **Chapman, H. C., Davidson, E. W., Laird, M., Roberts, D. W., and Undeen, A. H.**, Safety of microbial control agents to non-target invertebrates, *Environ. Conserv.*, 6, 278, 1979.
4. **Paillot, A.**, Le problème de l'équilibre natural chez les insectes phytophages, *Rev. Gen. Sci. Bull. Soc. Philomath*, 36, 206, 1925.
5. **King, K. M. and Atkinson, N. J.**, The biological control factors of the immature stages of *Euxoa ochrogaster* Gn. in Saskatchewan, *Ann. Entomol. Soc. Am.*, 21, 167, 1928.
6. **Jaques, R. P.**, The influence of physical stress on growth and nuclear-polyhedrosis of *Trichoplusia ni* (Hübner), *J. Insect Pathol.*, 3, 47, 1961.
7. **Beard, R. L.**, The toxicology of *Habrobracon* venom, a study of a natural insecticide, *Conn. Agric. Exp. Stn. New Haven Bull.*, 562, 27, 1952.
8. **Metalnikov, S. and Chorine, V.**, Du rôle joué par les hymenoptères dans l'infection de *Galleria mellonella*, *C. R. Acad. Sci. (Paris)*, 182, 729, 1926.
9. **Paillot, A.**, Sur une nouvelle flagellose d'insecte et un processus d'infestation naturelle non encore d'écrit, *C. R. Acad. Sci. (Paris)*, 177, 463, 1923.
10. **Payne, N. M.**, A parasitic hymenopteran as a vector of an insect disease, *Entomol. News*, 44, 22, 1933.
11. **Hamel, D. R.**, The effects of *Bacillus thuringiensis* on parasitoids of the western spruce budworm, *Choristoneura occidentalis* (Lepidoptera: Tortricidae), and the spruce coneworm, *Dioryctria reniculloides* (Lepidoptera: Pyralidae), in Montana, *Can. Entomol.*, 109, 1409, 1977.
12. **Dunbar, D. M., Kaya, H. K., Doane, C. C., Anderson, J. F., and Weseloh, R. M.**, Aerial application of *Bacillus thuringiensis* against larvae of the elm spanworm and gypsy moth and effects on parasitoids of the gypsy moth, *Conn. Agric. Exp. Stn. New Haven Bull.*, 735, 1973.
13. **Kaya, H., Dunbar, D., Doane, C., Weseloh, R. W., and Anderson, J.**, Gypsy moth: Aerial tests with *Bacillus thuringiensis* and pyrethroids, *Conn. Agric. Exp. Stn. New Haven Bull.*, p. 744, 1974.
14. **Wollam, J. D. and Yendol, W. G.**, Evaluation of *Bacillus thuringiensis* and a parasitoid for suppression of the gypsy moth, *J. Econ. Entomol.*, 69, 113, 1976.
15. **Weseloh, R. M. and Andreadis, G. G.**, Possible mechanism for synergism between *Bacillus thuringiensis* and the gypsy moth (Lepidoptera: Lymantriidae) parasitoid, *Apanteles melanoscelus* (Hymenoptera: Braconidae), *Ann. Entomol. Soc. Am.*, 75, 435, 1982.
16. **Weseloh, R. M., Andreadis, T. G., Moore, R. E. B., Anderson, J. F., Dubois, N. R., and Lewis, R. B.**, Field confirmation of a mechanism causing synergism between *Bacillus thuringiensis* and the gypsy moth parasitoid *Apanteles melanoscelus*, *J. Invertebr. Pathol.*, 41, 99, 1983.
17. **Wallner, W. E., Dubois, N. R., and Grinberg, P. S.**, Alteration of parasitism by *Rogas lymantriae* (Hymenoptera: Braconidae) in *Bacillus thuringiensis*-stressed gypsy moth (Lepidoptera: Lymantriidae) hosts *J. Econ. Entomol.*, 76, 275, 1983.
18. **Ahmad, S., O'Neill, J. R., Mague, D. L., and Nowalk, R. K.**, Toxicity of *Bacillus thuringiensis* to gypsy moth larvae parasitized by *Apanteles melanoscelus*, *Environ. Entomol.*, 7, 73, 1978.
19. **Salama, H. S., Zaki, Z. N., and Sharaby, A.**, Effect of *Bacillus thuringiensis* Berliner on parasites and predators of the cotton leafworm *Spodoptera littoralis*, *Z. Angew. Entomol.*, 94, 498, 1982.
20. **Salama, H. S. and Zaki, F. N.**, Interaction between *Bacillus thuringiensis* Berliner and the parasites and predators of *Spodoptera littoralis* in Egypt, *Z. Angew. Entomol.*, 95, 425, 1983.
21. **Temerak, S. A.**, Detrimental effects of rearing a braconid parasite on pink borer larvae inoculated by different concentrations of the bacterium, *Bacillus thuringiensis* Berliner, *Z. Angew. Entomol.*, 89, 315, 1980.

22. **Marchal-Segault, D.,** Larval development of Hymenoptera parasites *Apanteles glomeratus* L. and *Phaneratoma flavitestacea* F. in caterpillars infected with *Bacillus thuringiensis* Berliner, *Ann. Parasitol.* (Paris), 50, 223, 1975.

23. **Thomas, E. M. and Watson, T. F.,** Effect of (Dipel) *Bacillus thuringiensis* on the survival of immature and adult *Hyposoter exiguae* (Hymenoptera, Ichneumonidae), *J. Invertebr. Pathol.,* 47, 178, 1986.

24. **Dunbar, J. P. and Johnson, A. W.,** *Bacillus thuringiensis:* effects on the survival of a tobacco budworm parasitoid and predator in the laboratory, *Environ. Entomol.,* 4, 352, 1975.

25. **Hamed, A. R.,** Zur Wirkung von *Bacillus thuringiensis* auf Parasiten und Prädatoren von *Yponomeuta evonymellus, Z. Angew. Entomol.* 87, 294, 1979.

26. **Marchal-Segault, D.,** Effects of the spore-crystal complex on *Bacillus thuringiensis* Berliner on *Apanteles glomeratus* and *Phanerotoma flavitestacea, Ann. 2nd Ecol. Anim.,* 6, 521, 1974.

27. **Wysoki, M., Izhar, Y., Gurevitz, E., Swirski, E., and Greenberg, S.,** Control of the honeydew moth, *Cryptoblabes gnidiella* Mill with *Bacillus thuringiensis* Berliner in avocado plantations, *Phytoparasitica,* 3, 103, 1975.

28. **Bucher, G. E.,** Transmission of bacterial pathogens by the ovipositor of a hymenopterous parasite, *J. Insect Pathol.,* 5, 277, 1963.

29. **Bell, J., King, E., and Hamalle, R. J.,** Interactions between bollworms and braconid parasite and the bacterium *Serratia marcescens, Ann. Entomol. Soc. Am.,* 67, 712, 1974.

30. **Huger, A. M. and Krieg, A.,** Über eine fatale Bakteriose in einem Parasit/Wirt-System, *Jahresber. Biol. Bundesaust,* Lane-Forstwirtsch., Berlin Braunschweig, 1983, 80.

31. **Bracken, C. K. and Bucher, C.,** Mortality of hymenopterous parasite caused by *Serratia marcescens,* J. Invertebr. Pathol., 9, 130, 1967.

32. **King, E. G., Bell, J. V., and Martin, D. F.,** Control of the bacterium, *Serratia marcescens* in an insect host-parasite rearing program, *J. Invertebr. Pathol.,* 26, 35, 1975.

33. **Godwin, P. A. and Shields, K. S.,** Some interactions of *Serratia marcescens,* nucleopolyhedrosis virus and *Blepharipa pratensis* [Dip: Tachinidae] in *Lymantria dispar* [Lep: Lymantriidae] *Entomophaga,* 27, 189, 1982.

34. **Salama, H. S., Zaki, F. N., and Sharaby, A. F.,** Effect of *Bacillus thuringiensis* Berl. on parasites and predators of the cotton leafworm *Spodoptera littoralis* (Boisd.), *Z. Angew. Entomol.,* 94, 498, 1982.

35. **Ali, A. A. and Watson, T. F.,** Efficacy of Dipel and *Geocoris punctipes* (Hemiptera, Lygaeidae) against the tobacco budworm (Lepidoptera, Noctuidae) on cotton, *J. Econ. Entomol.,* 75, 1002, 1982.

36. **Salama, H. S. and Zaki, F. N.,** Impact of *Bacillus thuringiensis* Berl. on the predator complex of *Spodoptera littoralis* (Boisd.) in cotton fields, *Z. Angew, Entomol.,* 97, 485, 1984.

37. **Yousten, A. A.,** Effect of the *Bacillus thuringiensis* endotoxin on an insect predator which has consumed intoxicated cabbage looper larvae, *J. Invertebr. Pathol.,* 21, 312, 1973.

38. **Wicht, M. C., Jr., and Rodriquez, J. G.,** Integrated control of muscid flies in poultry houses using predator mites, selected pesticides and microbial agents, *J. Med. Entomol.,* 7, 687, 1970.

39. **Haverty, M. I.,** Sensitivity of selected nontarget insects to the carrier of Dipel 4L® in the laboratory, *Environ. Entomol.,* 11, 337, 1982.

40. **Franz, J. M.,** Influence of environment and modern trends in crop management on microbial control, in *Microbial Control of Insects and Mites,* Burgess, H. D. and Hussey, N. W., Eds., Academic Press, New York, 1971, 407.

41. **Tanada, Y.,** Epizootiology of infectious diseases, in *Insect Pathology: An Advanced Treatise,* Vol. 2, Steinhaus, E. A., Ed., Academic Press, New York, 1963, 423.

42. **Capinera, J. L. and Barbosa, P.,** Transmission of a nuclear polyhedrosis virus to gypsy moth larvae by *Calosoma sycophanta, Ann. Entomol. Soc. Am.,* 68, 593, 1975.

43. **Doane, C. C. and Redys, J. J.,** Characteristics of motile strains of *Streptococcus faecalis* pathogenic to larvae of the gypsy moth, *J. Invertebr. Pathol.,* 15, 420, 1970.

44. **Brooks, W. M.,** Protozoa: host-parasite-pathogen interrelationships, *Misc. Publ. Entomol. Soc. Am.,* 9, 105, 1973.

45. **Paillot, A.,** Sur *Thelohania mesnili,* microsporidie nouvelle parasite des chenilles de *Pieris brassicae L, C. R. Soc. Biol.,* 90, 501, 1924.

46. **Paillot, A.,** Sur la transmission des maladies à microsporidies chez les insectes, *C. R. Soc. Biol.,* 90, 504, 1924.

47. **Issi, I. V. and Maslennikova, V. A.,** Rol najezdnika *Apanteles glomeratus* (L.) (Hymenoptera, Braconidae) v transmissii mikrosporidii *Nosema polyvora* (Blunck) (Protozoa, Microsporidia), *Entomol Obozr.,* 45, 494, 1966.

48. **Blunck, H.,** Über die bei *Pieris brassicae* L., ihren Parasiten und Hyperparasiten schmarotozenden Mikrosporidien, *Trans. 9th Int. Congr. Entomol.,* 1, 432, 1952.

49. **Blunck, H.,** Mikrosporidien bei *Pieris brassicae* L., ihren Parasiten und Hyperparasiten, *Z. Angew. Entomol.,* 36, 316, 1954.

50. **Blunck, H., Krieg, R., and Scholtyseck, E.,** Weitere untersuchungen über die Mikrosporidien von Pieriden und deren Parasiten und Hyperparasiten, *Z. Pflanzenkr.*, 66, 129, 1959.

51. **Chlorine, V.,** Sur une microsporidie nouvelle *Thelohania vanessae* parasite des chenilles de *Vanessa urticae L., Zentralbl. Bakteriol. Parasitenkd. Infektionskr. Hyg. Abt. 1 Orig.*, 117, 86, 1930.

52. **Lipa, J. J.,** Observations on development and pathogenicity of the parasite of *Aporia crataegi* L. (Lepidoptera) *Nosema aporiae* n. sp, *Acta Parasitol Pol.*, 5, 559, 1957.

53. **Lipa, J. J.,** Studia inwazjologicizne i epizootiologiczne nad kilkoma gatunkami peirwotniaków z rzedu Microsporidia pasózytaujoacymi w owadach, *Pr. Nauk Inst. Ochr. Rosl.*, 5, 103, 1963.

54. **Naville, A.,** Recherches cytologiques sur les schizogregarines. I. Le cycle evolutif de *Mattesia dispora* n.g., n. sp, *Z. Zellf, Mikroskop. Ant.*, 11, 375, 1930.

55. **Smirnoff, W. A.,** Transmission of *Herpetomonas swanei* sp. n. by means of *Neodiprion swanei* (Hymenoptera, Tenthredinidae) parasites, *Can. Entomol.*, 103, 630, 1971.

56. **Tanada, Y.,** Field observations on a microsporidian parasite of *Pieris rapae L, Proc. Hawaii Entomol. Soc.*, 15, 609, 1955.

57. **Toumanoff, C.,** À propos d'une infection à protozoaire des fausses teignes et du rôle de *Dibrachys boucheanus* Ratzb. dans la destruction de ces insects, *Rev. Fr. Apiculture*, 2, 251, 1950.

58. **Laigo, F. M. and Tamashiro, M.,** Interactions between a microsporidian pathogen of the lawn-armyworm and the hymenopterous parasite *Apanteles marginiventris, J. Invertebr. Pathol.*, 9, 546, 1967.

59. **Brooks, W. M. and Cranford, J. D.,** Microsporidoses of the hymenopterous parasites, *Campoletis sonorensis* and *Cardiochiles nigriceps,* larval parasites of *Heliothis* species, *J. Invertebr. Pathol.*, 20, 77, 1972.

60. **Hamm, J. J., Nordlund, D. A., and Multinix, B. G., Jr.,** Interaction of the microsporidian *Vairimorpha* sp. with *Microplitis croceipes* (Cresson) and *Cotesia marginiventris* (Cresson) (Hymenoptera: Braconidae), two parasitoids of *Heliothis zea* (Boddie) (Lepidoptera: Noctuidae), *Environ. Entomol.*, 12, 1547, 1983.

61. **Larsson, R.,** Transmission of *Nosema mesnili* (Paillot) (Microsporida, Nosematidae), a microsporidian parasite of *Pieris brassicae* L. (Lepidoptera, Pieridae) and its parasite *Apanteles glomeratus* L. (Hymenoptera, Braconidae), *Zool. Anz.*, 203, 151, 1979.

62. **McNeil, J. N. and Brooks, W. M.,** Interactions of the hyperparasitoids *Catolacus aeneoviridis* (Hym:Pteromalidae) and *Spilochalcis side* (Hym:Chalcididae) with the microsporidans *Nosema heliothidis* and *N. campoletidis, Entomophaga.*, 19, 195, 1974.

63. **McLaughlin, R. E. and Adams, C. H.,** Infection of *Bracon mellitor* (Hymenoptera:Braconidae) by *Mattesia grandis* (Protozoa: Neogregarinida), *Ann. Entomol. Soc. Am.*, 59, 800, 1966.

64. **Laigo, F. M. and Pashke, J. D.,** *Pteromalus puparum* L. parasites reared from granulosis and microsporidiosis infected *Pieris rapae* L. chrysalids, *Philipp. Agric.*, 52, 430, 1968.

65. **Andreadis, T. G.,** *Nosema pyrausta* infection in *Macrocentrus grandii* a braconid parasite of the European corn borer, *Ostrinia nubilalis, J. Invertebr. Pathol.*, 39, 299, 1980.

66. **Nealis, V. C. and Smith, S. M.,** Interaction of *Apanteles fumiferanae* (Hymenoptera: Braconidae) and *Nosema fumiferanae* (Microsporidia) parasitizing spruce budworm, *Choristoneura fumiferana* (Lepidoptera: Tortricidae), *Can. J. Zool.*, 65, 2047, 1987.

67. **Masera, E.,** Rapporti fra *Apanteles glomeratus* Reinh. et *Pieris brassicae* L. infette di pebrina, in *Actes 7 Internat. Seric. Congr.*, Alès, France, 1948, 551.

68. **Hostounsky, Z.,** *Nosema mesnili* (Paill.) a microsporidian of the cabbageworm, *Pieris brassicae* (L.) in the parasites *Apanteles glomeratus* (L.), *Hyposoter ebenius* (Grav.) and *Pimpla instigator* (F.), *Acta Entomol. Bohemoslov.*, 67, 1, 1970.

69. **Allen, H. W. and Brunson, M. H.,** A microsporidian in *Macrocentrus ancylivorus, J. Econ. Entomol.*, 38, 393, 1945.

70. **Allen, H. W. and Brunson, M. H.,** Control of *Nosema* disease of potato tuber worm, a host used in the mass production of *Macrocentrus ancylivorus, Science*, 105, 394, 1947.

71. **McCoy, E. E.,** Elimination of a microsporidian parasite in the mass rearing of *Macrocentrus ancylivorus, J. N. Y. Entomol. Soc.*, 55, 51, 1947.

72. **Allen, H. W.,** *Nosema* disease of *Gnorimoschema opercullella* (Zeller) and *Macrocentrus ancylivorus* Rohmer, *Ann. Entomol. Soc. Am.*, 47, 407, 1954.

73. **Grosch, D. S.,** The relation of the midgut to growth and development of *Habrobracon* with a pertinent note on sporozoan infection, *J. Elisha Mitchell Sci. Soc.*, 65, 61, 1949.

74. **Thompson, H. M.,** The effect of a microsporidian parasite of the spruce budworm, *Choristoneura fumiferana* (Clem.) on two internal hymenopterous parasites, *Can. Entomol.*, 90, 694, 1958.

75. **Thompson, H. M.,** A list and brief description of the Microsporidia infecting insects, *J. Insect Pathol.*, 2, 346, 1960.

76. **Weiser, J.,** Die Mikrosporidien als Parasiten der Insekten, *Monogr. Angew. Entomol.*, 17, 1, 1961.

77. **Own, O. S. and Brooks, W. M.,** Interactions of the parasite *Pediobius foveolatus* (Hymenoptera: Eulophidae) with two *Nosema* spp. (Microsporida: Nosematidae) of the Mexican bean beetle (Coleoptera: Coccinellidae), *Environ. Entomol.*, 15, 32, 1986.

78. **Larsson, R.,** Transmission of *Nosema mesnili* (Paillot) (Microsporida, Nosematidae), a microsporidian parasite of *Pieris brassicae* L. (Lepidoptera, Pieridae) and its parasite *Apanteles glomeratus* L. (Hymenoptera, Braconidae), *Zool. Anz.,* 203, S., 151, 1979.

79. **Maddox, J. V.,** Protozoan diseases, in *Epizootiology of Insect Disease,* Fuxa, J. R., and Tanada, Y., Eds., John Wiley & Sons, New York, 1987, 417.

80. **Huger, A. M.,** Susceptibility of the egg parasitoid *Trichogramma evanescens* to the microsporidium *Nosema pyrausta* and its impact on fecundity, *J. Invertebr. Pathol.,* 44, 228, 1984.

81. **Andreadis, T. G.,** Impact of *Nosema pyrausta* on field populations of *Macrocentrus grandii* an introduced parasite of the European Corn Borer *Ostrinia nubilalis, J. Invertebr. Pathol.,* 39, 298, 1982.

82. **Kalavatti, C. and Narasimhammurti, C. C.,** A new microsporidian *Pleistophora eretesi* n. sp. from *Eretes sticticus* (L.) (Dytiscidae; Coleoptera), *Acta Protozool.,* 15, 139, 1976.

83. **Kalavatti, C. and Narasimhammurti, C. C.,** A new microsporidian parasite *Toxoglugea tillargi* sp. n. from an odonate *Tholymis tillarga, Acta Protozool.,* 17, 279, 1978.

84. **Lipa, J. J.,** *Nosema coccinellae* sp. n., a new microsporidian parasite of *Coccinella septempunctata, Hippodama tredecimpunctata* and *Myrrha octodecimguttata, Acta Protozool.,* 5, 369, 1968.

85. **Lipa, J. J. and Steinhaus, E. A.,** *Nosema hippodamiae* n. sp., a microsporidian parasite of *Hippodamia convergens* (Guerin) (Coleoptera, Coccinellidae), *J. Insect Pathol.,* 1, 304, 1959.

86. **Maddox, J. V. and Webb, D. W.,** A new species of *Nosema* from *Hylobittacus apicalis* (Insecta: Mecoptera: Bittacidae), *J. Invertebr. Pathol.,* 42, 207, 1983.

87. **Weiser, J.,** Die Mikrosporidien als Parasiten der Insekten, *Mongr. Angew Entomol.,* 17, 149, 1961.

88. **Young, O. P. and Hamm, J. J.,** The compatibility of two fall armyworm pathogens with a predaceous beetle, *Calosoma sayi* (Coleptera: Carabidae), *J. Entomol. Sci.,* 20, 212, 1985.

89. **Marti, O. G. and Hamm, J. J.,** Effect of *Vairimorpha* sp. on the survival of adult *Calleida decora* in the laboratory, *J. Agric. Entomol.,* 3, 242, 1986.

90. **Powell, W., Wilding, N., Brokyn, J., and Clark, S. J.,** Interference between parasitoids [Hym: Aphidiidae] and fungi [Entomophthorales] attacking cereal aphids, *Entomophaga,* 31, 293, 1986.

91. **Velasco, L. R. I.,** Field parasitism of *Apanteles plutellae* Kurdj. (Braconidae, Hymenoptera) on the diamond-back moth on cabbage, *Philipp., Entomol.,* 6, 539, 1983.

92. **Milner, R. J., Lutton, G. G., and Bourne, J.,** A laboratory study of the interaction between aphids, fungal pathogens and parasites, in *Proc. 4th Austr. Appl. Entomol. Res. Conf.,* Bailey, P. and Swincer, D., Eds., South Australian Dept. Agric., Adelaide, 1984, 375.

93. **King, E. G. and Bell, J. V.,** Interactions between a braconid, *Microplitis croceipes* and a fungus *Nomuraea rileyi,* in laboratory-reared bollworm larvae, *J. Invertebr. Pathol.,* 31, 337, 1978.

94. **Los, L. M. and Allen, W. A.,** Incidence of *Zoophthora phytonomi* [Zygomycetes; Entomophthorales] in *Hypera postica* (Coleoptera: Curculionidae) larvae in Virginia, *Environ. Entomol.,* 12, 1318, 1983.

95. **Fransen, J. J. and van Lenteren, J. C.,** Interaction between the parasitoid *Encarsia formosa* and the pathogen *Aschersonia aleyrodis* in the control of greenhouse whitefly, *Trialeurodes vaporariorum:* host selection and survival of the parasitoid in the presence of hosts infected with the fungus, in *Aschersonia aleyrodis as a Microbial Control agent of Greenhouse Whitefly,* Fransen, J. J., Ed., Dissertation, Landbouwenwersiteit te Wageningen, Netherlands, 1987, chap. 8.

96. **Fransen, J. J. and van Lenteren, J. C.,** Interaction between the parasitoid *Encarsia formosa* and the pathogen *Aschersonia aleyrodis* in the control of greenhouse whitefly; *Trialeurodes vaporariorum:* survival of the parasitoid after treatment of parasitized hosts with fungal spores, in *Aschersonia aleyrodis as a Microbial Control agent of Greenhouse Whitefly,* Fransen, J. J., Ed., Dissertation, Laudbouwenwersiteit te Wageningen, Netherlands, 1987, chap. 9.

97. **Ullyett, G. C., and Schonken, D. B.,** A fungus disease of *Plutella maculipennis* Curt., in South Africa, with notes on the use of entomogenous fungi in insect control, *Union South Afr. Dept. Agric. Sci. Bull.,* 218, 24, 1940.

98. **Keller, S. S.,** Histologische Untersuchungen an parasitierten, *Entomophthora*-infizierten Erbsenblattlausen, *Acrythosiphon pisum* Mitt. Schweiz *Entomol. Ges.,* 48, 247, 1975.

99. **Führer, E., El-Sufty, R., and Willers, D.,** Antibiotic effects of entomophagous endoparasites against microorganisms with the host body, *in 4th Int. Congr. Parasitol., Section F,* Warsaw, Poland, 1978, 100.

100. **El-Sufty, R. and Führer, E.,** Parasitäre Veränderungen der Wirtskuitikula bei *Pieris brassicae* und *Cydia pomonella* durch entomophage Endoparasiten, *Entomol. Exp. Appl.,* 30, 134, 1981.

101. **El-Sufty, R. and Führer, E.,** Wechselbeziehungen zwischen *Cydia pomonella* L. (Lep., Tortricidae), *Ascogaster quadridentatus* Wesm. (Hym., Braconidae) und dem Pilz *Beauveria bassiana* (Bals.) Vuill, *Z. Angew. Entomol.,* 99, 504, 1985.

102. **Führer, E. and El-Sufty, R.,** Produktion fungistatisches Metabolite durch Teratocyten von *Apanteles glomeratus L, Z. Parasitol.,* 59, 21, 1979.

103. **Willers, D., Lehmann-Danzinger, H., and Führer, E.,** Antibacterial and antimycotic effect of a newly discovered secretion from larvae of an endoparasitic insect, *Pimpla turionellae* L. (Hym.) *Arch. Microbiol.,* 133, 225, 1982.

104. **Willers, D. and Lehmann-Danzinger, H.,** Überleben endoparasitischer Hymenopteren in Puppen von Lepidopteren durch Hemmung der phenoloxidase, *Z. Parasitol.,* 70, 403, 1984.

105. **Führer, E. and Willers, D.,** The anal secretion of the endoparasitic larva *Pimpla turionellae:* sites of production and effects *J. Insect Physiol.,* 32, 361, 1986.

106. **Voukassovitch, P.,** Contribution à l'étude d'un champignon entomophyte *Spicaria farinosa* (Fries) var. *verticilloides* Fron, *Ann. Inst. Natl. Rech Agron.,* C., 2, 73, 1925.

107. **Teague, T. G., Horton, D. L., Yearian, W. C., and Phillips, J. R.,** Benomyl inhibition of *Cotesia marginiventris* in four lepidopterous hosts, *J. Entomol. Sci.,* 20, 76, 1985.

108. **Sewall, D. K. and Croft, B. A.,** Chemotherapeutic and nontarget side-effects of benomyl to the orange tortrix, *Argyrotaenia citrana,* (Lepidoptera: Tortricidae) and braconid endoparasite *Apanteles aristoteliae* (Hymenoptera: Braconidae) *Environ. Entomol,* 16, 507, 1987.

109. **Ishibashi, N., Young, Fah-Zu, Nakashima, M., Abiru, C., and Haraguchi, N.,** Effects of applications of DD-136 on silkworm *Bombyx mori,* predatory insect, *Agriosphodorus dohrni,* parasitoid, *Trichomalus apanteloctenus,* soil mites, and other non-target soil arthropods, with brief notes of feeding behavior and predatory pressure of soil mites, tardigrades, and predatory nematodes on DD-136 nematodes, in *Recent Advances in Biological Control of Insect Pests by Entomophagous Nematodes in Japan,* Ishibashi, N., Ed., Saga University, Saga, Japan, 1987, 158.

110. **Kaya, H. K.,** Interactions between *Neoaplectana carpocapsae* (Nematoda:Steinermatidae) and *Apanteles militaris* (Hymenoptera: Brachondiae), a parasitoid of the armyworm *Pseudaletia unipuncta, J. Invertebr. Pathol.,* 31, 358, 1978.

111. **Kaya, H. K.,** Infectivity of *Neoaplectana carpocapsae* and *Heterorhabditis heliothididis* to pupae of the parasite *Apanteles militaris, J. Nematol.,* 10, 241, 1978.

112. **Kaya, H. K. and Hotchkin, P. G.,** The nematode *Neoaplectana carpocapsae* Weiser and its effect on selected ichneumonid and braconid parasites, *Environ. Entomol.,* 10, 474, 1981.

113. **Triggiani, D.,** Influenza dei nematodi della famiglia *Steinernematidae* e *Heterorhabditidae* sue parassitoide *Apanteles ultor* RHD (Hymenoptera: Braconidae), *La Difesa delle Piante* 2, 293, 1985.

114. **Kaya, H. K.,** Entomogenous nematodes for insect control, in *Biological Control in Agricultural IPM Systems,* Hoy, M. A. and Herzog, D. C., Eds., Academic Press, New York, 1985, 283.

115. **Kaya, H. K.,** Effect of the entomogenous nematode *Neoaplectana carpocapsae* on the tachinid parasite *Compsilura concinnata* (Diptera: Tachinidae), *J. Nematol.,* 16, 9, 1984.

116. **Mrácek, Z. and Spitzer, K.,** Interaction of the predators and parasitoids of the sawfly, *Cephalcia abietis* (Pamphilidae: Hymenoptera) with its nematode *Steinernema kraussei, J. Invertebr. Pathol.,* 42, 397, 1983.

117. **Georgis, R. and Wojcik, W.,** The effect of entomogenous nematodes *Heterorhabditis heliothidis* and *Steinernema feltiae* on selected predatory soil insects, *J. Nematol.,* 19, 523, 1987.

118. **Stoltz, D. B. and Vinson, S. B.,** Viruses and parasitism in insects, *Adv. Virus Res.,* 24, 125, 1979.

119. **Stoltz, D. B., Krell, P., Summers, M. D., and Vinson, S. B.,** Polydnaviridae — a proposed family of insect viruses with segmented, double-stranded, circular DNA genomes, *Intervirology* 21, 1, 1984.

120. **Krell, P. J. and Stoltz, D. B.,** Unusual baculovirus of the parasitoid wasp *Apanteles melanoscelus,* isolation and preliminary characterization, *J. Virol.,* 29, 1118, 1979.

121. **Krell, P. J. and Stoltz, D. B.,** Virus-like particles in the ovary of an ichneumonid wasp, purification and preliminary characterization, *Virology,* 101, 408, 1980.

122. **Stoltz, D. B., Krell, P. J., and Vinson, S. B.,** Polydisperse viral DNAs in ichneumonid ovaries, a survey., *Can. J. Microbiol.,* 27, 123, 1981.

123. **Krell, P. J., Summers, M. D., and Vinson, S. B.,** Virus with a multipartite superhelical DNA genome from the ichneumonid parasitoid *Campoletis sonorensis, J. Virol.,* 43, 859, 1982.

124. **Blissard, G. W., Fleming, J. G. W., Vinson, S. B., and Summers, M. D.,** *Campoletis sonorensis* virus: expression in *Heliothis virescens* and identification of expressed sequences, *J. Insect Physiol.,* 32, 351, 1986.

125. **Edson, K. M., Vinson, S. B., Stoltz, D. B., and Summers, M. D.,** Virus in a parasitoid wasp; suppression of the cellular immune response in the parasitoid's host, *Science,* 211, 582, 1978.

126. **Davies, D. H., Strand, M. R., and Vinson, S. B.,** Changes in differential haemocyte count and *in vitro* behavior of plasmatocytes from host *Heliothis virescens* caused by *Campoletis sonorensis* polydna virus, *J. Insect Physiol.,* 33, 143, 1987.

127. **Dover, B. A., Davies, D. H., Strand, M. R., Gray, R. S., Keeley, L. L., and Vinson, S. B.,** Ecdysteroid titer reduction and developmental arrest of last-instar *Heliothis virescens* larvae by calyx fluid from the parasitoid, *Campoletis sonorensis, J. Insect Physiol.,* 33, 333, 1987.

128. **Theilmann, D. A. and Summers, M.D.,** Molecular analysis of *Campoletis sonorensis* virus DNA in the lepidopteran host *Heliothis virescens, J. Gen. Virol.,* 67, 1961, 1986.

129. **Hamm, J. J. and Styer, E. L.,** A new virus associated with the reproductive tract of the hymenopterous parasitoid *Cotesia marginiventris* and its replication in noctuid host larvae, *in Proc. 18th Annu. Mtg. Soc. Invertebr. Pathol.* (abstract), 1985.

130. **Styer, E. L., Hamm, J. J., and Nordlund, D. A.,** A new virus associated with the parasitoid *Cotesia marginiventris* (Hymenoptera:Braconidae): replication in noctuid host larvae, *J. Invertebr. Pathol.,* 50, 302, 1987.

131. **Stoltz, D. B. and Faulkner, P.,** Apparent replication of an unusual virus-like particle in both a parasitoid wasp and its host, *Can. J. Microbiol.,* 24, 1509, 1978.

132. **Stoltz, D. B., Krell, P., Cook, D., Mackinnon, E. A., and Lucarotti, C. J.,** An unusual virus from the parasitic wasp *Cotesia melanoscelus, Virology,* 162, 311, 1988.

133. **Irabagon, T. A. and Brooks, N. M.,** Interaction of *Campoletis sonorensis* and a nuclear polyhedrosis virus in larvae of *Heliothis virescens, J. Econ. Entomol.,* 67, 229, 1974.

134. **Bird, F. T.,** Transmission of some insect viruses with particular reference to ovarial transmission and its importance in the development of epizootics, *J. Insect Pathol.,* 3, 352, 1961.

135. **Raimo, B., Reardon, R. C., and Podgwaite, J. D.,** Vectoring gypsy moth nuclear polyhedrosis virus by *Apanteles melanoscelus* [Hym: Braconidae], *Entomophaga,* 22, 207, 1977.

136. **Hamm, J. J., Nordlund, D. A., and Marti, O. C.,** Effects of a nonoccluded virus of *Spodoptera frugiperda* (Lepidoptera:Noctuidae) on the development of a parasitoid, *Cotesia marginiventris* (Hymenoptera:Braconidae), *Environ. Entomol.,* 14, 258, 1985.

137. **Kelsey, J. M.,** Interaction of virus and insect parasites of *Pieris rapae L., in 11th Int. Cong. Entomol.,* Vol. 4, Vienna, 1962, 790.

138. **Versoi, P. L. and Yendol, W. G.,** Recognition of virus-diseased gypsy moth larvae by *Apanteles melanoscelus* Ratzeburg (Hymenoptera:Braconidae), *J. N.Y. Entomol. Soc.,* 86, 325, 1978.

139. **Versoi, P. L. and Yendol, W. G.,** Discrimination by the parasite, *Apanteles melanoscelus* between healthy and virus-infected gypsy moth larvae, *Environ. Entomol.,* 11, 42, 1982.

140. **Beegle, C. C. and Oatman, E. R.,** Differential susceptibility of parasitized and nonparasitized larvae of *Trichoplusia ni* to a nuclear polyhedrosis virus, *J. Invertebr. Pathol.,* 24, 188, 1974.

141. **Hotchkin, P. G. and Kaya, H. K.,** Pathological response of the parasitoid, *Glyptapanteles militaris,* to nuclear polyhedrosis virus-infected armyworm host, *J. Invertebr. Pathol.,* 42, 51, 1983.

142. **Hotchkin, P. G. and Kaya, H. K.,** Interactions between two baculoviruses and several insect parasites, *Can. Entomol.,* 115, 841, 1983.

143. **Federichi, B. A.,** Enveloped double-stranded DNA insect virus with novel structure and cytopathology, *Proc. Natl. Acad. Sci. U.S.A.,* 80, 7664, 1983.

144. **Kaya, H. K.,** Toxic factor produced by a granulosis virus in armyworm larva: effect on *Apanteles militaris, Science,* 168, 251, 1970.

145. **Kaya, H. K. and Tanada, Y.,** Response of *Apanteles militaris* to a toxin produced in a granolosis-virus-infected host, *J. Invertebr. Pathol.,* 19, 1, 1972.

146. **Kaya, H. K. and Tanada, Y.,** Hemolymph factor in armyworm larvae infected with a nuclear-polyhedrosis virus toxic to *Apanteles militaris, J. Invertebr. Pathol.,* 21, 211, 1973.

147. **Tower, D. G.,** Comparative study of amount of food eaten by parasitized and non-parasitized larvae of *Cirphis unipuncta, J. Agric. Res.,* 6, 455, 1916.

148. **Rahman, M.,** Effect on parasitism on food consumption of *Pieris rapae* larvae, *J. Econ. Entomol.,* 63, 820, 1970.

149. **Vinson, S. B. and Iwantsch, G. F.,** Host regulation by insect parasitoids, *Q. Rev. Biol.,* 55, 143, 1980.

150. **Reardon, R. C. and Podgwaite, J. D.,** Disease parasitoid relationships in natural populations of *Lymantria dispar* [Lep.:Lymantriidae] in the Northeastern United States, *Entomophaga,* 21, 333, 1976.

151. **Godwin, P. A. and Shields, K. S.,** Effects of *Blepharipa pratensis* [Diptera:Tachinidae] on the pathogenicity of nucleopolyhedrosis virus in stage V of *Lymantria dispar* [Lep.:Lymantriidae], *Entomophaga,* 29, 381, 1984.

152. **Weseloh, R. M.,** Host recognition behavior of the tachinid parasitoid *Compsilura concinnata, Ann. Entomol. Soc. Am.,* 73, 593, 1980.

153. **Elsey, K. D. and Rabb, R. L.,** Biology of *Voria ruralis* (Diptera:Tachinidae), *Ann. Entomol. Soc. Am.,* 63, 216, 1970.

154. **Levin, D. B., Laing, J. E., and Jaques, R. P.,** Transmission of granulosis virus by *Apanteles glomeratus* to its host *Pieris rapae, J. Invertebr. Pathol.,* 34, 317, 1979.

155. **Vail, P. V.,** Cabbage looper nuclear polyhedrosis virus-parasitoid interactions, *Environ. Entomol.,* 10, 517, 1981.

156. **Smirnoff, W. A.,** Predators of *Neodiprion swainei* Midd. (Hymenoptera: Tenthredinidae) larval vectors of virus diseases, *Can. Entomol.,* 91, 246, 1959.

157. **Vago, C., Fosset, J. and Bergoin, M.,** Dissemination des virus de polyèdres par les éphippigères predateurs d'insectes, *Entomphaga,* 11, 177, 1966.

158. **Hostetter, D. L.,** A virulent nuclear — polyhedrosis virus of cabbage looper, *Trichoplusia ni,* recovered from the abdomens of sarcophagid flies, *J. Invertebr. Pathol.,* 17, 130, 1971.

159. **Franz, J., Krieg, A., and Langenbuck, R.,** Untersuchungen über den Einfluss der Passage durch den Darm von Raubinsekten und Vögeln auf die Infektiosität Insektenpathogener Viren, *Z. Pflanzenkr. Pflanzenpathol. Pflanzenschutz Sonderh.,* 62, 721, 1955.

160. **Cooper, D. J.,** The role of predatory Hemiptera in disseminating a nuclear polyhedrosis virus of *Heliothis punctigera, J. Aust. Entomol. Soc.,* 20, 145, 1981.

161. **Burgess, A. F.,** *Calosoma sycophanta:* its life history, behavior, and successful colonization in New England, *Bull. Bur. Entomol. U.S. Dep. Agric.,* 101, 1911.

162. **Beekman, A. G.,** The infectivity of polyhedra of nuclear polyhedrosis virus (NPV) after passage through gut of an insect-predator, *Experientia,* 36, 858, 1980.

163. **Young, S. W. and Yearian, W. C.,** *Nabis roseipennis* adults (Hemiptera:Nabidae) as disseminators of nuclear polyhedrosis virus to *Anticarsia gemmatalis* (Lepidoptera:Noctuidae) larvae, *Environ. Entomol.,* 16, 1330, 1987.

164. **Abbas, M. S. T. and Boucias, D. G.,** Interaction between nuclear polyhedrosis virus-infected *Anticarsia gemmatalis* (Lepidoptera:Noctuidae) larvae and predator *Podisus maculiventris* (Say) (Hemiptera:Pentatomidae), *Environ. Entomol.,* 13, 599, 1984.

165. **Biever, K. D., Andrews, P. L., and Andrews, P. A.,** Use of a predator, *Podisus maculiventris* to distribute virus and initiate epizootics, *J. Econ. Entomol.,* 75, 150, 1982.

166. **Young, D. P. and Hamm, J. J.,** The effect of the consumption of NPV-infected dead fall armyworm larvae on the longevity of two species of scavenger beetles, *J. Entomol. Sci.,* 20, 90, 1955.

167. **Milner, R. J.,** Pathogen importation for biological control—risks and benefits: Pests and parasites as migrants: an Australian perspective, *in Proc. Aust. and New Z. Assoc. Adv. of Sci./Aust. Acad. Sci. Joint Symp. of Toxic Disease,* Canberra, May 1984, 1985, chap. 18.

168. **Tamashiro, M.,** Effect of insect pathogens on some biological control agents in Hawaii, *in Proc. Joint U.S. Japan Sem. Microbiol. Control Insect Pests,* Fukuoka, April 21-23, 1967, 1968, 147.

169. **Kevan, P. G. and Baker, H. G.,** Insects as flower visitors and pollinators, *Annu. Rev. Entomol.,* 28, 407, 1983.

170. **Fuxa, J. R. and Tanada, Y.,** *Epizootiology of Insect Diseases,* John Wiley & Sons, New York, 1987.

171. **Kaya, H. K.,** *Steinernema feltiae* use against foliage feeding insects and effect on nontarget insects, *in Fundamental and Applied Aspects of Invertebrate Pathology,* Samson, R. A., Vlak, J. M., and Peters, D., Eds., Found. 4th Int. Colloq. Invertebr. Pathol., Wageningen, The Netherlands, 1986, 268.

172. **Cantwell, G. E., Lehnert, T., and Fowler, J.,** Are biological insecticides harmful to the honey bee?, *Am. Bee J.,* p. 255, 1972.

173. **Dutky, S. R. and Hugh, W. S.,** Note on a parasite nematode from codling moth larvae, *Carpocapsa pomonella, Entomol. Soc. Wash. Proc.,* 57, 244, 1955.

174. **Kaya, H. K., Marston, J. M., Lindegren, J. E., and Peng, Y. S.,** Low susceptibility of the honey bee, *Apis mellifera* L. (Hymenoptera:Apidae), to the entomogenous nematode *Neoplectana carpocapsae* Weiser, *Environ. Entomol.,* 11, 920, 1982.

175. **Hackett, K. J. and Poinar, G. O., Jr.,** The ability of *Neoplectana carpocapsae* Weiser (Steinernematidae: Rhabditoidea) to infect adult honeybees *Apis mellifera,* Apidae, Hymenoptera), *Am. Bee J.,* 113, 100, 1973.

176. **Cantwell, G. E., Knox, D. A., Lehnert, T., and Michael, A. S.,** Mortality of the honey bee, *Apis mellifera,* in colonies treated with certain biological insecticides, *J. Invertebr. Pathol.,* 8, 228, 1966.

177. **Knox, D. A.,** Tests of certain insect viruses on colonies of honey bees, *J. Invertebr. Pathol.,* 16, 152, 1970.

178. **Haragsim, O. and Vankova, J.,** Effet pathogène comparé de 12 souches de *Bacillus thuringiensis* Berl. sur l'abeille domestique et son couvain, *Ann.Abeille (Paris),* 11, 31, 1968.

179. **Krieg, A. and Herfs, W.,** Über die wirkung von *Bacillus thuringiensis* auf Bienen, *Entomol. Exp. Appl.,* 6, 1, 1963.

180. **Celli, G. and Giordani, G.,** Richerche sull'attivita del *Bacillus thuringiensis* Berliner in riguardo all' *Apis mellifera,* L. *Boll. Ist. Ento. Univ. Studi. Bologna* 28, 141, 1966.

181. **Leskova, A. Y. and Kulikov, N. S.,** Action of Entobacterin-3 and Thuricide on bees (in Russian), *Pchelovodstvo,* 40, 3233, 1963.

182. **Johansen, C.,** Impregnated foundation for waxmoth control, *Glean. Bee Cult.,* 90, 682, 1962.

183. **Wilson, W. T.,** Observations on the effects of feeding large quantities of *Bacillus thuringiensis* Berliner to honey bees, *J. Insect Pathol.* 4, 269, 1962.

184. **Burges, H. D.,** *Bee World,* 59, 129, 1984.

185. **Bailey, L.,** *Honey Bee Pathology,* Academic Press, New York, 1982.

186. **Applebaum, S. W.,** Biochemistry of digestion, *in Comprehensive Insect Physiology, Biochemistry and Pharmacology.,* Vol. 4, Kerkut, G. A. and Gilbert, L. I., Eds., Pergamon Press, Oxford, 1985, 279.

187. **Evans, H. G. and Entwistle, P. F.,** Viruses, *in Epizootiology of Insect Diseases,* Fuxa, J. R. and Tanada, Y., Eds., John Wiley & Sons New York, 1987, 258.

188. **Lipa, J. J.,** Miscellaneous observations on protozoan infections of *Nepa cinerea* Linnaeus including descriptions of two previously unknown species of microsporidia, *Nosema bialoviesianae* sp. n. and *Thelohania nepae* sp. n., *J. Invertebr. Pathol.,* 8, 158, 1966.

189. **Andreadis, T. G.,** Transmission, in *Epizootiology of Insect Diseases,* Fuxa, J. R. and Tanada, Y., Eds., John Wiley & Sons, New York, 1987, 159.

190. **Lysenko, O.,** Principles of pathogenesis of insect bacterial diseases as exemplified by the nonsporeforming bacteria, in *Pathogenesis of Invertebrate Microbial Diseases,* Davidson, E. W., Ed., Allanheld, Osmun Publishers, Totowa, NJ, 1981, 163.

191. **Sweetman, H. L.,** *Principals of Biological Control,* Wm. C. Brown Publishing, Bubridge, IA, 1958.

192. **Askew, R. R.,** *Parasitic Insects,* American Elsevier, New York, 1971.

193. **Stairs, G. R.,** Artificial initiation of virus epizootics in forest tent caterpillar populations, *Can. Entomol.,* 97, 1059, 1965.

194. **Starr, C. K.,** Enabling mechanism in the origin of sociality in the Hymenoptera: the sting's the thing, *Ann. Entomol. Soc. Am.,* 78, 836, 1985.

195. **Kurstak, E. S.,** Étude des relations entre l'infection à *Bacillus thuringiensis* Berliner et le parasitism par *Nemeritis canescens* (Gravenhorst) (Ichneumonidae) chez *Ephestia kuhniella* Zeller (Pyralidae) *Ann. Epiphyties,* 17, 451, 1966.

196. **Tanada, Y.,** Epizootiology and microbial control, in *Comparative Pathiobiology. Biology of the Microsporidia* Vol. 1, Bulla., L. A., Jr. and Cheng, T. C., Eds., Plenum Press, New York, 1976, 247.

197. **Stoltz, D. B., Guzo, D., and Cook, D.,** Studies on polydna-virus transmission, *Virology,* 155, 120, 1986.

198. **Vinson, S. B. and Iwantsch, G. F.,** Host suitability for insect parasitoids, *Annu. Rev. Entomol.,* 25, 397, 1980.

199. **Vinson, S. B. and Barbosa, P.,** Interrelationships of nutritional ecology of parasitoids. *in: Nutritional Ecology of Insects, Mites, Spiders and Related Invertebrates,* Slansky, F., Jr. and Rodriquez, J. G., Eds., Wiley-Interscience, New York, 1987, 673.

200. **Martin, G., Juchault, P., and Legrand, J. J.,** Mise en évidence d'un micro-organisme intracytoplasmique symbiote de l'oniscoide *Armadillidium vulgare* Latr. dont la présence accompagne l'intersexualité ou la féminisation totale des mâles génétiques de la lignée thélygène, *C.R. Acad. Sci. (Paris),* 276, 2313, 1973.

201. **Bulnheim, H. P.,** Interaction between genetic, external and parasitic factors in sex determination of the crustacean amphipod *Gammarus duebeni, Helgol. Wiss. Meeresunters.,* 31, 1, 1978.

202. **Bull, J. J.,** *Evolution of Sex Determining Mechanisms,* Benjamin and Cummings, Menlo Park, CA, 1983.

203. **Skinner, S. W.,** *Son-killer:* a third extrachromosomal factor affecting the sex ratio in the parasitoid wasp, *Nasonia (= Mormoniella) vitripennis, Genetics,* 109, 745, 1985.

204. **Huger, A. M., Skinner, S. W., and Werren, J. H.,** Bacterial infections associated with the son-killer trait in the parasitoid wasp *Nasonia (= Mormoniella) vitripennis* (Hymenoptera:Pteromalidae), *J. Invertebr. Pathol.,* 46, 272, 1985.

205. **Walter, G. N.,** Differences in host relationships between male and female heteronomous parasitoids: a review of host location, oviposition and pre-imaginal physiology and morphology, *J. Entomol. Soc. S. Afr.,* 46, 261, 1983.

206. **Luck, R. F., Podaler, H., and Kfir, R.,** Host selection and egg allocation behavior by *Aphytes metinus* and *A. lingnanensis:* comparison of two facultatively gregarious parasitoids, *Ecol. Entomol.,* 7, 397, 1982.

207. **Nozato, K.,** The effect of host size on the sex ratio of *Itoplectis cristitae* Momoi, a pupal parasite of the Japanese pine shoot moth, *Petrova (= Evetria) cristata* (Walsingham), *Kontyu.,* 37, 134, 1969.

Chapter 6

POTENTIAL IMPACT OF MICROBIAL INSECTICIDES ON THE FRESHWATER ENVIRONMENT, WITH SPECIAL REFERENCE TO THE WHO/UNDP/WORLD BANK, ONCHOCERCIASIS CONTROL PROGRAMME

C. Dejoux and J.-M. Elouard

TABLE OF CONTENTS

I. INTRODUCTION

Inland waters, both flowing (lotic) and still (lenitic), are potential sites of collection, and at times concentration, of the multifarious byproducts of human activity. An excessive use of fertilizer on lakes or ponds, followed by runoff, leads to deleterious eutrophication. Equally familiar is the contamination of rivers by industrial wastes. More insidious, though very common, is the slow accumulation in continental waters of products, or compounds resulting from them, used tens or even hundreds of kilometers away and then windborne before being dropped to the earth's surface by rain (acid rains, drift of insecticidal dusts following aerial application, etc.).

Again, the control of disease vectors having aquatic developmental stages may necessitate introducing insecticides directly into waters, where their selectivity for the species under attack is never absolute. In all such cases it is essential to minimize the level of contamination so induced in aquatic ecosystems.

For many years the practical employment of microbial insecticides was restricted to agriculture and forestry, and little attention was paid to the effect of their toxicity to freshwater fauna. Only a few organisms, such as *Daphnia* spp., were tested in the laboratory, and records of the LD_{50} established for these crustaceans were simply added to other data required for the documentary support for the registration of relevant microbial control products. With growing appreciation of the hazards of transportation of pesticide molecules toward the aquatic biota, and of the direct addition of compounds with antivectorial properties into larval habitats, need has been highlighted for a better understanding of insecticide impact on aquatic organisms and the hydrosphere at large.

The nature of the particular toxicological tests to which microbial insecticides must be submitted in assessing their safety, depends on the manner in which these agents are to be employed in practice. It is obvious, for example, that the registration criteria for the use of *Bacillus moritai* Aizawa and Fujiyoshi (essentially applied against the housefly, *Musca domestica,* in domestic environments) will be far different from those required for *B. thuringiensis* spp. *israelensis* de Barjac if it must be directly introduced into lenitic or lotic waters for control of Culicidae or Simuliidae, respectively.

The consequences are that toxicity studies are often restricted to some laboratory tests carried out on one or two organisms (in general *Daphnia magna* and *Gambusia affinis*). In the case of insecticides introduced directly into water, the research has to concern different biological or ecological levels, right up to the observation of toxin impact on entire ecosystems.

II. WAYS OF EVALUATING THE IMPACT OF MICROBIAL INSECTICIDES ON NONTARGET AQUATIC ORGANISMS (NTOs)

The concept of laboratory evaluation of the potential toxicity of new insecticides to aquatic NTOs is not new. It was even urged as an absolute necessity by some of the experts participating in the meetings of WHO scientific working groups. For example, one such meeting recommended that, " . . . all products of natural origin, able to replace chemical insecticides, living or dead, have to be tested in depth in order to discard any risk of toxicity either to man or to any organisms other than the target species."[1]

More recently the same idea has been reiterated and enlarged upon in another WHO report, "An important consideration for the realization of vector control products is the study of the innocuity spectrum of the candidate compounds to the beneficial and non-target fauna which coexist with the vectors and exercise regulatory pressure upon them. This is true for chemical compounds as well as for biological ones. The ecological innocuity of biological compounds will be measured not only by the immediate response of isolated individuals or groups of organisms, but also by the research of effects on the population of variate organisms (predators as well as detrivors) which are living in the same biotopes as the target species."[2]

If these principles are valid, the toxicity tests necessary for the registration of a microbial insecticide will differ from one product to another, taking into account the future field of utilization. However, it has to be borne in mind that a product initially conceived for a specific type of use may later, as circumstances demand, be employed for a purpose other than that originally intended, following due modification of its original presentation or formulation.

Nevertheless, considerations of time and money often lead to use of compounds which are not intended to be introduced directly into aquatic biota, being tested only against certain "standard" laboratory organisms. Products destined for use in aquatic vector control must, however, undergo a variety of specific and graduated tests. It is also clear that only those products whose effectiveness against a specific target group has been clearly demonstrated will be submitted to a systematic and complete toxicity screening against aquatic fauna and flora.

III. ECOLOGICAL IMPACT OF THE MAIN MICROBIAL INSECTICIDES ON THE FRESHWATER BIOTA

A. *BACILLUS THURINGIENSIS* BERLINER

It was not until the 1950s, some 4 decades after its isolation (from the Mediterranean flour moth, *Ephestia kuehniella*) and description, that *B. thuringiensis* entered into use as a biological insecticide. Today, it is the sporulating bacterium most commonly used for crop protection and (via its subspecies *israelensis*[3,4]) vector control. First employed in its primary powder form and subsequently as formulations, *B. thuringiensis* has become generally accepted as an environmentally safe control agent.

Active against larvae of both Culicidae and Simuliidae, *B. thuringiensis* ssp. *israelensis* is directly introduced into water bodies, whether lenitic or lotic. This circumstance has occasioned many toxicity studies, first in the laboratory and later in the field, of its possible impact on freshwater ecosystems. The following pages present a general survey of relevant trials at different levels.

1. Laboratory Tests
a. Influence of Abiotic Factors

It is important to understand that some environmental factors capable of influencing the degree of efficacy of this agent against target vectors can also play an important role with respect to certain NTOs.

For example, the influence of temperature on *B. thuringiensis* ssp. *israelensis* has been extensively studied.[5-9] All concerned have concluded that this agent's activity declines rapidly when water temperature decreases. This important finding is potentially negative for the environment, insofar as tropical regions are those chiefly exposed to heavy applications for mosquito or blackfly control. Tropical aquatic NTOs may thus be more vulnerable than those of cold or temperate regions. It has been claimed, also, that storage of the more commonly used formulations (Vectobac®, Bactimos®, Teknar®) under tropical field conditions does not alter their efficacy. Variations in hydrogen-ion activity, at least between pH 6 and pH 10, have no influence on the activity of *B. thuringiensis*.[5] No information is available on the possible adverse effects of high electrical conductivity or dissolved salts, although the efficacy of *B. thuringiensis* ssp. *israelensis* against salt marsh mosquito larvae[10] indicates that sea-salt content is not specifically deleterious. However, a high ferrous content has been shown to negate the latter subspecies' larvicidal effect against *Aedes detritus*.[11] Similar consequences have been noted in the case of chlorine, which appears to destroy the delta-endotoxin.[12,13]

A high content of suspended matter in natural waters is held to contraindicate larval mosquito control,[14] leading as it does to a chelation of *B. thuringiensis* ssp. *israelensis* spores with organic particles; these agglomerations rapidly sedimenting, and thus ceasing to be available to a diversity of mosquito larvae.[15] Such a phenomenon must also be considered as negative for benthic detrivorous NTOs, for these would clearly have far greater access to the microbial agent's spores than were the latter to remain in suspension.[9,16]

Heavy exposure to sunlight/UV radiation leads to a rapid decrease in viable spore content, but does not reduce the larvicidal efficacy of *B. thuringiensis* ssp. *israelensis*.[17,18] Thus, UV action cannot be considered a basic factor of toxicity reduction for filter-feeding NTOs.

b. Toxicity in Bioassays

Other references to laboratory studies in the present context, indicating no adverse effects of *B. thuringiensis* ssp. *israelensis* to NTOs, include those of Colbo and Undeen,[19] Gallagher,[11] Dunn,[20] Garcia and colleagues,[21-25] Mastri,[26] Larget and de Barjac,[27,28] Miura et al.,[29] Pantuwatana,[30] Prasertphon,[31] and Rajagopalan.[32] Garcia et al.,[33] who noted that Dixidae were as susceptible to *B. thuringiensis* as were Culicidae, detected no adverse effects to chironomid larvae below 10^7 cells per ml. Working with Teknar at 25 and 100 ppm, Dejoux et al.[34] reported adverse effects to *Hydra* sp. only, while Schnetter et al.,[35] using a locally produced culture of ssp. *israelensis*, recorded 100% mortality in two species of Chironomidae, two of Chaoboridae, and Ceratopogonidae, at dosages ranging from 1.6 to 160 mg/1, but only minimal adverse effect on larval Tipulidae.

Globally, over the past 20 years, numerous trials have been conducted on *B. thuringiensis* toxicity. Many of these were not scientifically comparable with one another. In some cases, too, experimental conditions were imprecise. Nevertheless, the overall results clearly suggest that the introduction of *B. thuringiensis* ssp. *israelensis* into water affects only a handful of taxa. A broad spectrum of invertebrates has been investigated in the present connection. Garcia and colleagues[21-25,33] furnish preliminary evidence of vulnerability of species beyond the Nematocera to this overwhelmingly mosquito/simuliid entomopathogen. Burges[36] and WHO[37] summarize other initial field trials as well as laboratory ones. These findings permit us to correlate the absence of toxicity with defined testing conditions, particular formulations or specific concentrations. They suggest, too, that the aquatic stages of Diptera, primarily Culicidae and Simuliidae and their near relatives (Dixidae, Chaoboridae), and (some) Chironomidae are the target families and NTOs most vulnerable to *B. thuringiensis* ssp. *israelensis*.

2. Field Trials
a. Short-Term Studies

Investigations of the rather ephemeral effects of *B. thuringiensis* ssp. *israelensis* on NTOs, have been more numerous than those conducted on a longer-term basis. They are covered in detail by WHO[37] and in Chapter 12 herein, and clearly indicate a less-than-drastic adverse impact upon freshwater ecosystems; despite the results of these studies often having been achieved via methodologies not always well adapted to the purpose. Of course, in the case of a catastrophic effect, such immediate consequences as the mortality of aquatic organisms are self-evident.

When the impact is less drastic, only quantitative and comparative studies before and after treatment are able to demonstrate any adverse effect. In this case the methods generally used consist of collecting series of Surber samples and/or core samples in suitable places immediately before the insecticide reaches the aquatic biota, and obtaining other series from the same location at regular intervals after treatment. The use of artificial substrates in the same way[6] can also be considered a useful method in the case of short-term trials, as well as the study of 24-h *in situ* drift variations, covering an equal period (24 h, 48 h, etc.) before and after insecticide application in the case of treatment of running waters. This method provides useful indication of the immediate response of aquatic ecosystems to insecticides, microbiological as well as chemical. This method has been in regular use since the earlier days of the World Health Organization/United Nations Development Programme/World Bank/Onchocerciasis Control Programme (WHO/OCP) in the Volta River Basin,[34,38-40] these field results from the largest long-sustained practical usage of *B. thuringiensis* ssp. *israelensis* to date, revealing only the slightest of hazards to any of the NTOs tested. The two- or threefold increase in the diurnal or nocturnal drift intensity that was sometimes registered, represents only a minimal adverse environmental impact by comparison with the 20- to 40-fold such increase of day-drift regularly occurring in WHO/OCP streams after the application of different chemical compounds employed under identical ecological conditions.

The one drawback to this method is the impossibility of ascertaining precisely what part of the ecosystem participates in the drift as a result of the insecticidal application (whether chemical or microbiological). To explore this question we designed a gutter system able to reveal the drift pattern before and after the application. This allowed in-depth analysis of the participation of each NTO taxon or group of taxa in the increased drift, permitting precise assessment, at the end of each experiment, of the effects of the control agent on the various faunal elements assayed.[34,41,42] These and related assessments from elsewhere have been summarized by WHO.[37] Key papers with respect to WHO/OCP include those of Lacey et al.,[6] Dejoux et al.,[34] Gibon et al.,[38] Dejoux,[39] Yameogo,[40] and Elouard and Fairhurst.[43] The general outcome may be summed up in the customarily very strong mortality of Simuliidae accompanied by no significant increase of drift-rate for NTOs except for Chironomidae/Orthocladiinae,[39] and associated with a lack of adverse effect to the insect fauna (notably predators of simuliids such as Hydropsychidae located on stones in the full current previously thickly inhabited by immature *Simulium damnosum* s.l.).[34] The only noninsects detected as being particularly sensitive to *B. thuringiensis* ssp. *israelensis* were molluscs of the family Ancylidae.[39]

Relevant observations from places other than West Africa that have reported no adverse effects of *B. thuringiensis* ssp. *israelensis* on Chironomidae and a wide range of other NTOs include some from small streams, e.g., in Newfoundland, Canada, treated with a locally produced culture[19] and in New York, treated with Roger Bellon powder R-153-78 and Sandoz SAN 402/WDC;[8] and others from ponds, e.g., in Camargue, France, treated with Vectobac.[44] Reporting on trials with Teknar in a stream in Quebec, Back et al.[45] found no significant elevation of drift rate for Chironomidae, Ephemeroptera, Plecoptera, and Trichoptera, but reductions of 26 and 39% for larval chironomids of the genera *Eukiefferella* and *Polypedilum*, respectively, on artificial substrates. They also noted a high level of drift for Diptera/Blepharoceridae over the 2 d following application of the microbial preparation. However, in trials with Sandoz SAN 402/WDC in 0.02-ha experimental plots in California, Miura et al.[29] noted that while a wide range of other NTOs were unaffected, chironomid larvae collected immediately after treatment showed 100% mortality after 2 d. Severe adverse effects to Chironomidae resulting from the use of Abbott powder ABG-6108 in Florida, were recorded by Ali,[46] his mortality figures ranging from 23 to 61% (2 kg/ha) to 53 to 88% (10 kg/ha) in 4 × 6-m × 45 to 50-cm deep experimental ponds over a 4-week period; while in the same period there was 27 to 65% control of larval chironomids subjected to 3 kg/ha in a 1-ha golf course pond.

b. Medium-Term Studies

Investigations under this subheading have been fewer than those concerning short-term impacts of microbials on aquatic ecosystems, perhaps because of their costlier nature and the need for participation of qualified personnel over a relatively long period. Also, if we consider several months of regular observation to constitute a medium-term study, a further difficulty becomes apparent; namely that of distinguishing between the eventual impact of the tested compound and the natural evolution of the populations observed.

TABLE 1
Comparison of the Density of Invertebrates Living on Stones in the Current Between the Section Treated with *Bacillus thuringiensis* ssp. *israelensis*[a] and the Untreated Section

Taxa	Control zone		Treated zone		Difference of average density %
	N/m²	%	N/m²	%	
Chironomidae	4402	45.05	4205.5	84.0	−4.5
Simuliidae	94	0.96	39.5	0.79	−58.0
Ceratopogonidae	23.5	0.24	70.0	1.40	+66.4
Tipulidae	284.5	2.92	127.5	2.55	−55.2
Hydropsychidae	5028	51.46	562	11.23	−88.8
Baetidae	38.5	0.39	2	0.04	−94.8
Total	9770.5		5006.5		−48.8%

[a]   Weekly applications at a dosage of 1.6 mg/l/10′ for 9 weeks

It is necessary to know how these populations evolve naturally in order to correlate, for example, an eventual decrease in number of certain taxa to the insecticidal impact, without allowing the data to be distorted by seasonal variations which belong to a natural pattern.

A typical example is provided by the study carried out by Yameogo[40] on a small stream in central Ivory Coast. Weekly applications of *B. thuringiensis* ssp. *israelensis* (Sandoz 402 I.W. DC formulation) were carried out over 2 months. The evolution of the treated populations of invertebrates during this period was studied and compared to that of a similar population in an untreated section of the same stream.

Traditional methods such as core and Surber samples, utilization of artificial substrates, and *in situ* measurements of drift intensity were used.

Yameogo's conclusions were as follows: ''The Chironomidae (Orthocladiinae) demonstrated an immediate response to the first treatment (significant drift increase), but did not appear to be greatly affected by later applications, and their density on artificial substrates and on the stones in the current increased regularly. The Trichoptera/Hydroptilidae demonstrated the same pattern of evolution. However, the Trichoptera/Hydropsychidae and Philopotamidae populations reduced significantly in number during the two months of observation.''[40] We have studied Yameogo's data in more detail in order to take his results still further.[34]

If we examine the invertebrate fauna living on the stones in the current, it appears that the general pattern of increasing density of invertebrates during the observation period occurs in both the treated and the untreated section of the stream, but with a lower intensity in the treated section, a finding which is not statistically significant for $p = 0.01$.

The Chironomidae were completely unaffected by the treatment and remained numerous in both sections. The Ceratopogonidae were found to be more numerous in the treated section but there was a marked decrease in the density of Hydropsychidae and Tipulidae in the treated area (see Table 1). Similar results were found for the artificial substrates (Table 2), but in the case of sandy bottoms the results differed slightly (Table 3). There, a reduction in the numbers of Chironomini and Tanytarsini was apparent. This could be due to higher sedimentation of the microbial product on the sandy bottom than on the stones which were scoured regularly by the current.

In conclusion, it would appear that a modification of the populations occurs only after 2 months of regular treatments. This conclusion is based mainly on the decrease in density of groups which had never shown a particular sensitivity in short-term studies. Conversely, the Chironomidae which had appeared sensitive in short-term trials were apparently unaf-

TABLE 2
Comparison of the Average Density of Invertebrates Collected on Artificial Substrates Over a 7-Week Period, in the Treated and Control Areas

Taxa	% in control zone (7 samples)	% in treated zone (7 samples)	Difference
Chironomini	18.4	18.3	− 0.1
Tanytarsani	8.2	12.3	+ 4.1
Tanypodinae	1.7	2.5	+ 0.8
Orthocladiinae	8.4	35.3	+ 26.9
Simulium adersi	4.5	1.5	− 2.5
Simulium hargreavesi	0.2	0.5	+ 0.3
Simulium ruficorne	1.2	0	− 1.2
Baetidae	3.2	8.4	+ 5.2
Caenidae	0.3	1.7	+ 1.4
Hydropsychidae	51.8	14.7	− 37.1
Hydroptilidae	1.0	1.1	+ 0.1
Philopotamidae	0.4	0.9	+ 0.5
Libellulidae	0.1	0	− 0.1
Zygoptera	0.03	0.2	+ 0.17
Potadoma sp.	0.3	1.5	+ 1.2
Oligochaeta	0.27	1.0	+ 0.73

TABLE 3
Comparison of the Density of Invertebrates Living on Sandy Bottoms in the Current Between the Section Treated with *B. thuringiensis* ssp. *israelensis*[a] and the Untreated Section

Taxa	Control zone		Treated zone		Difference of average density %
	N/dm³	%	N/dm³	%	
Chironomini	51	46.4	32.9	39.0	
Tanytarsini	47.8	43.5	8.8	10.4	− 12.1[b]
Tanypodinae	2.2	2.0	8.3	9.8	
Orthocladiinae	1.7	1.5	19.3	22.9	
Ceratopogonidae	0.5	0.5	2.2	2.6	+ 77.3
Caenidae	1.4	1.3	0	0	—
Potadoma sp.	5.4	4.9	9.4	11.1	+ 42.6
Hydropsychidae	0	0	3.5	4.1	—
Total	110		84.4		− 23.3%

[a] Weekly applications at a dosage of 1.6 mg/l/10′ for 9 weeks.
[b] Figure represents difference between treated and control population for all chironomids.

fected by medium-term treatments. In this specific case there is a strong possibility that the decrease in numbers of predacious Hydropsychidae was beneficial to the chironomid population and compensated for any possible direct impact of *B. thuringiensis* ssp. *israelensis* on these midges.

c. Long-Term Studies

At the start of WHO/OCP in 1974, control of the onchocerciasis vector, *Simulium damnosum* s.l., was achieved via weekly applications of Abate® (temephos, an organophosphate compound) to production sites throughout the Volta River Basin. Eventual widespread resistance to temephos and cross-resistance to the first alternative chemical pesticide, chlorphoxim, in key cytospecies of the *S. damnosum* complex, necessitated the search for and use of other suitable control agents, including *B. thuringiensis* ssp. *israelensis*.[47-51] Long-term environmental impact studies in WHO/OCP have mainly concerned insecticide impact. There has been a strong component for the microbial agent under consideration during the 1980s, and Teknar has lately seen increasingly widespread use at a concentration of 1.2 mg/l/10′ via regular weekly applications by helicopter or fixed-wing aircraft. Small rivers are being so treated throughout the year, the larger ones being alternately treated with Teknar during the dry season (low-discharge period) and organic chemical compounds (notably temephos and chlorphoxim) during the season of high water and maximum flow. Unfortunately, alternation of the microbial insecticide with chemical ones (which are less selective) makes it impossible to ascribe adverse effects on NTOs to one or the other control agent.

The dry-season use of *B. thuringiensis* ssp. *israelensis* in streams where the rate of discharge is less than 50 to 75 m^3/s.[49,51] reflects the physical bulk of Teknar by comparison with that of temephos. While the fact that the invertebrate communities exposed to insecticides change somewhat between the wet and dry seasons does not permit determination of the microbial agent's seasonal impact on specific NTOs as well as the aquatic community structure, interesting results have been obtained by one of us (J.-M. E.) from comparing the fauna of rivers following treatment with Teknar for several years, with the fauna known to have existed prior to control.

From mathematical analysis of data from the Maraoué River, which was treated with *B. thuringiensis* ssp. *israelensis* for over a year in 1983 to 1984, the structure of the invertebrate community encountered during the treatment phase can be well characterized. The results indicate that it differs in some respects from that found during both untreated periods and periods of application of temephos or chlorphoxim (Figures 1 and 2). Differences related to diurnal and nocturnal drift are evident, as well as others concerning the invertebrate fauna of the vegetation growing on stones in the current. However, the chief results are similar. For example, it is clear that Trichoptera of the family Ecnomidae (which some regard as a subfamily of the Psychomyiidae) actually benefit from Teknar treatment. Moreover, hydropsychid predators on simuliids do not demonstrate any particular sensitivity, a finding that contradicts those obtained earlier from the medium-term study undertaken on the Kan River by Yameogo.[40] Two possible explanations for this inconsistency are some difference between the formulations used in the two cases, and a specific resistance on the part of Hydropsychidae having been induced by the alternation of the microbial and chemical insecticides. Overall, though, such alternation as practiced in WHO/OCP on the lower Maraoué River (where resistance to temephos first appeared), does not appear to disrupt NTO populations any more than does each insecticide individually,[52] while there is no evidence to date of any long-term deleterious effect of *B. thuringiensis* ssp. *israelensis* on the ecosystems of streams receiving weekly applications throughout the dry season.[43] The graphs of Figures 3a through 3d exhibit the change of population density of selected groups of invertebrates on stones in rapids after treatments with *B. thuringiensis* ssp. *israelensis* following temephos and chlorphoxim applications. Among the Chironomidae the Tanytarsini seem to be the most sensitive NTOs, a finding in agreement with various short-term studies. It can also be stated that the results of environmental monitoring in WHO/OCP have not revealed any significant direct or indirect effects of operational dosages of Teknar on lotic fish populations.[49]

To conclude, *B. thuringiensis* ssp. *israelensis* is certainly both the most valuable microbial control agent yet commercialized for use against major nematoceran pests and disease vectors, and a conspicuously "safe" insecticide insofar as freshwater NTOs and whole ecosystems are concerned.

Nevertheless, it is urged that international standardization of field as well as laboratory testing, based on a short list of widely distributed freshwater test organisms, is necessary at this time.

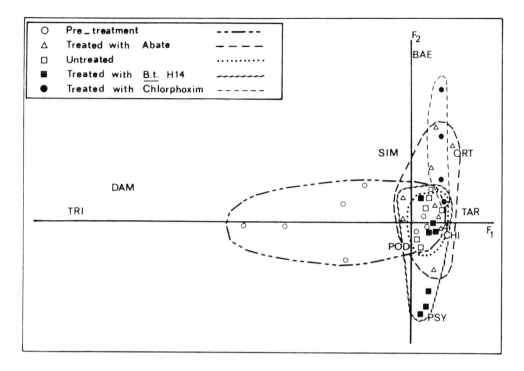

FIGURE 1.   Factorial analyses of correspondence were applied to data concerning the densities of benthic fauna living on stones in the current in the Maraoué River during the dry season. This river was treated successively with temephos, chlorphoxim, and *Bacillus thuringiensis* ssp. *israelensis*. The axis F_1 draws a clear separation between pretreatment samples associated with a dominance of *Simulium damnosum* and Trichorythidae and all the other samples collected under insecticide treatment conditions. The axis F_2 draws a separation between the period with chlorphoxim treatment associated here with an abundance of Baetidae and the period when *Bacillus thuringiensis* ssp. *israelensis* was used associated with an abundance of Hydropsychidae. BAE = Baetidae; CHI = Chironomini; DAM = *Simulium damnosum*; ORT = Orthocladiinae; PSX = Hydropsychidae; TAR = Tanytarsini; TRI = Trichorythidae; POD = Tanypodinae.

B. *BACILLUS SPHAERICUS* NEIDE 1904

When considering aquatic NTOs, the ability of *B. sphaericus* 1593 (see Chapter 12) to survive and recycle in polluted waters must be taken into account. Survival can be up to 9 months under certain conditions, without decrease of toxicity of the toxin.[53] On the other hand, *B. sphaericus* is somewhat susceptible to UV radiation; and it has also been found that the presence of large amounts of suspended material in the water (particularly organic matter) induces a reduction of larvicidal activity against mosquitoes. Such a factor may be considered positive for NTOs if the sedimented organic particles, after chelation with *B. sphaericus*, are not eaten by benthic fauna.

Finally, it has been demonstrated that the toxicity of *B. sphaericus* does not increase significantly when the temperature increases, unlike that of *B. thuringiensis* ssp. *israelensis*.[9] The same authors have also shown that too much agitation of *B. sphaericus*, during transportation for example, can induce destruction of the spore cells. This leads to a reduction of efficacy. In the same way, this phenomenon can also reduce any toxic effect on NTOs. It is also important to consider that the difficulty remains of standardizing fermentation of the spores and producing batches with equal efficiency towards target fauna. By the same token, the toxicity of each batch produced is subject to variation.

Toxicity studies for *B. sphaericus* are not so numerous as those dealing with *B. thuringiensis* ssp. *israelensis*; but the general results obtained from laboratory and field experiments lead to a similar conclusion, indicating the high innocuity of the different strains commonly used in pest control.

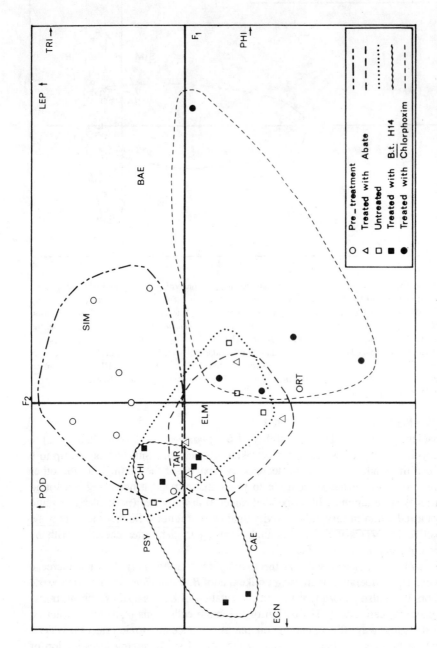

FIGURE 2.   The method used in Figure 1 was applied to day drift data and a clear opposition appears again between the period under chlorphoxim treatments and the period under *B. thuringiensis* ssp. *israelensis* treatments here characterized by high densities of Ecnomidae. BAE = Baetidae, LEP = Leptophlebiidae, PHI = Philopotamidae, ECN = Ecnomidae, CAE = Caenidae, ORT = Orthocladiinae, PSY = Hydropsychidae, ELM = Elmidae, SIM = other Simuliidae, TAR = Tanytarsini, CHI = Chironomini, POD = Tanypodinae, TRI = Trichorythidae.

Emphasis has been placed mainly on NTOs living in mosquito larval habitats. Going up in the taxonomic classification, studies made on planktonic crustaceans have shown that neither copepods nor cladocerans are affected by concentrations of 10^5 cells per ml. This is the case, e.g., for natural populations of *Acanthocyclops vernalis, Moina* spp., *Cypris* spp. and *Ceriodaphnia* spp., tested in California ponds, as well as for *Moina rectirostris, Cyprinotus* sp. or *Cypridopsis* sp.[54] In these experiments, strains 1593 and 2362 were used. Turning to Malacostraca/Decapoda, the North American crayfish, *Orconectes rusticus,* proved similarly unaffected.[55]

Similarly, no adverse effects were detected on Odonata, Ephemeroptera (*Callibaetis* spp.), Heteroptera (Corixidae, Notonectidae), or Coleoptera (Dytiscidae, Hydrophilidae).[56]

Among the Diptera no effects have been noted on *Chironomus stigmaterus* (we have also seen that many chironomid species are not sensitive to *B. thuringiensis* spp. *israelensis*) and it seems that larvae of the genus *Culicoides* (Ceratopogonidae) are affected only at *B. sphaericus* dosages much higher than those needed to kill mosquitoes.[2]

In a more recent study, Sinègre et al.[44] tried to control chironomid populations in some shallow brackish-water ponds of the Camargue (France) where mass outbreaks were considered a nuisance. Using BSP 2, a liquid containing the strain 2362 of *B. sphaericus,* at a dosage between 1 to 9 1/ha (i.e., about 3 ppm), the treatment had no discernible effect on the following species: *Chironomus salinarius, C. halophilus, C. plumosus,* nor on some unidentified Tanytarsini and Tanypodinae. Such a result suggests that *B. sphaericus* could be less toxic to chironomids than is *B. thuringiensis* spp. *israelensis*, although a similar result was obtained by the same authors, using the latter under the same conditions.

Toxicity for fish appears to be nonexistent at normal dosages. No deleterious effect on fry of *Gambusia affinis* was noted, for example, in a trial using the Stauffer wettable powder formulation of *B. sphaericus* strain 1593, after the fish had been in contact with a solution containing 10^4 and 5×10^4 spores per ml for 96 h.[54] Neither were *Epilatys bifasciatus* and *Aphyosemion gardneri* damaged by exposure to comparable dosages.[57]

Similar results have been reported by Chapman[58] from research carried out in the laboratory. The organisms concerned were again *Gambusia affinis*, crayfish, tadpoles, and different aquatic insects. However, the formulations used in these experiments were " . . . poor and shelf life tenuous, so little faith can be put in the results" as the author himself points out.

More interesting are the following results quoted by WHO:[2] "Although field treatments with *B. sphaericus* (1593 and 2362) caused no noticeable effects, a laboratory trial was carried out, in which *B. sphaericus* infected mosquito larvae were offered as the only source of food to several predacious organisms. *Culex* larvae (L4) exposed to 1000 mg/l (100 times the larvicidal rate) were offered for several days to predators, such as dragon-flies, damselfly naiads and the notonectid *Notonecta unifasciata*: the full daily food requirement of these predators was provided by larvae whose guts were filled with a lethal dose of *B. sphaericus* (2362, BSP-1). Predation on treated larvae did not induce any acute adverse effects nor alter the developmental rates of the predacious organisms."[2]

Such results are of primary importance when we know that the main, if not the unique mode of entry of *B. sphaericus* to an aquatic animal is by ingestion. The lack of effect of such high concentrations on all tested organisms demonstrates the highly selective toxicity of this bacterial agent. Nevertheless, if no effects of the different strains of *B. sphaericus* such as 1593, 1593 M, SSII-1, 2362, or 2297 have been noted at normal dosages required for effective larviciding, the fact remains that pending the production of normalized formulations more studies of toxicity are still required.

C. INFECTIVITY OF BACULOVIRUSES

Mainly effective against defoliant insects, the nuclear polyhedrosis viruses (NPVs) of the genus *Baculovirus* (Baculoviridae) are becoming more and more widely employed since

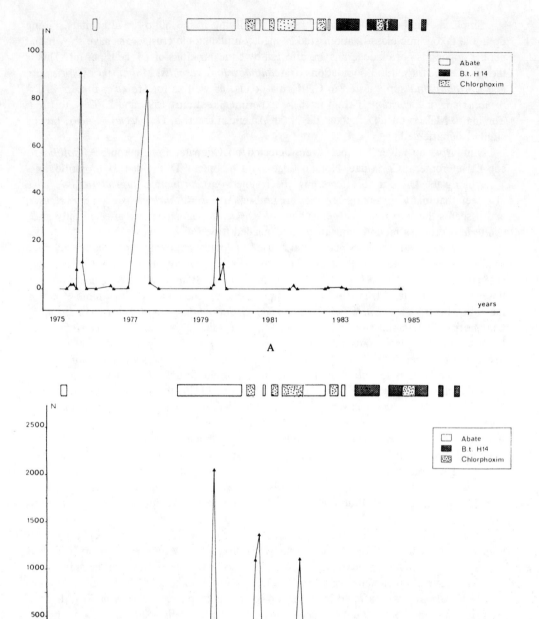

FIGURE 3. Evolution of the average densities of selected benthic organisms on stones in the current (collected by Surber Sampling), in the case of *B. thuringiensis* var. *israelensis* treatments following temephos and chlorphoxim applications. (A) Trichorythidae; (B) Tanytarsini; (C) Hydropsychidae; and (D) Ecnomidae.

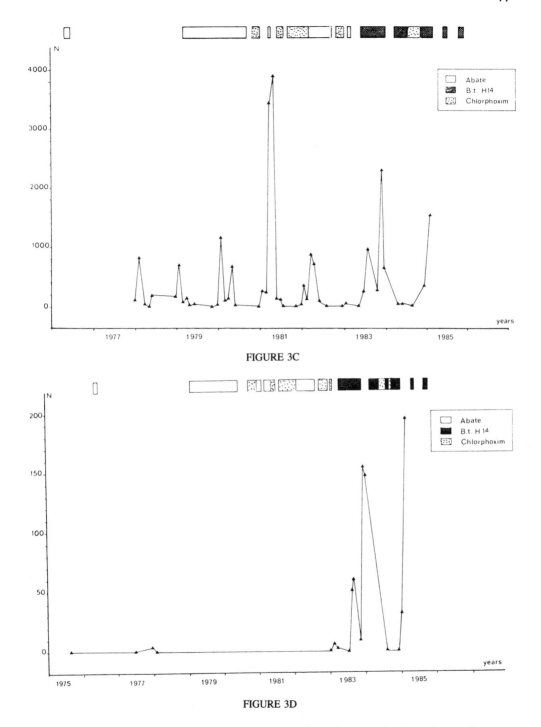

FIGURE 3C

FIGURE 3D

their relatively recent registration, although their pathogenic action has been known for more than 30 years.

Aerial applications inevitably lead to baculoviral preparations reaching rivers or lakes within treated forest areas, and for this reason their pathogenicity among some aquatic organisms has been studied. Also, despite their relative specificity, they prove fatal to many insects (mainly Lepidoptera) which may fall into water and be eaten by fish. Finally, some target lepidopteran species (such as various Pyralidae) represent, during their larval stage, a significant component of the benthic fauna in certain aquatic habitats. An important reduction of their biomass can then induce local disturbances of the food chain, thus indirectly affecting the fish population.

Before we consider the eventual pathogenicity of baculoviruses to aquatic freshwater fauna, it has to be mentioned — as pointed out by Gröner et al.[59] — that high concentrations (up to 10/20 units/ha) can frequently be present in the environment without detectable prejudice to the natural populations (vertebrates as well as invertebrates) commonly present in such habitats.

Where the toxicity of aquatic organisms is concerned, it is apparent that once again more research has been carried out on adverse effects to fish than to aquatic invertebrates.

As regards the latter, the most relevant investigations have been carried out by Streams.[60] His results were presented in an unpublished report cited by Doane and McManus.[61]

Frequently abundant species of aquatic invertebrates such as *Daphnia magna, Notonecta undulata,* and other waterboatmen (Hemiptera) and *Chironomus thummi* (Diptera) were tested. These organisms were exposed to high but vaguely specified concentrations of gypsy moth NPV. (i.e., "approximately the concentration which would occur when a very shallow pond is sprayed at the rate of 3.75 × 10/13 PIB's per hectare".[61] No direct mortality of any of these organisms was observed by the authors. In addition, they state that no harmful effects were observed throughout the development stages of the tested species, a particularly important result. It is also important to note that *Daphnia magna* as well as *Notonecta undulata* and the waterboatmen did not accumulate gypsy moth NPV, as shown by bioassays conducted on tested specimens.

Similar results were obtained by Geraci and Hicks[62] working on *Daphnia pulex* held in water contaminated by the red-headed pine sawfly (*Neodiprion lecontei*) NPV and kept under observation for 14 d. Their conclusions were later confirmed by Hicks et al.[63] using concentrations of 2.4 × 10 PIB of the same NPV per ml. As well as the absence of direct mortalities, no adverse effects were observed as far as fecundity, lesions or body abnormalities were concerned. Some other invertebrates have also been tested, such as young instars of the penaeid shrimps, *Penaeus setiferus* and *Penaeus aztecus.* In this case, the NPV of the alfalfa looper (*Autographa californica*) was tested, introduced via injections or by feeding. In the case of muscular injection the NPV concentration used about 4.7 × 10 virus rods per shrimp. After a 30-d observation, shrimp mortality attributable to viral infection did not occur, neither from the injected nor the orally introduced virus. A careful examination of tissues contaminated by NPV injections did not lead to the discovery of any nuclear polyhedra.[64]

Fish toxicity has been the subject of more extensive research, and special studies have been carried out on the physiological effects of a direct contamination by Baculoviridae (incubation, injection, ingestion, etc.). All results indicate an absence of histopathological toxicity for these organisms.[65-67] Some 12 species of fish, mainly Salmonidae, have been tested in this way and similar studies carried out on amphibians produced an identical result.

Fish cell cultures exposed to *Baculovirus* (Douglas fir tussock moth NPV) for 24 h showed no pathological changes and no alteration occurred in their growth rate.[65] In these experiments, cells from chinook salmon (*Oncorhynchus tshawytscha*) and rainbow trout (*Salmo gairdneri*) were used. Fingerlings of coho salmon (*Oncorhynchus kisuch*), chinook salmon, and rainbow trout exposed to the same *Baculovirus* by three different routes (i.e., i.p. injection, waterborne exposure, and by feeding) were not adversely affected by any of these contamination methods. No histopathological troubles were detected one month after exposure. This result confirms Wolf's[68] conclusions, which also established that *Dendroctonus pseudotsugae* NPV had no deleterious effect on amphibian and fish cell lines.

All these initial results have been confirmed by subsequent research; e.g., Moore[69] established a total absence of harmful effects after the 96-h exposure of 240 juvenile bluegills

(*Lepomis macrochirus*) and 240 brown trout (*Salmo trutta*) to gypsy moth NPV concentrations 100 times higher than those normally effective against that insect.

Similarly, rainbow trout and white suckers (*Catostomus commersoni*) inoculated with spruce budworm NPV were not significantly affected by the virus.[70] Rainbow trout reactions to red-headed pine sawfly (*Neodiprion lecontei*) NPV incubations or topical applications were also found to be nonexistent by Hicks et al.[63]

These results all lead to the general conclusion that viruses of the Baculoviridae are nontoxic to aquatic invertebrates as well as fish, although possible long-term effects have not been studied. It is also regrettable that we have no observation relating to global ecosystem modifications which might occur in the case of regular applications of Baculoviridae over several years in the same forest area.

D. INFECTIVITY OF MICROSPORIDIANS

To our knowledge, very little work has been carried out on protozoan pathogenicity to aquatic NTOs, as far as their insecticidal properties are concerned. Some microsporidians which are effective against mosquito larvae or Simuliidae are known to survive in dead insects or in viable infected eggs for more than 10 months.[71] *Nosema stegomyiae* and *Nosema algerae* can be found in their dormant stage in the eggs of *Aedes* for over a year, and are then able to contaminate eventual predators. Spores of *N. algerae* remain viable for 1 to 2 months at 20°C. Weiser also states that "Under conditions of man-guided distribution of spores, *Nosema algerae* or *Vavraia culicis* are able to impact different hosts including mollusks and crayfish."[71]

In a study undertaken by Van Essen and Anthony,[72] nine nontarget aquatic predators were fed with mosquito larvae previously heavily infected with *N. algerae*. Of these NTOs, 50% of the *Notonecta undulata* (Heteroptera) tested developed infections. The other predators (dragonfly, hydrophilid, nepid, megalopteran, decapod, and the larvivorous fish, *Gambusia affinis*) were in no way affected. If we consider that feeding organisms directly with a high dosage of pathogen is much more dangerous for them than a single contact in contaminated water, one can only conclude that *Nosema algerae* will be safe for many other aquatic organisms.

On the other hand, the low mortality rate of mosquito larvae challenged with *N. algerae*, as well as with *V. culicis* (except in the case of anopheline mosquitoes), does not favor the greatly extended use of these Microspora.[73]

IV. CONCLUSION ON THE HAZARD OF MICROBIAL INSECTICIDES TO FRESHWATER FAUNA

At this point it is perhaps worth quoting various conclusions drawn by other researchers concerning aquatic biotas.

"There is no danger of ill-effect of microbial insecticides on nontarget organisms . . . ".[71]

"I believe that a pathogen should be registered as safe when there is reasonable evidence that it is so and in the absence of concrete evidence that it is not. A "no risk" situation does not exist, certainly not with chemical pesticides, and even with biological agents one cannot absolutely prove a negative."[36]

It is now clear that control of disease vectors as well as of agricultural and forest pests cannot be based purely on the use of chemicals because of their generally low specificity, the risk of environmental contamination and also the increasing resistance of insects following intensive applications. Microbial pathogens have great potential, but this necessitates, at least as far as freshwater ecosystems are concerned, more attention to their possible impact.

Despite the relatively large amount of work related to *B. thuringiensis* ssp. *israelensis*, it appears that a lot of questions have only received partial answers. These questions were addressed by Forsberg et al.[74] in their review of *B. thuringiensis* toxicity, and can be summarized as follows. They remain relevant today, although we have changed the original wording slightly.

1. What are the toxic effects of the components of commercial formulations of microbial insecticides, in the medium and the long-term?
2. What quantities of formulation enter the environment and how persistent are the toxic components in each of their different modes of application?
3. What effects on ecosystems are related to long-term and large-scale field applications of microbial insecticides?
4. Can intensive use of entomopathogens lead to their mutation? Are the microbial agents susceptible to modification by genetic transfer into other forms? Are these new forms capable of producing toxic components, or components with modified toxicity or host specificity?

These questions leave open a rich field of investigation, both at a fundamental and a more practical level, linked to application campaigns. We can conclude that whereas further *B. thuringiensis* ssp. *israelensis* laboratory tests may be superfluous, there is an overall lack of standardized laboratory and field studies covering all microbial insecticides. The setting-up of a well-defined and systematic screening process will enable comparable data to be obtained, and will lead to a better evaluation of toxicity for different conditions of application. In this respect, the safety-testing procedures for bacterial and fungal agents proposed by WHO[75] could be enlarged upon in order to include more specifications related to aquatic environmental safety.

REFERENCES

1. **WHO,** Securité d'Emploi des Pesticides, OMS, Technical report, World Health Organization, 1967, 356.
2. **WHO,** Informal Consultation on the Development of *Bacillus sphaericus* as a Microbial Larvicide, WHO/TDR/BCB/*sphaericus*/85.3, mimeographed document, World Health Organization, 1985.
3. **de Barjac, H.,** Une nouvelle variété de *Bacillus thuringiensis* très toxique pour les moustiques: *Bacillus thuringiensis* var. *israelensis* sérotype 14, *C. R. Acad. Sci. Ser. D.,* 286, 797, 1978.
4. **WHO,** Data Sheet on the Biological Control Agent *Bacillus thuringiensis* serotype H.14 (de Barjac, 1978), WHO/VBC/79-750, mimeograph, World Health Organization, 1979.
5. **Mulla, M. S.,** Potential of Some New Insecticides, Pyrethroids and Insect Growth Regulators, WHO/OCP/SWG/78.22, mimeograph, Geneva, 1978.
6. **Lacey, L. A., Escaffre, H., Philippon, B., Seketeli, A. and Guillet, P.,** Large river treatment with *Bacillus thuringiensis* (H-14) for the control of *Simulium damnosum* s.l. in the Onchocerciasis Control Programme, *Z. Tropenmed. Parasitol.,* 33, 97, 1982.
7. **Molloy, D. and Jamnback, H.,** Screening and Evaluation of *Bacillus sphaericus* and *B. thuringiensis* serotype H-14 as Black Fly Control Agents, Annu. Rep. from New York State Museum, Albany, N.Y., 1981.
8. **Molloy, D. and Jamnback, H.,** Field evaluation of *Bacillus thuringiensis* var. *israelensis* as a black fly biocontrol agent and its effect on nontarget stream organisms, *J. Econ. Entomol.,* 74, 314, 1981.
9. **Wraight, S. P., Molloy, D., and McCoy, P.,** A comparison of laboratory and field tests of *Bacillus sphaericus* strain 1593 and *Bacillus thuringiensis* var. *israelensis* against *Aedes stimulans* larvae (Diptera: Culicidae), *Can. Entomol.,* 114, 55, 1982.
10. **Purcell, B. H.,** Effects of *Bacillus thuringiensis* var. *israelensis* on *Aedes taeniorhynchus* and some non-target organisms in the salt marsh, *Mosq. News,* 41, 476, 1981.

11. **Gallagher, R.,** Assessment of *Bacillus thuringiensis* var. *israelensis* as a Biological Agent for the Control of the Pest Mosquito *Aedes detritus*, unpublished report, Entomol. Dept. London Sch. Hyg. Trop. Med., 1981.

12. **Sinègre, G., Gaven, B., and Jullien, J. L.,** Evaluation de l'Activité Larvicide de *Bacillus thuringiensis* var. *israelensis* sur les Culicides. Performances Comparées des Formulations Commerciales. Impact du Produit sur la Faune Non-cible, EID Document, Montpellier, 1979.

13. **Sinègre, G., Gaven, B., and Vigo, G.,** Contribution à la normalisation des épreuves de laboratoire concernant des formulations expérimentales et commerciales du sérotype H-14 de *B. thuringiensis*. II. Influence de la témperature, du chlore résiduel, du pH et de la profondeur de l'eau sur l'activité biologique d'une poudre primaire, *Cah. O.R.S.T.O.M. Ser. Entomol. Med. Parasitol.*, 19, 149, 1981.

14. **Lebrun, P. and Vlayen, P.,** Etude de la bioactivité comparée et des effets secondaires de *Bacillus thuringiensis H, Z. Angew. Entomol.*, 91, 15, 1981.

15. **Standaert, J. Y.,** Persistance et efficacité de *Bacillus thuringiensis* H14 sur les larves d' *Anopheles stephensi, Z. Angew. Entomol.*, 91, 292, 1981.

16. **Yousten, A. A. and Benoit, R.,** The Stability of the Toxin and Spores of *Bacillus thuringiensis* serovar. *israelensis* (H-14) and *Bacillus sphaericus* 1593 in Dialysis Bags under Field Conditions in Pond Water, WHO/VBC/82.844, mimeograph, Geneva, 1982.

17. **Garcia, R. and Des Rochers, B.,** Toxicity of *Bacillus thuringiensis* var. *israelensis* to some California mosquitoes under different conditions, *Mosq. News*, 39, 541, 1979.

18. **Krieg, A., Engler, S., and Rieger, M.,** Produktion von Praparaten auf der Basis von *Bacillus thuringiensis* mit UV-inaktivierten Sporen zur biologischen Bekampfung von Mückenlarven, *Anz. Schaedlingskd. Pflanz. Umweltschutz*, 53, 129, 1980.

19. **Colbo, M. H. and Undeen, A. H.,** Effect of *Bacillus thuringiensis* var. *israelensis* on non-target insects in stream trials for control of Simuliidae, *Mosq. News*, 40, 368, 1980.

20. **Dunn, P. H.,** Effects of Ingestion of Biotrol on Fish, Experiment 7, information supplied by Nutrilite Products, Inc., Buena Park, CA, 1980.

21. **Garcia, R., Federici, B. A., Hall, I. M., Mulla, M. S., and Schaeffer, C. H.,** BTI, a potent new biological weapon, *Calif. Agric.*, 34, 18, 1980.

22. **Garcia, R., Des Rochers, B., and Tozer, W.,** Further Studies on *Bacillus thuringiensis* var. *israelensis* against Mosquito Larvae and Other Organisms, Annu. Rep. Mosq. Control Res., University of California, Berkeley, 1980, 54.

23. **Garcia, R., Des Rochers, B., and Tozer, W.,** Further Studies on *Bacillus thuringiensis* var. *israelensis* against Mosquito Larvae and Other Organisms, Mosquito Control Res., unpublished report to World Health Organization, University of California, Berkeley, 1980.

24. **Garcia, R., Des Rochers, B., and Tozer, W.,** Studies on *Bacillus thuringiensis* var. *israelensis* against organisms found in association with mosquito larvae, *Proc. Calif. Mosq. Vector Control Assoc.*, 48, 33, 1980.

25. **Garcia, R., Tozer, W., and Des Rochers, B.,** Effects of *BTI* on aquatic organisms other than mosquitoes and black flies, Annu. Rep. Mosq. Control Res., University of California, Berkeley, 1981, 68.

26. **Mastri, C.,** Four-day fish toxicity study on *Bacillus thuringiensis*, Industrial Biotest Laboratories Ltd., information supplied by Abbott Laboratories, North Chicago, Ill., 1970.

27. **Larget, I. and de Barjac, H.,** The serotype H-14 of *Bacillus thuringiensis, Parasitology*, 82, 117, 1981.

28. **Larget, I. and de Barjac, H.,** Specificité et principe actif de *Bacillus thuringiensis* var. *israelensis, Bull. Soc. Pathol. Exot.*, 74, 216, 1981.

29. **Miura, T., Takahashi, R., and Mulligan, F. S., III,** Effects of the bacterial mosquito larvicide *Bacillus thuringiensis* serotype H-14 on selected aquatic organisms, *Mosq. News*, 40, 619, 1980.

30. **Pantuwatana, S.,** Laboratory evaluation of various formulations of B.t. H-14, personal communication to World Health Organization, 1980.

31. **Prasertphon, S.,** A brief summary of biological control activities at WHO/VBC/RU-I, report to World Health Organization, Kaduna, Nigeria, 1979.

32. **Rajagopalan, P.,** personal communication (from Vector Control Research Centre, ICMR, Pondicherry, India) to World Health Organization, 1982.

33. **Garcia, R., Des Rochers, B., Voight, W., and Goldberg, L.,** Studies on the toxic effect of the bacterial spore ONR 60A on nontarget organisms, Annu. Rep. Mosq. Control. Res., University of California, Berkeley, 1977.

34. **Dejoux, C., Gibon, F. M., and Yameogo, L.,** Toxicité pour la faune non-cible de quelques insecticides nouveaux utilisés en milieu aquatique tropical. IV. Le *Bacillus thuringiensis* var. *israelensis, Rev. Hydrobiol. Trop.*, 18, 31, 1985.

35. **Schnetter, W., Engler, S., Morawcsik, J., and Becker, N.,** Wirksamkeit von *Bacillus thuringiensis* var. *israelensis* gegen Stechmückenlarven und Nontarget-Organismen, *Mitt. Dtsch. Ges. Angew. Entomol.*, 2, 195, 1981.

36. **Burges, H. D.**, Safety, safety testing and quality control of microbial pesticides, in *Microbial Control of Pests and Plant Diseases 1970—1980*, Burges, H. D., Ed., Academic Press, New York, 1981, 737.

37. **WHO**, Data Sheet on the Biological Control Agent *Bacillus thuringiensis* serotype H-14 (deBarjac, 1978), WHO/VBC/79.750, mimeograph, revised version, Geneva, 1982.

38. **Gibon, F.-M., Elouard, J.-M., and Troubat, J.-J.**, Action du *Bacillus thuringiensis* var. *israelensis* sur les invertebres aquatiques. Effects d'un traitement experimental sur la Maraoue, *Rapp. Lab. Hydrobiol. O.R.S.T.O.M. Bouake*, 38, 1, 1980.

39. **Dejoux, C.**, Recherches Préliminaires Concernant l'Action de *Bacillus thuringiensis israelensis* de Barjac sur la Faune des Invértebrés d'un Cours d'Eau Tropical, WHO/VBC/79.721, mimeograph, Geneva, 1979.

40. **Yameogo, L.**, Modification des Entomocenoses d'un Cours d'Eau Tropical Soumis à un Traitement Antisimulidien avec *Bacillus thuringiensis* var. *israelensis*, Mem. d'ingenieur de l'Université de Ouagadougou, mimeograph, 1980.

41. **Dejoux, C.**, Nouvelle technique pour tester in situ l'impact de pesticides sur la faune aquatique non-cible, *Cah. O.R.S.T.O.M. Ser. Entomol. Med. Parasitol.*, 13, 75, 1975.

42. **Troubat, J.-J.**, Dispositif à gouttières multiples destiné à tester in situ la toxicité des insecticides vis-à-vis des invertébrés benthiques, *Rev. Hydrobiol. Trop.*, 14, 149, 1981.

43. **Elouard, J.-M., and Fairhurst, C. P.**, Ten years of surveillance of the rivers treated with antiblackfly insecticides by the Onchocerciasis Control Programme: medium and long-term impact on the invertebrates, *Chemosphere*, in press.

44. **Sinègre, G., Gaven, B., Jullien, J. L., Vigo, G., and Tourenq, J. N.**, Evaluation de Terrain de Quelques Larvicides sur les Chironomes des Etangs du Sud de la France, Comm. 2nd Meet. European Soc. Vector Ecol., mimeograph, Heidelberg, 1987.

45. **Back, C., Boisvert, J., Lacoursiere, J. O., and Charpentier, G.**, High-dosage treatment of a Quebec stream with *Bacillus thuringiensis* serovar. *israelensis*: efficacy against black fly larvae (Diptera: Simuliidae) and impact on non-target insects, *Can. Entomol.*, 117, 1523, 1985.

46. **Ali, A.**, *Bacillus thuringiensis* serovar. *israelensis* (ABG-6108) against chironomids and some nontarget aquatic vertebrates, *J. Invertebr. Pathol.*, 38, 264, 1981.

47. **Guillet, P.**, La lutte contre l'onchocercose humaine et les perspectives d'intégration de la lutte biologique, *Entomophaga*, 29, 121, 1984.

48. **Kurtak, D. C.**, Insecticide resistance in the Onchocerciasis Research Programme *Parasitol. Today*, 2, 20, 1986.

49. **Kurtak, D. C., Grünewald, J., and Baldry, J. A. T.**, Control of black fly vectors of onchocerciasis in Africa, in *Black Flies: Ecology, Population Management, and Annotated World List*, Kim, K. C. and Merritt, R. W., Eds., Proc. Int. Conf. Ecology and Population Management of Black Flies, Pennsylvania State University Press, University Park, PA, 1987.

50. **Kurtak, D., Jamnback, H., Meyer, R., Ocran, M., and Renaud, P.**, Evaluation of larvicides for the control of *Simulium damnosum* s.l. (Diptera: Simuliidae) in West Africa, *J. Am. Mosq. Control Assoc.*, 3, 201, 1987.

51. **Kurtak, D., Meyer, R., Ocran, M., Ouedraogo, M., Renaud, P., Sawadogo, R. O., and Tele, B.**, Management of insecticide resistance in control of the *Simulium damnosum* complex by the Onchocerciasis Control Programme, West Africa: potential use of negative correlation between organophosphate resistance and pyrethroid susceptibility, *Med. Vet. Entomol.*, 1, 137, 1987.

52. **Elouard, J.-M. and Gibon, F.-M.**, Incidence on Nontarget Insect Fauna of the Alternate Use of Three Insecticides (Temephos, Chlorphoxim and B.t. H-14) for the Control of the Larvae of *Simulium damnosum* s.l., WHO/OCP/VCU/HYBIO/84.4, mimeograph, Geneva, 1984.

53. **Hertlein, B. C., Levy, R., and Miller, T. W., Jr.**, Recycling potential and selective retrieval of *Bacillus sphaericus* from soil in a mosquito habitat, *J. Invertebr. Pathol.*, 33, 217, 1979.

54. **Mulligan, F. S., III, Schaeffer, C. H., and Miura, T.**, Laboratory and field evaluation of *Bacillus sphaericus* as a mosquito control agent, *J. Econ. Entomol.*, 71, 774, 1978.

55. **WHO**, Data Sheet on the Biological Control Agent *Bacillus sphaericus*, strain 1593, WHO/VBC/80.777, mimeograph, World Health Organization, Geneva, 1980.

56. **Chapman, H. C.**, What mosquito control districts might want to know about biological control, *Mosq. News*, 38, 479, 1978.

57. **Singer, S.** Potential of *Bacillus sphaericus* and related sporeforming bacteria for pest control, in *Microbial Control of Pests and Plant Diseases 1970—1980*, Burges, H. D., Ed., Academic Press, New York, 1981, 283.

58. **Chapman, H. C.**, Safety of nontarget species for ten of the most promising biological control agents, unpublished document, Gulf Coast Mosquito Research, U.S. Department of Agriculture, Washington, D.C., 1979.

59. **Gröner, A., Huber, J., and Krieg, A.**, Anwendung von Baculoviren im Pflanzenschutz: Unbedenklichkeit fur aquatische Organismen, *Z. Binnenfisch, DDR, Z. Bienenforsch.*, 31, 25, 1981.

60. **Streams, F. A.**, Effect of gypsy moth NPV on selected aquatic invertebrates, unpublished report, Forest Service, U.S. Department of Agriculture, Hamden, CT, 1977.

61. **Doane, C. C. and McManus, M. L.,** The Gypsy Moth: Research Toward Integrated Pest Management, Tech. Bull. 1584, Doane, C. C. and McManus, M. L., Eds., Forest Service, U.S. Department of Agriculture, Science and Education Agency, Animal and Plant Health Inspection Service, Washington, D.C., 1981, 475.

62. **Geraci, J. R. and Hicks, B. D.,** A Study on the Effect of Red-headed Pine Sawfly, *Neodiprion lecontei,* Nuclear Polyhedrosis Virus to Rainbow Trout, *Salmo gairdneri,* and *Daphnia pulex,* Final Report on Contract No. DSS 01SU.KL 013-8-0013, Environment Canada, mimeograph, Can. Forest Serv., Sault Ste. Marie, Ontario, Canada, 1979, 302.

63. **Hicks, B. D., Geraci, J. R., Cunningham, J. C., and Arif, B. M.** 1981. Effects of red-headed pine sawfly, *Neodiprion lecontei,* nuclear polyhedrosis virus on rainbow trout, *Salmo gairdneri,* and *Daphnia pulex, J. Environ. Sci. Health,* B16, 493, 1981.

64. **Lightner, D. V., Proctor, R. R., Sparks, A. K., Adams, J. R., and Heimpel, A. M.,** Testing penaeid shrimp for susceptibility to an insect nuclear polyhedrosis virus, *Environ. Entomol.,* 2, 611, 1973.

65. **Banowetz, G. M., Fryer, J. L., Iwai, P. J., and Martignoni, M. E.,** Effects of the douglas-fir tussock moth nucleo-polyhedrosis virus (baculovirus) on three species of salmonid fish, Paper PNW-214, Forest Service Res., U.S. Department of Agriculture, Washington, D.C., 1976.

66. **Ignoffo, C. M.,** Effects of entomopathogens on vertebrates, *Ann. N.Y. Acad. Sci.,* 217, 141, 1973.

67. **McIntosh, A. H.,** *In vitro* specificity and mechanism of infection, in *Baculoviruses for Insect Pest Control: Safety Considerations,* Summers, M., Engler, R., Falcon, L. A., and Vail, P., Eds., American Society for Microbiology, Washington, D.C., 1975, 63.

68. **Wolf, F. E.,** Evaluation of the exposure of fish and wildlife to nuclear polyhedrosis and granulosis viruses, in *Baculoviruses for Insect Pest Control: Safety Considerations,* Summers, M., Engler, R., Falcon, L. A., and Vail, P., Eds., American Society for Microbiology, Washington, D.C., 1975, 109.

69. **Moore, R. B.,** Determination of the Effects of Nuclear Polyhedrosis Virus in Trout and Bluegill Sunfish under Laboratory Conditions, Final Report (unpublished), Forest Service, U.S. Department of Agriculture, Hamden, CT, 1977.

70. **Savan, M., Budd, J., Rend, P. W., and Darley, S.,** A study of two species of fish inoculated with spruce budworm nuclear polyhedrosis virus, *J. Wildl. Dis.,* 15, 331, 1979.

71. **Weiser, J.,** Microbial insecticides in the environment, in *Basic Biology of Microbial Larvicides of Human Diseases,* Michal, F., Ed., UNDP/World Bank/WHO Report, 1982, 69.

72. **Van Essen, F. W. and Anthony, D. W.,** Susceptibility of non-target organisms to *Nosema algerae* (Microsporida:Nosematidae), a parasite of mosquitoes, *J. Invertebr. Pathol.,* 28, 77, 1976.

73. **WHO,** Report of the Seventh Meeting of the Scientific Working Group on Biological Control of Vectors, WHO/TDR/BCV/SWG-7/84.3, mimeograph, World Health Organization, Geneva, 1984.

74. **Forsberg, C. W., Henderson, M., Henry, E., and Roberts, J. R.,** *Bacillus thuringiensis:* Its Effects on Environmental Quality, Report 15385, National Research Council of Canada, ACSCEQ, 1976.

75. **WHO,** Report of the Fifth Meeting of the Scientific Working Group on Biological Control of Vectors, WHO/TDR/VEC/SWG (5)/81.3, mimeograph, World Health Organization, Geneva, 1981.

Chapter 7

POTENTIAL IMPACT OF MICROBIAL INSECTICIDES ON THE ESTUARINE AND MARINE ENVIRONMENTS

J. A. Couch and S. S. Foss

TABLE OF CONTENTS

I. INTRODUCTION

Estuarine and coastal regions are particularly significant impact zones for the products and other consequences of human enterprise. For decades these regions have absorbed the effects of industrial and agricultural chemicals. For example, chemical pesticides have been entering the freshwater, estuarine, and coastal regions of the world since before World War II. In the last 20 years, new living pest control agents (viral, bacterial, fungal, and protozoal) have been developed for use in the integrated control of arthropod pests and disease vectors. These microbial agents offer promise as alternatives and substitutes for hazardous synthetic organic pesticides. Because microbials must be shown to have high levels of safety to nontarget organisms (NTOs) and ecosystems before being used widely, considerable recent interest has focused on their possible environmental effects in coastal regions, particularly estuaries, which are inhabited by hundreds of NTO species. Fish and invertebrates (particularly bivalve Mollusca and Crustacea), have the potential of being exposed to microbial agents that enter the estuarine mixing zone for freshwater and the sea. Of particular interest is the fact that taxa in which most such agents are found (e.g., viruses, bacteria, fungi, and protozoa) contain congeneric species that are natural parasites and/or pathogens in estuarine and coastal nontarget invertebrate and vertebrate hosts.[1-6]

Species of both insects and crustaceans, all definitive arthropods, are often common hosts of microbial pathogens from the same microbial higher taxa (genera and above). Not only are these microbial genera (found in Insecta and Crustacea) closely related taxonomically, but their species also share similar modes of infectivity, pathogenicity, and occasionally toxicity. Because most of them are both parasites and pathogens, their parasitic evolutionary adaptations have resulted in many similar effects on their different hosts. Along with shared pathogenic characteristics, however, are stringent differences that evoke the phenomenon of host specificity. In most cases, because of established host specificity, one cannot expect parasites with long evolutionary relationships, in either insectan or crustacean hosts, to suddenly accept and utilize a new host group in nature, even when the potential hosts may occur in the same phylum. Therefore, much of the concern about potential risks of microbial pest control agents to nontarget species focuses on the possible exceptions to the expected that may occur, rather than on the expected or the "norm", particularly when nontarget species closely related to the normal host(s) may suddenly be exposed to large doses of the agent.

These theoretical points must be considered when studies are planned to determine "safety" of microbial control agents to NTOs in any environment or habitat type. Such studies have been and are underway at university, private, and government laboratories worldwide. They are usually based on the null hypothesis that infection and related effects will not occur in NTOs. To date, this hypothesis has not been rejected based on results of experimental exposure of nontarget estuarine or marine species via natural exposure routes to relatively high concentrations of infective stages of microbial insecticides.

II. MICROBIAL CONTROL AGENTS AND SAFETY STUDIES

To date, relatively few microbials have been tested for safety in selected nontarget estuarine and marine species. The purpose of this section will be to review the most relevant of these studies, including approaches and experimental design, species of microbials tested, systems used, and endpoints and results evaluated for determination of risks of these agents to nontarget marine species. The order of this review will be microbial agents as follow: (1) viruses; (2) bacteria; (3) fungi; and (4) protozoa.

A. VIRUSES (EXAMPLE, *BACULOVIRUS* OF *AUTOGRAPHA CALIFORNICA*, = [ACB])

The entity selected as our example has a much broader host range among Lepidoptera than do other equally well known baculoviruses, for example, that of *Heliothis zea*. A number of baculoviruses isolated from insects have now been developed for practical use in the control of agricultural pests.[7] Recent findings of at least three enzootic species of baculoviruses in shrimps and crabs have spurred interest in determining the safety of those from insects to Crustacea and other estuarine NTOs.[3,8-13]

1. Presumed Mode of Action in Natural Hosts

Baculovirus inclusion bodies contain many rod-shaped virions in a protein matrix. Inclusion bodies are the vehicles in which the infectious virion resides until it reaches the proper microhabitat, the gut of a vulnerable insect. Insects obtain inclusion bodies by feeding on materials contaminated by debris from cadavers of diseased insects or, in the case of microbial pesticides, after mass production in the laboratory and application in the field.

Once in the gut of a target insect and at higher pH (pH 10 to 12), the inclusion bodies' protein matrix dissolves. Virions are released into the lumen of the gut in proximity to vulnerable epithelial cells. Receptors on the virion envelope or cell apparently permit attachment to the tips of epithelial cell microvilli. The nucleocapsid of the virus enters the apical cytoplasm of the cell by unknown mechanisms and migrates to the nuclear envelope. There the nucleocapsid aligns itself at a nuclear pore and injects (uncoats) its DNA into the nucleoplasm. Integration of the viral DNA occurs in the host cell genome and viral replication and inclusion body production begins within hours of integration.

Cytopathic effects are produced at this time in the infected insect cells, including lysis or breakdown of the cell due to growth of inclusion bodies, loss of structural integrity of the cell, and possible irreversible chemical changes in the latter.

An understanding of these events in the natural hosts of baculoviruses, such as insects and crustaceans, provides clues to endpoints in the infectious cycle and diseased state that should be used in studies of potential nontarget test animals. Obviously certain techniques involving light microscopy, electron microscopy (EM), serology, and genetic probe methods would be useful in measuring selected endpoints.

2. Safety Studies

Lightner et al.[14] carried out the first test of an insect baculovirus in noninsect arthropods — the white shrimp, *Penaeus setiferus* and brown shrimp, *Penaeus aztecus*. Early and late juvenile stages were exposed to baculovirus from *Autographa californica*, the alfalfa looper, by intramuscular injection and by feeding virus polyhedra. Detailed light and EM examination of key tissues from the exposed shrimps did not reveal infections or pathogenesis related to the virus. Exposure-related mortality did not occur.

The only other extensive studies on possible effects of insect virus on marine organisms were reported by Couch et al.[15] These investigators determined possible infectivity, pathogenicity, or toxicity of the *A. californica* baculovirus in two species of estuarine grass shrimps (*Palaemonetes vulgaris* and *Palaemonetes pugio*) and their studies will be reviewed in detail.

3. Test Animals and Test Conditions

Adult Grass shrimp were collected from grass beds and acclimated for 30 d in a large outdoor aquarium. No preexisting inclusion body virus or other pathogen was found in routine histological baseline samples. Tissues examined included stomach, midgut, hindgut, hepatopancreas, muscle, and gonads.

Glass tanks with a substrate of clean oyster shell over filtration systems were filled with filtered seawater (25 °/$_{oo}$). During the static test, evaporation loss was replaced with deionized water.

Shrimp were selected from the acclimated source group for size similarity and nongravid status and placed in cylindrical holding cups constructed of mesh screen. Holding cups were then placed on a glass tray in experimental and control aquaria where the water level was maintained just below the top of the holding cups.

Except for injection, exposure of test animals via feeding appeared to be the most efficient, natural and certain method to ensure internal exposure to the virus. During a second 30-d period of acclimation in the test aquarium, different feeding regimens were attempted to maximize survival and ensure statistical validity of the test. Inadequate food supply caused cannibalism, whereas excessive food caused bacterial contamination which also endangered shrimp survival. We found that virus exposure probably was most assured if shrimp were fed virus-contaminated food pellets twice a week (one pellet per holding cup per feeding) and excess food was removed from the cups the next day.

During the test, the shrimp were checked twice daily to ensure that weakened or dead specimens were recovered in time for histological examination. All animals that died during the test were fixed in a combination fixative for light microscopy and EM.[16]

4. Virus — Preparation and Feeding

A pure preparation of freeze-dried *A. californica* nuclear polyhedrosis virus, a pathogenic baculovirus produced in the alternate host insect, *Spodoptera exigua*, was fed to the shrimp. Virus-containing food was prepared by grinding standard flaked fish food into a fine powder, adding a known quantity of the virus, and then pressing the powder into pellets with a commercial pellet maker. Control pellets were virus free. Each pellet weighed approximately 0.2 g.

Shrimp were fed twice weekly, with one pellet per holding cup. Control animals received virus-free pellets; experimental shrimp were fed pellets that contained 15×10^6 PIBs (polyhedral inclusion bodies) per pellet. This concentration far exceeded the amount needed to cause significant mortality in the host insect (more than 3×10^3 times the LD_{90} concentration).

The day after feeding, all holding cups were removed to plastic trays, excess food was pipetted into a large glass flask and the cups returned to their tanks. Virus-contaminated refuse was decontaminated by adding sodium hypochlorite to the glass flask, plastic tray, and utensils. Waste from the control aquarium was merely washed into a sink with tap water.

5. Bioassay of Food

Commercial food pellets used to feed the shrimp were bioassayed by incorporating 2 g of the shrimp food into semisynthetic insect medium and placing one neonatant cabbage looper (*Trichoplusia ni*) larva into each assay cup.[17] For the control, uncontaminated food and food that contained the LD_{50} dose of polyhedra (experimental), 36 larvae were used in each of two replicates; thus 72 *T. ni* larvae were exposed to the control food and 72 were exposed to polyhedra-dosed food. There were also 72 *T. ni* and *Spodoptera exigua* that served as controls, with no control food or polyhedra added to the semisynthetic insect diet.

6. Serological Testing

The test was terminated after 30 d exposure, and all surviving animals were killed and processed. Frozen specimens of exposed and control shrimp (both whole body and visceral masses) were prepared for serological testing.

Antisera were prepared against whole AcB polyhedra, AcB polyhedral protein, and *T. ni* baculovirus polyhedral protein according to the method of Tompkins.[18] The antisera were precipitated with ammonium sulfate and stored at − 20°C. Immunodiffusion was performed against these three antisera and polyhedral protein of AcB produced in beet armyworm, using 1% Bio-Rad agarose in Tris-barbital buffer, pH 8.6. The slides were incubated for 72 h at 4°C in a moist chamber, overlaid with 1.0 m NaCl for 3 h, rinsed with distilled water for 10 min, dried, stained in a solution of 2% Coomassie brilliant blue R-250, destained, and photographed.

Immunoelectrophoresis was performed in 10 × 10-cm glass slides that contained a medium of 15 ml of 1% Bio-Rad agarose in Tris-barbital buffer, pH 8.6. Wells were punched in the medium and appropriate antigens added. Electrophoresis was in Tris-barbiturate buffer at 8.3 V/cm, 9°C, for 1 h and 45 min. After electrophoresis, appropriate antisera were added and the slides incubated for 73 h at 4°C in a moist chamber, washed, stained, and destained as previously described with the immunodiffusion slides.

All test and control supernates from the homogenized shrimp and the pellets after addition of 0.1 M Na_2CO_3 to 0.1 m NaCl were subjected to immunodiffusion and immunoelectrophoresis using antisera against *T. ni* virions, polyhedra protein, and/or antisera to AcB protein.

7. Histology

Surviving shrimp were killed at the end of the test and preserved in Davidson's fixative.[19] Complete visceral masses were removed and processed for light microscopy. Following embedding in paraffin, 7-μm longitudinal sections were made and stained with either routine Harris' hematoxylin and eosin or with a modified bromo-phenol blue technique.[20] Sections were then examined for inclusion bodies, with major emphasis on nuclei of hepatopancreas and midgut cells.

8. EM

Hepatopancreas specimens of control and treated animals reserved for EM were fixed, processed, and embedded in epon. Silver sections, cut on an ultramicrotome, were stained with uranyl acetate and lead citrate. Nuclei and cytoplasm of hepatopancreas cells from random samples were searched at 2500 × to 5000 × in a Zeiss EM10. All sections on each grid were examined for presence or absence of inclusion body, virus, or cytopathic effect.

9. Bioassay Results of Agent in Natural Insect Hosts

There was less than 5% mortality in the insects *T. ni* larvae or *S. exigua* larvae fed synthetic shrimp diet without virus. Mortality was 47% among *T. ni* larvae fed synthetic shrimp diet containing LD_{50} dose of AcB, and 54.6% among *S. exigua*.

10. Gross Observations of Nontarget Hosts During Test

Control and experimental shrimp, after the introduction of the virus, did not exhibit differences in feeding behavior, equilibrium, or activity. Mortality during acclimation to the holding cups was 35 to 40%. During the test, cumulative mortality (for 30 d) was 30% among control animals and 28% among experimental animals.

11. Serology Results

Immunodiffusion patterns showed similar precipitin bands between antisera against *Autographa californica* polyhedral protein and antisera against *T. ni* polyhedra protein to *A. californica* polyhedral protein. Although whole *A. californica* polyhedral protein gave no visible reaction by immunodiffusion, it did give similar precipitin bands by immunoelectrophoresis.

Since antisera against *T. ni* polyhedrin and *A. californica* polyhedrin gave the most prominent bands, either was used for the rest of the immunodiffusion and immunoelectrophoresis tests.

No serological evidence of presence of the polyhedral protein or virion protein was found by immunodiffusion or immunoelectrophoresis utilizing internal tissues of exposed shrimp. Complement fixation tests were also negative.

12. Histopathology Results

Light microscopical examination revealed no difference between tissues of control shrimp and those exposed to the viral PIBs. No evidence of patent infection (inclusion bodies) or other pathogenic response was seen in the hepatopancreas or midgut, nor was there indication of altered nuclei or cytoplasmic change usually indicative of viral infection.

13. EM Results

Careful examination of representative fields in ultra-thin sections of hepatopancreas revealed no indications of viral infection or cytopathologic activity characteristic of viral infections.

B. BACTERIA (EXAMPLE, *BACILLUS THURINGIENSIS* SSP. *KURSTAKI*)
1. Presumed Mode of Action in Natural Hosts

B. thuringiensis preparations consist of spores and parasporal bodies (crystals). The toxic effect occurs when the parasporal bodies are dissolved and activated by enzymes of the gut lumen, releasing protein endotoxin, and causing swelling and eventual destruction of the gut epithelium. An infective stage may occur when the gut epithelium is damaged. It is thought that hemolymph leakage into the gut lumen, which lowers the pH, stimulates germination of spores; which, however, happens only in certain groups of insects, and is unimportant with respect to pathology in most. This is followed by passage of the bacillus or vegetative stages through the damaged gut into the hemocoel. Within the hemocoel the bacillus is carried throughout the animal. Critical damage occurs to the gut and internal tissues by endotoxin and vegetative growth of *B. thuringiensis*[21] and *B. thuringiensis* ssp. *israelensis*.[22,23]

2. Test Animals and Test Conditions

The grass shrimp, *Palaemonetes pugio*, was selected for this test. Adults of similar size, including males and gravid and nongravid females, were tested. Two methods of exposure to *B. thuringiensis* ssp. *kurstaki* were used: (1) food pellets and (2) direct gavaging. For the shrimp exposed by food pellets, test conditions, including schedule of feeding, cleaning, and decontamination were the same as described for the viral agent. Gavage was done by injecting a solution of the test material into the oral cavity. Histological samples were taken at the time of gavage and at 3-, 6-, and 16-h intervals afterwards.

3. Bacterium — Preparation and Feeding

A dry powder preparation of technical grade *B. thuringiensis*, ssp. *kurstaki* (potency of 4000 IU/mg) was used. Food pellets were prepared by grinding flaked fish food into a fine powder and adding a known quantity of the technical grade powder. Food was formed into pellets, each weighing approximately 0.1 g, with a potency of approximately 160,000 IU per pellet. A concentrated suspension of *B. thuringiensis* and seawater was mixed and injected into the oral cavity of each shrimp, which had not been fed for the previous 24 h, by use of a microliter syringe. Approximately 8 μl of the suspension, a potency of approximately 2560 IU, was injected into the oral cavity of each shrimp. The suspension traversed the entire length of the digestive tract in less than 2 h, indicated by appearance of new fecal strands.

4. Bioassay of Food

Experimental and control food pellets were tested for potency and virulence in natural insect hosts. The bioassays were performed on *Trichoplusia ni*, *Spodoptera exigua*, and *S. frugiperda*, and were done by surface feeding and incorporation into the diet. One pellet was used for 24 cups, each containing one insect. The pellet was suspended in 5-ml distilled water, mixed, and either pipetted onto the surface of the cup and allowed to dry, or blended into the diet and poured into cups, before placing a neonate insect in the cup.

5. Histology — Exposure by Food Pellets

All surviving exposed and control shrimp were killed at the end of the 30-d test and preserved in Davidson's fixative. Carapaces were removed and whole shrimps were embedded in paraffin. Seven-micron parasagittal and cross-sections were stained with routine Harris' hematoxylin and eosin.

6. Direct Exposure by Gavaging

All exposed and control shrimp were killed at the end of each sample period of 0, 3, 6, and 16 h. The entire digestive tract was dissected from the shrimp, preserved in Davidson's fixative, and processed for light microscopy. Parasagittal and cross-sections (7 μm) were stained with routine Harris' hematoxylin and eosin.

7. EM

Shrimp from the feeding exposure were selected for examination. Only the digestive tracts of experimental and control shrimps were processed. Tissue samples were taken from three areas of each digestive tract; the foregut, midgut, and hindgut. Samples were fixed, sectioned, stained, and examined with a Zeiss EM10. Special attention for possible toxic effects was given to cells of the epithelial lining of the midgut.

8. Bioassay Results of Agent in Insect Hosts

The experimental food pellets containing *B. thuringiensis* proved to be very effective against three species of natural insect hosts at this dose, with control pellets causing mortalities within acceptable limits.

9. Gross Observations of Nontarget Host During Test

There were no noticeable differences in behavior between the control and experimental shrimp, before, during, and after feeding. Cumulative mortality was 14% among control and 16% among experimental groups. We felt that the mortality percentages were within acceptable limits for this test system and that mortality was caused by reasons other than exposure to *B. thuringiensis*. There was no mortality among shrimp that were gavaged with *B. thuringiensis* powder and saltwater.

10. Histopathology and EM Results

Light microscopy and EM revealed no difference between tissues of control shrimp and those exposed to *B. thuringiensis*. No evidence of the bacterium was present, nor were any toxic or pathogenic effects seen in the hepatopancreas or digestive tract (see Reference 58 in Chapter 12).

C. FUNGI (EXAMPLE, *LAGENIDIUM GIGANTEUM* — CALIFORNIA STRAIN)

This fungus, originally isolated from its natural insect host, the mosquito (*Culex tarsalis*), is a potential registrant as a microbial insecticide. Although *L. giganteum* is a freshwater organism, we believed it is a good candidate for initial fungal studies for three reasons. First, it provides a good prototype organism for initial methods development; second, a member of the same genus, *L. callinectes*, is a known marine/estuarine pathogen of crab eggs and crustacean larvae, and third, we believe investigation of its possible effects on estuarine organisms, particularly those that occur in low salinity, is necessary and valid because of its tolerance of low salinity.[24,25]

1. Presumed Mode of Action in Natural Hosts

The zoospore of *L. giganteum*, a motile flagellated cell, is the infective stage which attaches to the first, second, third, or earlier fourth instar of a mosquito larva (*Culex* spp., *Aedes* spp., *Anopheles* spp.), and encysts on the cuticle. It then penetrates and begins to grow inward as a hypha. This process of penetration has yet to be elucidated. One theory proposed by Domnas et al.[26] suggests a process involving either a protolytic or lipolytic enzyme in conjunction with mechanical pressure. After the cuticle is breached, hyphal strands begin to grow within the larva, eventually filling the hemocoel. The hyphae then develop septa, forming sporangia. These swell, become spherical, and begin to form exit or discharge tubes which push through the cuticle and form exit vesicles external to the larva's body. Zoospores are cleaved out within the vesicles and released into the aqueous environment surrounding the larval remains when vibrations caused by the maturing zoospores rupture the vesicle's membrane, thus beginning a new cycle.

2. *Lagenidium* Culture and Zoospore Production

A subtaxon of *L. giganteum* (California strain, a reisolate of the North Carolina strain) on an agar medium was transferred to a liquid medium (standard medium) for growth and maintenance. For experimental purposes, multiple cultures were made similarly from a single parent culture.

Zoospore production was induced by adding distilled water to a culture and blending the mixture with a ultrasonic blender at low speed. Zoospore production began 1 to 2 h following this procedure, with peak production at approximately 14 to 18 h. The diluted culture or an aliquot thereof was added to test situations 2 to 6 h following addition of the distilled water. To prevent possible release to the environment, fungal cultures were then destroyed.

3. Test Animals and Test Conditions

Preliminary studies with larval grass shrimp (*P. pugio*) were done in salinities of 3°/$_{oo}$ and 5°/$_{oo}$. Three groups of ten larvae each were used at each salinity: a control group, a group exposed to induced *Lagenidium* culture (standard medium), and a group exposed to induced cultures from medium at the appropriate test salinity (either 3°/$_{oo}$ or 5°/$_{oo}$ salinity). Each group was held at 25°C in individual compartments, each containing 10 ml of seawater at the appropriate salinity. Daily inoculations of 1 ml of induced culture were made each afternoon for 4 consecutive days. Each morning following inoculation, the compartments were cleaned and seawater changed. Observations for mortality or stress were made regularly, and suspect organisms removed and evaluated for fungal infection by microscopic examination of fresh preparations.

Preliminary studies of susceptibility of *Palaemonetes pugio* eggs and adults to *L. giganteum* were performed. Three gravid adult females were tested, each in a large culture dish containing 500 ml of 3°/$_{oo}$ seawater (25°C) to which 10 ml of induced *Lagenidium* culture (standard media) was added. Dosing was repeated for 4 consecutive days with cleaning and observation between each dosing. Three control gravid adults were tested under identical conditions without dosing. Adult shrimp and egg masses were processed for histological examination at test termination. Slides were stained with hematoxylin and eosin and with PAS and hematoxylin for evaluation.

4. Fungal Results

During the test, some mortality was noted in experimental and control larval shrimp. However, examination of dead shrimp during the test and those remaining at termination revealed no fungal infection or fungus-related death in experimental organisms. Probable cause of the mortalities was salinity stress or some other nonexposure-related cause.

Gross and histological evaluation of adult shrimp and egg masses revealed no fungal infection or attachment of zoospores to either adults or developing embryos.

D. PROTOZOA (EXAMPLE, *NOSEMA CUNEATUM*)
1. Presumed Mode of Action in Natural Hosts

The infectious stage for microsporidian protozoans is a spore. The spores of *Nosema* are ingested in the diet of the target insect host. In the midgut of the host, the spores come into proximity with the epithelial cells of the gut. Germination occurs when the spore releases a discharge tube that attaches to the plasmalemma of the epithelial or other primary cell type. The sporoplasm leaves the spore via the everted polar tube, to enter the primary host target cell.[27] Not a great deal is known about routes of early infectious stages of microsporidia in their hosts. However, "wandering cells", especially undifferentiated mesenchymal cells, hemocytes, macrophages (depending on whether hosts are invertebrates or vertebrates), and body fluids probably aid in their distribution. In suitable cells, the sporoplasms became meronts. The microsporidia are certainly disseminated to final cell types such as gonadal cells, muscle, nerve, or connective tissue where they cause massive infections and much tissue destruction and loss of vital function (e.g., muscle lysis). It is at these ultimate tissue sites that sporulation begins, usually following massive vegetative proliferation of the schizont stages. Spores are released into the environment or enter a new host when the diseased insect is decomposed or eaten by scavengers/predators.

2. Test Animals and Test Conditions

Test conditions were identical to those designed for exposure to the baculovirus from *A. californica*, except for use of *P. pugio* only as the test animal. Also, no attempt was made to separate gravid from nongravid female shrimp for test purposes.

3. Protozoa — Preparation and Feeding

Concentrated spores of *N. cuneatum*, a microsporidian pathogen of the North American migratory grasshopper, *Melanoplus sanguinipes*, were used in this test. Contaminated food pellets were prepared weekly. Nine grams of a flaked fish food was ground to a fine powder. The spores, in a suspension of distilled water, were sprayed over this preparation of powdered food using an atomizer. The water was allowed to evaporate at room temperature and the dry powder pressed into 0.1 g food pellets. This provided a large dose of spores (approximately 440,000 spores per pellet) to the experimental shrimp at each feeding. Control shrimp were fed protozoan-free pellets made of flake fish food only.

Exposure was by food pellets, and the schedule of feeding, cleaning, and decontamination was as for the virus test agent.

4. Bioassay of Food

Several contaminated and control food pellets were selected from each week's supply and stored at approximately 4°C for bioassay purposes. All food pellet samples were sent to the U.S. Department of Agriculture Rangeland Insect Laboratory, Bozeman, MT, to be tested for efficacy and pathogenicity against *M. sanguinipes*.

5. Histology

The test was terminated after 30 d and all surviving shrimp were killed, injected with Davidson's fixative, and left overnight in a vacuum oven (20 psi) to insure complete penetration of the fixative. They were processed for histopathology after treatment with a decalcifyer. Whole shrimp were processed histologically, enabling good examination of the entire digestive tract and hepatopancreas (digestive gland).

6. Bioassay Results of Agent

Preliminary results indicate that the microsporidians from contaminated food pellets were pathogenic and viable in *M. sanguinipes*.

7. Gross Observations of Nontarget Host During Test

Control and experimental shrimp, after introduction of the test agent, showed no difference in feeding behavior or activity. Cumulative mortality for the 30-d test was 20% among control animals and 14% among experimental animals. These mortalities were within acceptable limits for this testing procedure and in accord with control animal expected mortality.

8. Histopathology Results

No difference between tissues of the control shrimp and those exposed to the protozoan agent were seen by light microscopic examination. There was no indication of any pathogenic response in any of the tissues of the hepatopancreas or midgut. Histological examination revealed no evidence of persisting spores within the digestive tract of the exposed shrimp, and elimination of ingested spores was indicated.

III. CONCLUSIONS

This chapter is concerned chiefly with research that is in an early state of development. Some of the methods reported here are still being evaluated and validated. Results, to date, with agents representative of the four major microbial pesticidal groups exposed to NTOs via natural routes (Table 1), have been negative; i.e., neither infections, severe pathogenesis, or toxicoses have been induced in the experimental, nontarget hosts tested. This is hardly surprising since the null hypothesis was that there would be no effect of the agents in nontarget species. Thus far the null hypothesis (based on experimental, estuarine nontarget animal *in vivo* results) has not been rejected.

Not reported herein, but worthy of note at this time, is ancillary research on the *in vitro* use of cell cultures, and use of genetic probe techniques to determine if baculoviruses can enter the nucleus of nontarget vertebrate cell lines. Cell lines of rainbow trout (*Salmo gairdneri*) and eastern box turtle (*Terrapene carolina*) heart were exposed under controlled conditions to the baculovirus of *Autographa californica*, the same microorganism evaluated in *in vivo* tests. EM techniques and genetic probe methods were applied to both qualitatively and quantitatively assess the fate and effects of this virus in the vertebrate cells. The results

TABLE 1
Studies of Selected Microbial Control Agents in Estuarine NTOs

Agent	Type of test	Method of exposure and concentration	Endpoints studied	Ref.
Virus (*Baculovirus* of *Autographa californica*)	Single species; static	Food pellets: mixture of polyhedra virus and flaked fish food (15 × 10^6 PIBs per pellet)	L,B,S,H,U	15
	Single species; static	Liquid mixture of polyhedra virus and artificial diet at three different concentrations	L,B,H,U	14
	Single species; static	Intramuscular injection of a virus suspension (4.7 × 10^{11} virus rods per injection)	L,B,H,U	14
Bacteria (*Bacillus thuringiensis*)	Single species; static	Food pellets: mixture of technical grade powder formulation and flake fish food (150,000 IU per pellet)	L,B,H,U	28
Fungi (*Lagenidium giganteum*)	Single species; static	Direct addition of a standard medium *Lagenidium* culture to seawater	L,B,H	29
Protozoa (*Nosema cuneatum*)	Single species; static	Food pellets: mixture of spores and flaked fish food (440,000 spores per pellet)	L,B,H	28

Note: L — lethality; B — behavioral changes; S — serological (determination of antigens); H — histopathology of selected tissues and organs (infectivity and pathogenicity); and U — ultrastructure of exposed cells (infectivity and pathogenicity).

were (1) EM revealed entry of the insect baculovirus into heart cells of rainbow trout and eastern box turtle heart cells; further EM demonstrated the alignment of the nucleocapsid of the virus at the nuclear pores of the nontarget cells and (2) genetic probe methods (DNA hybridization) revealed amounts of viral DNA that entered the nuclei and how long it persisted after entry. Viral replication did not occur in the nontarget cells.[28]

These studies have raised several questions on what should be done next in regard to determining possible long-term significance (if any) of entry of viral DNA into nontarget cells. Chief among these questions are (1) does subtle, low-level integration of insect virus DNA with host cell DNA occur?, (2) what effects, if any, are possible if genetic integration occurs?, and (3) what, if any, research should be done to follow up this observation?

Further *in vitro* test methods should involve attempts to establish crustacean cell lines for use in screening microbial pest control agents for infectivity, pathogenicity, or toxicity.

In contrast to the single species systems[29,30] described herein, new multispecies test systems are being developed in which to determine safety of both natural and genetically altered microbial control agents.[31] By multispecies systems, we mean laboratory systems that include several species of nontarget animals and plants that may be exposed simultaneously to test the safety of a microbial agent in the laboratory. These systems provide a more complex exposure scenario with opportunity for at least limited interaction of the agent with several nontarget species representing different phyletic and/or trophic groups. These systems may provide keener insights into more natural exposure situations and potential effects.

REFERENCES

1. **Sindermann, C. J.,** *Principal Diseases of Marine Fish and Shellfish,* Academic Press, New York, 1970.
2. **Sprague, V. and Couch, J.,** An annotated list of protozoan parasites, hyperparasites, and commensals of decapod crustacea, *J. Protozool.,* 18, 526, 1971.
3. **Couch, J. A.,** Diseases, parasites, and toxic responses of commercial penaeid shrimps of the Gulf of Mexico and South Atlantic coasts of North America, *Fish. Bull.,* 76, 1, 1978.
4. **Overstreet, R. M.,** Marine Maladies? Worms, Germs and other Symbionts from the Northern Gulf of Mexico, No. MASGP-78-021, Mississippi-Alabama Sea Grant Consortium, Ocean Spring, MS, 1978.
5. **Lightner, D. V.,** A review of the diseases of cultured penaeid shrimps and prawns with emphasis on recent discoveries and developments, in Proc. 1st Int. Conf. Culture of Penaeid Prawns/Shrimps, 1984 SEAFDEC Aquaculture Department, Iloilo City, Philippines, 1985, 79.
6. **Sparks, A. K.,** *Synopsis of Invertebrate Pathology, Exclusive of Insects,* Elsevier, Amsterdam, 1985.
7. **Jaques, R. P.,** Application of viruses to soil and foliage for control of the cabbage looper and imported cabbage worm, *J. Invertebr. Pathol.,* 15, 328, 1970.
8. **Couch, J. A.,** An enzootic nuclear polyhedrosis virus of pink shrimp: ultrastructure, prevalence, and enhancement, *J. Invertebr. Pathol.,* 24, 311, 1974.
9. **Couch, J. A., Summers, M. D., and Courtney, L.,** Environmental significance of baculovirus infections in estuarine and marine shrimp, *Ann. N. Y. Acad. Sci.,* 266, 528, 1975.
10. **Couch, J. A.,** Viral diseases of invertebrates other than insects, in *Pathogenesis of Invertebrate Microbial Diseases,* Davidson, E. W., Ed., Allanheld Osmun Publishing, Totowa, NJ, 1981, 127.
11. **Lightner, D. V. and Redman, R. M.,** A baculovirus-caused disease of the penaeid shrimp, *Panaeus monodon, J. Invertebr. Pathol.,* 38, 299, 1981.
12. **Sano, T., Nishimura, T., Oguma, K., Momoyama, K., and Takeno, N.,** Baculovirus infection of cultured Kuruma shrimp, *Penaeus japonicus* in Japan, *Fish. Pathol.,* 15, 185, 1981.
13. **Lightner, D. V., Redman, R. M., and Bell, T. A.,** Observations on the geographic distribution, pathogenesis and morphology of the baculovirus from *Penaeus monodon* Fabricius, *Aquaculture,* 32, 209, 1983.
14. **Lightner, D. V., Proctor, R. R., Sparks, A. K., Adams, J. R., and Heimpel, A. M.,** Testing penaeid shrimp for susceptibility to an insect nuclear polyhedrosis virus, *Environ. Entomol.,* 2, 611, 1973.
15. **Couch, J. A., Martin, S. M., Tompkins, G., and Kinney, J.,** A simple system for the preliminary evaluation of infectivity and pathogenesis of insect virus in a nontarget estuarine shrimp, *J. Invertebr. Pathol.,* 43, 351, 1984.
16. **Hayat, M. A.,** *Principles and Techniques of Electron Microscopy: Biological Applications,* Vol. 1, University Park Press, Baltimore, 1981, 128.
17. **Thomas, E. D., Reichelderfer, C., and Heimpel, A.,** Accumulation and persistence of a nuclear polyhedrosis virus of the cabbage looper in the field, *J. Invertebr. Pathol.,* 20, 157, 1972.
18. **Tompkins, G. T.,** Characterization of Multiple Embedded Nuclear Polyhedrosis Viruses Isolated from the Cabbage Looper, Ph.D. thesis, University of Maryland, College Park, 1979.
19. **Shaw, B. L. and Battle, H. I.,** The gross and microscopic anatomy of the digestive tract of the oyster *Crassostrea virginica* (Gmelin), *Can. J. Zool.,* 35, 325, 1957.
20. **Johnson, P.,** *Histology of the Blue Crab,* Callinectes sapidus: *A Model for the Decapoda,* Praeger, New York, 1980, 395.
21. **Percy, J. and Fast, P. G.,** *Bacillus thuringiensis* crystal toxin: Ultrastructural studies of its effect on silkworm midget cells, *J. Invertebr. Pathol.,* 41, 86, 1983.
22. **Lacey, L. A.,** *Bacillus thuringiensis* serotype H-14, in *Biological Control of Mosquitoes,* Bull. No. 6, Chapman, H. C., Ed., American Mosquito Control Association, Fresno, CA, 1985, 132.
23. **Aly, C., Mulla, M. S., and Federici, B. A.,** Sporulation and toxic production by *Bacillus thuringiensis* var. *israelensis* in cadavers of mosquito larvae (Diptera:Culicidae), *J. Invertebr. Pathol.,* 46, 251, 1985. 1985.
24. **McCray, E. M., Jr., Umphlett, C. J., and Fay, R. W.,** Laboratory studies on a new fungal pathogen of mosquitoes, *Mosq. News,* 33(1), 54, 1973.
25. **Merriam, T. L. and Axtell, R. C.,** Salinity tolerance of two isolates of *Lagenidium giganteum* (Oomycetes: Lagenidiales), a fungal pathogen of mosquito larvae, *J. Med. Entomol.,* 19, 388, 1982.
26. **Domnas, A. J., Srebro, J. P., and Hicks, B. F.,** Sterol requirements for zoospore formation in the mosquito-parasitizing fungus *Lagenidium giganteum, Mycologia,* 69, 875, 1977.
27. **Canning, E. U. and Lom, J.** (with a contribution by Dykova, I.), *The Microsporidia of Vertebrates,* Academic Press, Orlando, 1986.
28. **Brusca, J., Summers, M., Couch, J., and Courtney, L.,** *Autographa californica* nuclear polyhedrosis virus efficiently enters but does not replicate in poikilothermic vertebrate cells, *Intervirology,* 26, 207, 1986.

29. **Couch, J. A., Foss, S. S., and Courtney, L. C.,** Evaluation for Risks of an Insect Virus, Bacterium, and Protozoan to a Nontarget, Estuarine Crustacean, Report EPA/600/X-85/290, U.S. Environmental Protection Agency, Washington, D.C., 1985.
30. **Foss, S. S., Courtney, L. A., and Couch, J. A.,** Evaluation of a Fungal Agent *(Lagenidium giganteum)* Under Development as an MPCA for Nontarget Risks, Report EPA/600/X-86/229, U.S. Environmental Protection Agency, Washington, D.C., 1986.
31. **Couch, J. A., Duke, T. W., Foss, S. S., and Perez, K. T.,** Enclosed systems for testing microbial pest control agents, Proc. Workshop at ERL/Gulf Breeze, U.S. Office of Pesticide Programs and Office of Research and Development, Environmental Protection Agency, Washington, D.C., 1986.

III. Safety of Microbial Insecticides to Vertebrates

Chapter 8

SAFETY OF MICROBIAL INSECTICIDES TO VERTEBRATES — HUMANS

J. P. Siegel and J. A. Shadduck

TABLE OF CONTENTS

I. INTRODUCTION

A. OVERVIEW

Development and use of microbial insecticides has increased as chemical insecticides have become less popular because of problems associated with the resistance of target insect species, nontarget toxicity, and environmental persistence. The first microbial insecticide registered in the U.S. was *Bacillus popillae* (1948). By 1985 there were 14 microbial pest control agents in over 100 products registered in the U.S. for use in agriculture, forestry, mosquito control, and the home.[1,2] From the outset, microbial insecticides underwent the same safety studies as their chemical counterparts, including chronic carcinogenicity studies as well as infectivity studies. For example, *B. thuringiensis* was fed to avian species for as long as 690 d and to rats for 730 d. Also, parenteral injection, inhalation, and short-term feeding studies were conducted on mice, rats, guinea pigs, and human volunteers. Humans ate 10^{10} spores per d for 5 d and inhaled 10^9 spores with no ill effects.[1,3-5] However, it became evident that the body of tests available for assessing the safety of chemical insecticides were not directly applicable to microbial agents for a variety of reasons.

Chemical safety tests assume that a measurable biological effect can be achieved if the dose administered is high enough. Typically this is expressed as a median lethal dose or LD_{50}. In contrast, it is often impossible to achieve an LD_{50} using microbial agents if they are administered to nonhost species by conventional means such as aerosol, dermal, or oral application. The quantities of material needed to produce death are so large as to suffocate the test animal or block its gastrointestinal tract.[6] Chemical tests also assume that a compound is metabolized and/or excreted and that metabolic products may be as toxic or more toxic than the original compound. There is no evidence to date that microbial insecticides are metabolically activated in mammals. Although alkaline-solubilized δ-endotoxin of *Bacillus thuringiensis* ssp. *israelensis* is lethal when injected into mice and is cytolytic to human erythrocytes, mouse fibroblasts, and primary pig lymphocyte cultures *in vitro*, the activation step cannot occur in mammals because gut conditions are unsuitable.[7-9] Microbial agents differ from chemical toxicants because they possess the ability to multiply, while some chemical agents are sequestered in host body fat and exert an effect over time, necessitating both acute and chronic studies. Finally, chemical safety tests assume that a knowledge of a compound's chemical structure and information on related compounds can give insight into a compound's potential hazard to man, while there can be a wide disparity within the same genus of microorganism with respect to host range and pathogenicity. For instance, *B. thuringiensis* is lethal to Lepidoptera while *B. anthracis* is a virulent pathogen of ruminants and man.

In summary, while both chemical and microbial insecticides share safety concerns with respect to toxicity, irritancy, and allergenicity, microbial agents are unique in their ability to multiply within a suitable host. In this chapter, we will summarize the guidelines for conducting mammalian safety tests and discuss several key issues concerning mammalian infectivity and toxicity of entomopathogens, with particular emphasis on bacterial, fungal, and protozoan entomopathogens tested in our laboratory.

B. PHILOSOPHY OF HAZARD EVALUATION

Interest developed in establishing new safety testing protocols that recognized the unique characteristics of entomopathogens. The philosophy of testing known as maximum challenge advocated the use of extreme test conditions such as intraocular and intracerebral injection in order to achieve detectable biological effects. The approach used is parallel to the LD_{50} concept but the route of exposure rather than the dose is varied. In a maximum challenge test, the highest dose possible (dependent on the physical nature of the material) is given by the route that most severely compromises an animal's natural defenses.[6] Selection of the route most vulnerable to the animal is based on two parameters. First, a literature search is conducted to locate instances where the organism was isolated from humans or animals or where lesions were produced by the candidate organism or species closely related to it. Tests are then designed to target the appropriate organ system in an attempt to reproduce the infection. Second, some mammalian organs such as the brain or eye are considered to be highly vulnerable to infection because of their limited range of responses and the devastating effect of even moderate lesions on the functions of these tissues. These principles are illustrated in the safety tests of the following entomopathogens. *B. sphaericus* was involved in a fatal case of meningitis in man, and the maximum challenge safety tests therefore included intracerebral injection studies.[10,11] Likewise, the microsporidian *Encephalitozoon cuniculi* causes brain lesions in rabbits and primates, so the entomopathogenic microsporidia *Nosema algerae* and *N. locustae* were injected intracerebrally in rabbits as part of an effort to assess their hazard to mammals.[6,12]

The primary difficulty associated with maximum challenge testing involves the interpretation of mortality data and extending the results to human safety. Premature rejection of a candidate is a major concern because even nonpathogenic organisms can cause death when injected in large quantities into a vulnerable site such as the brain; the ability to cause mortality should not automatically lead to a candidate's rejection until other routes of exposure are evaluated. The value of these tests lies in negative results, which provide convincing evidence of an organism's safety to mammals and by extension to man. Additionally, information associated with what is clearly a "worst case" scenario such as intracerebral injection could prove to be invaluable in instances in which the entomopathogen is isolated from humans.

C. TESTING GUIDELINES

Current safety testing protocols combine elements of maximum challenge testing with conventional methods of exposure. In 1981, the World Health Organization (WHO) published a memorandum advocating a three-tier mammalian safety testing scheme for entomopathogens.[12] The first tier employed short-term tests (4 weeks or less) to evaluate infectivity, toxicity, irritancy, and allergenicity by means of a single oral, inhalation, i.p., dermal, and ocular exposure. Doses were not to exceed 5 g/kg and the minimum i.p. dose was 10^6 organisms per mouse and 10^7 organisms per rat. Successful passage through Tier I led to a candidate entomopathogen being declared safe with no limitations, and Tier I clearly departed from the chemical insecticide legacy of costly, long-term carcinogenicity studies. However, if difficulties arose in the first tier of tests, Tiers II and III provided the development of exposure data as well as longer-term studies involving both single and multiple exposure to quantify toxicity, infectivity, and irritancy. In addition, intracerebral injection was specified in Tier I for protozoa because of the mammalian central nervous system vulnerability to *Encephalitozoon cuniculi*. This three-tier concept was subsequently adopted by the U.S. Environmental Protection Agency (EPA) in 1982, the tests serving as the foundation for the biorational pesticide assessment guidelines of Subdivision M.[13] These guidelines specify the species of mammals to be tested as well as the number of animals per treatment group and in addition, stipulate that immunocompromised animals be included in Tier I testing.

Evaluation of the role played by an intact host immune system in clearance of an entomopathogen, through the use of partially immunodeficient athymic mice as test animals or the use of chemical immune suppressants such as hydrocortisone and cyclophosphamide, can be regarded as a logical extension of the maximum challenge philosophy. This issue was not addressed in the WHO memorandum and was felt by some to be controversial because of the difficulty associated with relating positive findings such as infection or mortality to human health. Some researchers noted that the variety of immune suppressants available and their varying modes of action would make comparisons between safety studies difficult. Thus the view has been expressed that immunosuppressed humans might suffer from natural vertebrate diseases before entomopathogens could have time to act, but the question remains open.[1] Acquired Immunodeficiency Syndrome (AIDS) is prevalent in regions where entomopathogens are an important component of vector control programs. In North America and Western Europe outpatient treatment of cancer patients undergoing immunosuppressive therapies is common, leading to concern on the part of the EPA about the possible hazard posed to these individuals by entomopathogens.

Currently, toxicity is an issue with bacteria and fungi but it has not been demonstrated with protozoa or viruses; infectivity is a concern with all four groups. Candidates that pass Tier I are considered nonhazardous to mammals and by extension to man, but one must be wary about claiming absolute safety for any entomopathogen. Burges[1] summed this up when he noted that "A "no-risk" situation does not exist, certainly not with chemical pesticides, and even with biological agents one cannot absolutely prove a negative. Registration of a chemical is essentially a statement of usage in which the risks are acceptable. The same must apply to biological agents."

II. INFECTIVITY OF ENTOMOPATHOGENS TO MAMMALS

A. INFECTION VS. PERSISTENCE

Infection can be defined as the invasion of the body by pathogenic microorganisms and the subsequent reaction of tissues to their presence. The term can also refer to the simple presence of microorganisms within tissues, whether or not this results in pathological effects.[14] For our purposes in this chapter, the term *infection* will refer to multiplication of an organism in mammals, as measured by the presence of vegetative stages in tissue or the recovery of increased numbers of viable organisms from the subject over time. The term *persistence* will refer to survival of viable organisms without multiplication. It can be demonstrated by simple recovery of the organism. This distinction is crucial because an agent that can replicate in mammals poses a greater hazard to man than one that cannot multiply. In practice, it can be difficult to distinguish between infection and persistence using standard microbial culture techniques such as swabbing suspect lesions, because this technique can only demonstrate that viable organisms were present and cannot determine the life-history stage present. Unfortunately, some researchers regard microbial recovery as synonymous with infection. This has given rise to several controversial reports of entomopathogens infecting animals and man. In the following paragraphs, we will summarize data concerning infectivity and persistence of entomopathogenic bacteria, fungi, and protozoa in immune intact and immunocompromised animals, and will use these results to determine the hazard posed by these agents to man.

B. ENTOMOPATHOGENIC BACTERIA

Bacillus thuringiensis was reported by Samples and Buettner[15] to have caused an ocular ulcer in an 18-year-old farmer accidentally splashed in the face by a commercial suspension of this microorganism. Their claim was based on recovery of *B. thuringiensis* from a swab of the eye taken 13 d after exposure, although the ulcer was not examined directly for the presence of vegetative forms of the bacterium. The authors did not acknowledge the possibility that spores might persist in the eye and be recovered, although not necessarily causatively related to the ulcer. Experiments conducted in our laboratory with two entomopathogenic bacilli, *B. thuringiensis* ssp. *israelensis* and *B. sphaericus,* indicated that both bacteria could be recovered from rabbit eyes as long as 8 weeks after instillation into the conjunctival cul-de-sac, although there was no evidence of multiplication. Flushing the treated eyes with water for 30 s following exposure did not remove all of the inoculum. Therefore, the recovery of *B. thuringiensis* spores from a patient 13 d after exposure more than likely represents persistence. In addition, this microorganism can be recovered from the spleen for as long as 8 weeks after i.p. injection and from the brain for as long as 4 weeks after intracerebral injection. We thus consider recovery of its spores from an eye 13 d after exposure to be unremarkable.[16]

There have been unsubstantiated claims that *B. thuringiensis* incorporated into diet might produce pneumonia, myocarditis, and hepatic lesions in sheep, but a recently published[17] 5-month diet incorporation study using several commercial formulations of this entomopathogen found no evidence of disease. Hadley et al.[17] noted in their sheep investigations that *B. thuringiensis* spores not only remained viable in the rumen but were also recovered from the blood. While this finding may suggest that this microorganism has invasive properties, it has been shown that scratches and wounds in the mouth of sheep caused by food are a portal of entry into the bloodstream for a variety of bacteria such as *B. anthracis* and *Actinomyces lignierisi*.[18,19] Additionally, bacteria such as *Clostridium novyi,* can passively migrate through an abraded intestinal wall and enter the bloodstream. The authors concluded that finding *B. thuringiensis* present in sheep blood was not cause for concern and that the species is nonpathogenic to this host.[17,20]

Partially immunodeficient mice were not susceptible to infection by *B. thuringiensis* when tested in our laboratory.[21] Athymic (nude) mice, which lack T-lymphocytes and thus the capacity to mount T-lymphocyte-dependent antibody response, successfully cleared the microorganism from the spleen after i.p. injection. *B. thuringiensis* was recovered for as long as 5 weeks after injection but it was also recovered from immune intact inbred mice (BALB/c) for a similar amount of time. However, the rate of clearance differed and this experiment suggests that immune status may influence clearance dynamics.[16,21] This experiment indicates that a completely intact immune system is not required for a successful defense against *B. thuringiensis* and by analogy, this may extend to man.

There have been no claims that *B. thuringiensis* ssp. *israelensis* is infectious to man, possibly because it has not been used for as long as the lepidopteran strains, hence there has been less human exposure. This subspecies has been subjected to extensive safety testing by several laboratories. Oral, ocular, i.p., aerosol, and intracerebral administration of this bacterium to mice, rats, and rabbits in our laboratory have not provided any evidence of infection. Nevertheless, *B. thuringiensis* ssp. *israelensis* persisted for as long as 7 weeks in the spleens of mice.[16,21,22] There was no evidence of multiplication and no illness following administration. The microorganism disappeared completely from the lungs of rats within 1 week after aerosol exposure and disappeared completely from the brains of rats within 3 weeks after intracerebral injection.

Immunodeficient and immunosuppressed mice were not susceptible to infection by *B. thuringiensis* ssp. *israelensis*. Athymic mice injected i.p. with 2.6×10^7 spores and inbred mice (BALB/c) immune suppressed with hydrocortisone acetate (60 mg/kg) and injected i.p. with 3.4×10^7 spores, successfully cleared this microorganism from their spleens over a 7-week period. Clearance rates differed between athymic and corticosteroid-treated mice as well as untreated BALB/c mice and immunocompromised mice (both athymic mice and corticosteroid-injected), but there was no evidence of infection.[16,21]

B. sphaericus has been recovered from humans and was implicated in nine cases of meningitis as well as a fatal case of leptomeningitis and has also been recovered from a pseudotumor of a human lung.[10,11,23] These human isolates were not pathogenic to young rabbits injected i.v. and i.p. or mice inoculated i.v., i.p., and intracerebrally. Several insect isolates of *B. sphaericus* obtained from WHO/CCBC (accession numbers 1321-1, 1404, and 1593-4) underwent maximum challenge and conventional exposure safety testing.[10,11,24] The central nervous system of rabbits and mice did not appear vulnerable to infection by as many as 4.7×10^8 colony-forming units (cfu) of this species. Rats injected intracerebrally cleared 5×10^5 such units within 17 d. Intraocular injection of as many as 1×10^9 cfu of *B.*

TABLE 1
Maximum Persistence of Bacterial, Fungal, and Protozoan
Entomopathogens in Mammals

Organism	Route	Persistence (days)	Species
Bacillus thuringiensis ssp. *kurstaki*	i.p.	≥35	Mouse
B. thuringiensis ssp. *israelensis*	i.p.	≥70	Mouse
B. sphaericus	i.p.	≥67	Mouse
Metarhizium anisopliae	i.p.	18	Rat
Lagenidium giganteum[a]	i.p.	28	Mouse
Nosema algerae[b]	s.d.	27	Mouse
N. locustae[c]	i.p.	42	Rabbit

Note: i.p. = intraperitoneal; s.d. = subdermal.

[a] Oospores observed, viability undetermined.
[b] Presporal stages — some multiplication.
[c] Spores observed, viability undetermined — accidental intrahepatic deposition.

sphaericus in rabbits produced no evidence of infection, although the entomopathogen could be recovered 14 d after injection. A human isolate of this species, NCTC #11025, injected s.c. and i.p. into mice and rats, did not cause infection. Immunodeficient athymic mice received as many as 9.5×10^8 cfu of NCTC #11025 s.c. without developing infection, and none of this isolate was recovered from tissue homogenate.

Experiments currently conducted in our laboratory with a commercial formulation of *B. sphaericus* provided by WHO indicate that the microorganism can be recovered from the spleen of mice for as long as 10 weeks after i.p. injection, and from the conjunctival cul-de-sac of rabbits 8 weeks after instillation. *B. sphaericus* appears comparable to both *B. thuringiensis* and its subspecies *israelensis* in terms of prolonged persistence. The maximum persistence of these bacilli is summarized in Table 1. If bacterial spores do not germinate, they may behave as inert particles and their clearance dynamics should be similar to latex particles or colloids. However, it may be argued that prolonged bacterial recovery or persistence, in fact, represents slow multiplication with constant replacement of bacteria destroyed by host defenses. If only spores are recovered following injection and the number of organisms recovered decreases with time, this scenario can be ruled out. Alternatively, recovery of vegetative forms would suggest that the bacterium has at least the potential for multiplication.

An experiment was conducted in our laboratory using mice injected with spores of *B. sphaericus* to determine if germination did in fact occur in the spleens of mice, as well as to determine the rate of clearance. Spleen homogenates from 40 mice were heated in a water bath at 65°C for 30 min to kill bacilli (vegetative stages); these conditions should not affect spores.[17] Comparison of the number of cfu recovered from heat-treated and nontreated spleen aliquots indicated that heat-treated *B. sphaericus* aliquots contained significantly fewer cfu than nontreated ones. This suggest that spore germination had occurred. Control experiments indicated that spores were not affected by this temperature regime, and that heated spleen homogenate did not agglutinate spores, which could result in fewer colonies. The above experiment indicates that there may be a potential for multiplication by *B. sphaericus* in mammals, although no bacilli were observed in histological sections of spleen tissue. The number of cfu of this species recovered over time declined as a power function according to the formula $\ln Y = \ln 12.761 - 0.111 X$, with Y as the number of cfu per gram spleen and X as the number of days after injection.

C. ENTOMOPATHOGENIC FUNGI

Metarhizium anisopliae has never been reported as infecting humans. Rats, mice, and rabbits given this entomopathogen in our laboratory by inhalation, oral administration, s.c. injection, i.p. injection, and topical administration had no signs of infection or illness. Viable conidia were recovered from the spleens of rats for as long as 18 d after i.p. injection (Table 1) and *M. anisopliae* was recovered from the stomach, spleen, and lungs of mice for as long as 14 d after inhalation of conidia. However, recovery of this fungus declined and it disappeared from the lungs after 14 d, suggesting that no multiplication occurred. Immuno-compromised animals were not tested in our laboratory but researchers had earlier reported[25] that immunosuppressed rats were not susceptible to infection by *M. anisopliae*.

Following dermal, oral, intratracheal, intraperitoneal, and ocular instillation, *Lagenidium giganteum* proved not infective to rats, mice, and rabbits. Oospores were observed in abscesses of the liver 28 d after i.p. injection. Although their viability was not assessed due to germination difficulties on culture media, germination and growth must have taken place prior to injection (it is noted that oospores normally require 30 or more days for their development in water). Mycelia were also observed in mesenteric abscesses and these abscesses were sterile when cultured. Although *L. giganteum* triggered an inflammatory response in mammals, there was no evidence of infection; temperatures above 30°C killed its vegetative mycelial stage.[26] This fungus was not administered to immunocompromised animals. Based upon the upper temperature limit of 30°C for mycelia as well as its non-pathogenicity in our tests, it seems unlikely that *L. giganteum* poses substantial risk to humans.

D. ENTOMOPATHOGENIC PROTOZOA

Undeen and Alger[27] reported that s.c. injection of *Nosema algerae* into the footpads and tails of mice resulted in infection. Vegetative stages were observed up to 10 d after injection and presporal stages were found as long as 27 d after injection (Table 1). Heat-killed spores persisted in tissue up to 12 d.[27] *N. algerae* died within 3 d after infecting pig kidney cells maintained at 37°C, and no infected cells were found in flasks incubated at 38°C.[28]

In our laboratory, i.p., s.c., and intracerebral injection of *N. algerae* into mice and rabbits failed to substantiate the above results, producing no illness or infection as defined by the presence of vegetative forms. Outbred mice immune-suppressed by hydrocortisone acetate (25 mg per mouse) were not vulnerable to infection. Athymic mice were not tested because the inoculum of this microsporidian was contaminated with bacteria and fungi from the mosquito homogenate. *N. algerae* does not appear to pose a significant threat to humans because of its upper temperature limit of 35°C and its innocuousness when injected by a variety of challenging routes into immune-intact and immunosuppressed mammals.

Nosema locustae has been used in grasshopper control for almost 10 years without any instance of human or animal infection. This microsporidian underwent conventional and maximum challenge testing without causing illness or infection. Rabbits were injected in-tracerebrally, intraocularly, and i.p. without incident. Microscopical examination subse-quently revealed *N. locustae* spores in the liver of one rabbit 6 weeks after i.p. administration. In this instance, the spores were accidentally injected intrahepatically and their viability was not determined (Table 1). Mice were injected intracerebrally without becoming infected. *N. locustae* elicited an inflammatory response in mammals and the spores were often present in granulomatous foci in the spleen and liver. However, there was a large quantity of insect material in the inoculum and this contaminant probably triggered the inflammatory response, since grasshopper fragments without *N. locustae* elicited a granulomatous response as well. Oral administration of spores to rats produced no lesions and this entomopathogen was not a primary skin irritant. The spores of the latter species appear to persist longer in mammals than do those of *N. algerae*. It would be interesting to determine if they remain viable for as long as the spores of entomopathogenic bacteria. The test results indicate that *N. locustae* was not pathogenic to the test mammals and that it should not be a significant threat to humans.

E. INFECTIVITY — CONCLUSION

Entomopathogens possess a variety of life stages that can survive when environmental conditions are unsuitable for normal growth and development (spores, conidia, oospores). It should thus come as no surprise that these microorganisms remain viable in mammalian tissue for various lengths of time. Their persistence in tissue and subsequent recovery may give rise to spurious claims of infection; therefore, increased emphasis must be placed on determining the life stage recovered if true infections are to be recognized. We believe that recovery of entomopathogens from humans will be primarily determined by the frequency of human exposure. As use of microbial insecticides increases, isolation of commercially produced entomopathogens from humans is inevitable. The body of mammalian safety data available as well as a greater understanding of the maximum persistence time of these microbial agents in mammals, should aid in assessing the significance of human isolations of entomopathogens, if they do occur. Based upon the available mammalian data, immunocompromised humans do not appear to face a greater risk of infection from entomopathogens than do the general population. However, the bulk of the work conducted to date has evaluated the T-lymphocyte component of the immune system.

III. TOXICITY OF BACTERIAL AND FUNGAL ENTOMOPATHOGENS TO MAMMALS

A. BACTERIA

In our laboratory, toxicity is considered a cause for concern when it is associated with exposure to 10^6 or fewer bacteria per mouse or 10^7 or fewer bacteria per rat (these are the single dose parenteral limits suggested by Burges[1]). We note that other authors[4] suggest testing at 10 to 100 times the average field dose per acre with a conversion ratio of the weight of test animal to weight of man. We consider mortality in laboratory animals at concentrations above the limits of Burges acceptable for human safety, although an LD_{50} should be determined when practical. Exotoxins are a separate issue and should be evaluated as chemicals, using the standard chemical safety protocols, and their safety will not be discussed in this chapter. It is important to note that while bacteria themselves may not be toxic when grown in pure culture, the culture components and fermentation byproducts may be, so that registration requirements include both active ingredient and end use-product testing.[13]

The original test results published for *Bacillus thuringiensis* indicated that it was toxic to mice at a concentration of 3×10^8 cfu when injected i.p. and was also toxic to guinea pigs at an unspecified concentration. The toxic factor to mice was heat stable since autoclaved cultures caused mortality (5 of 5 mice within 16 h). Guinea pigs were also affected as 7 of 10 died following i.p. injection from a slant culture of undetermined concentration. This study concluded that toxicity was a characteristic of the carrier because it was not associated with living organisms.[5] Lamanna and Jones[29] evaluated the i.p. toxicity and infectivity of several strains of *B. thuringiensis as well as B. anthracis* and *B. cereus* to mice and noted that vegetative forms were more toxic than spores. However, the *B. thuringiensis* LD_{50} for vegetative stages per mouse was 8.4×10^6 cells and for spores was 3.7×10^8 cfu per mouse. The toxicity was therefore within acceptable limits, since it occurred above 10^6 organisms. As a comparison, the LD_{50} for one strain of *B. anthracis* was two spores per mouse, which is a difference of 100 million-fold between these two species of *Bacillus* with respect to toxicity to mice.[29] It should be noted that the results of these two studies may have been confounded by the presence of β-exotoxin, which does have mammalian activity and strains containing this exotoxin have been banned from use in the U.S. since 1973.[4]

TABLE 2
Toxicity Associated with Exposure to Bacterial and Fungal
Entomopathogens in Our Laboratory

Organism	Route	No. dead/ no. exposed	Dose	Species
Bacillus thuringiensis ssp. *kurstaki*	i.c.	5/6	10^{7a}	Rat
B. thuringiensis ssp. *kurstaki*	i.c.	2/6	10^{6a}	Rat
B. thuringiensis ssp. *israelensis*	i.c.	5/6	10^{7a}	Rat
B. thuringiensis ssp. *israelensis*	i.c.	1/6	10^{6a}	Rat
B. thuringiensis ssp. *israelensis*	i.p.	26/42	10^8	Mouse[b]
B. thuringiensis ssp. *israelensis*	i.p.	3/42	10^8	Mouse[c]
B. sphaericus	i.c.	5/5	10^7	Mouse
B. sphaericus	i.p.	42/49	10^8	Mouse
Lagenidium giganteum	i.t.	6/9	10^5	Rat

Note: i.c. = intracerebral; i.p. = intraperitoneal; and i.t. = intratracheal.

[a] No mortality when inoculum autoclaved.
[b] Athymic mice.
[c] BALB/c mice injected with hydrocortisone acetate.

Intracerebral injection experiments using rats, conducted in our laboratory with a strain that does not produce β-exotoxin — *B. thuringiensis* 3a,3b — revealed a central nervous system vulnerability to concentrations of 10^6 and 10^7 cfu of the microorganism. Mortality was dose dependent and did not occur below 10^6, and the toxic factor was heat labile. Culture medium affected toxicity, as *B. thuringiensis* grown on brain heart infusion (BHI) broth was more toxic than that grown on BHI agar (Table 2).[21] In contrast to the earlier work cited, a heat-labile toxic factor was present. Our results also confirm that culture medium or carrier may influence toxicity. However, toxicity remained within our guidelines for safety, especially when the highly challenging route of administration is considered.

B. thuringiensis ssp. *israelensis* produced mortality in rats following intracerebral injection and its toxicity was similar to lepidopteran strains, with no mortality occurring below 10^6 cfu per rat. The toxic factor was heat labile and toxicity was affected by growth medium composition as well as incubation temperature. BHI broth at 30°C produced the most toxic inoculum. Intraperitoneal injection of *B. thuringiensis* spp. *israelensis* also produced mortality in one experiment designed to determine the susceptibility of immunodeficient and immunocompromised mice to infection with this subspecies. A dose of 2.6×10^7 cfu was given to 40 athymic mice and one of 3.4×10^7 cfu to 40 BALB/c mice injected with corticosteroids, and 42 BALB/c control mice. Within a 10-h period, 62% of the athymic mice were dead. Seven percent of the corticosteroid-treated mice had died too, but all control mice survived (Table 2); there were no discernible lesions in the dead mice. The inoculum consisted of a washed suspension produced by one manufacturer, wheras a commercial suspension produced elsewhere and given by the same route did not cause mortality. It was unlikely that bacterial contaminants were present, since the culture was pure. Therefore, the toxicity may have resulted from the bacterium itself or fermentation products that adhered to the spores of *B. thuringiensis* ssp. *israelensis*. However, it is important to emphasize that the toxicity did not occur below 10^7 spores.

We will not deeply explore the issue of the toxicity of alkali-solubilized (pH 12) δ-endotoxin of *B. thuringiensis* ssp. *israelensis* to mammals, because the protoxin is not cleaved into active fragments in the latter. It has been reported recently[9] that the 25 kDa polypeptide is responsible for the hemolytic action of this toxin, and that it must bind to phospholipid receptors on the cell membrane in order to increase cell permeability.

B. sphaericus isclates 1321-1, 1404, and 1593 grown on Sphaericus Synthetic Medium broth and injected intracerebrally as whole broth cultures, were not toxic to rats at a concentration of 6×10^6 cfu per rat while mice injected intracerebrally with 1.2×10^7 cfu experienced cerebral hemorrhages and died (Table 2). Rats experienced no mortality following i.p. injection with as many as 2.4×10^8 cfu while s.c. injection of 6.9×10^9 cfu in mice produced no adverse effects.[10,11]

A commercial suspension of *B. sphaericus* produced mortality following i.p. injection, with 86% of the mice dying within 24 h (Table 2). Subsequent experiments indicated that toxicity was dose dependent and that no death occurred when 10^7 or fewer cfu were injected. The toxic factor was not present in filtered supernatant, so that toxicity was associated with the spores of this microorganism or products that adhered to the spore coat, and the toxic factor was heat stable. It does not seem likely that the insecticidal toxin caused the mice to die because crude toxin preparations of *B. sphaericus* 1593 and 2362 were not cytotoxic to baby hamster kidney cells or to human red blood cells, and injection of crude toxin in distilled water (0.75 mg/ml) failed to kill mice.[30] The toxicity of this commercial preparation falls within our stated safety limits. It may result from the use of a different strain of *B. sphaericus,* or as with the toxicity associated with the single commercial preparation of *B. thuringiensis* ssp. *israelensis* tested, may illustrate differences arising when laboratory cultures are compared to end-use products.

B. FUNGI

Metarhizium anisopliae was not toxic to laboratory animals. Routes of administration included aerosol, s.c. injection, gavage (forcible feeding via a force-pump and a tube passed into the stomach), oral, ocular instillation, and i.p. injection. There were lesions at the site of s.c. injection (s.c. granulomas). Granulomas were observed following i.p. injection, but these probably resulted from the presence of foreign particulate material since autoclaved M. *anisopliae* produced lesions of equal severity.

Lagenidium giganteum caused significant mortality in rats (67%) following intratracheal instillation of 1.2×10^5 oospores in 0.5 ml distilled water (Table 2). Some rats died within 4 min of receiving the dose while the remainder died of acute pneumonia within 24 h. The deaths resulted from obstruction of airways and/or the host inflammatory reaction to large quantities of foreign material. Reduction of the inoculum to 2.9×10^4 oospores in 0.5 ml distilled water per rat eliminated mortality, although lesions were present in the lungs due to host inflammatory response to foreign material.[24] No mortality was associated with dermal, oral, ocular, and i.p. exposure to *L. giganteum* and we conclude that this organism is not toxic to man.

C. TOXICITY — CONCLUSION

The toxicity of the bacterial entomopathogens tested by our laboratory fell within acceptable limits and mortality occurred only following invasive routes of administration such as i.p. and intracerebral injection. We note that commercial products may contain substances that are toxic to laboratory animals in high concentrations since discrepancies were noted between material grown in our laboratory and commercially produced products. The differential mortality may result from strain and isolate differences as well; after all, "*B. sphaericus*" may eventually prove to be a species complex, the members of which are indistinguishable from one another on morphological grounds. This is hinted at by current knowledge of serotypes and their differential insecticidal activity. Our studies did not rigorously examine this possibility. We note that *B. sphaericus* strain 2362 produced mortality following i.p. injection of 10^9 cfu (6 of 6 mice died), but we must emphasize that no mortality was associated with doses below 10^8 cfu.

Mortality in animals exposed to fungi was mechanical, i.e., blockage of the airways.

Both *Lagenidium giganteum* and *Metarhizium anisopliae* produced lesions in the lungs and peritoneum of rats and mice, but since autoclaved material produced lesions of comparable severity, we concluded that the lesions resulted from host reaction to heat stable foreign material. Laboratory animal data support the conclusion that the entomopathogenic bacteria and fungi tested in our laboratory do not pose a significant threat, except under extreme conditions when massive quantities are injected. Human vulnerability to exotoxins is a separate issue and must be evaluated according to established chemical safety protocols.

IV. OVERALL CONCLUSION

About 4 decades ago a chicken farm owner in Maryland lost some birds following treatment of her county with *B. popillae* and as reported by Heimpel,[3] she subsequently took the county commissioners to court in order to seek damages. Dr. George Langford, the scientist in charge of the program, testified in court about the vertebrate safety of *B. popillae* and, while doing so, "ate a heaping spoonful of the spore dust, washing it down with a glass of water. The magistrate immediately suspended the case following further observation of Dr. Langford," who according to Heimpel,[3] was (more than 20 years after the event), alive, well and — as State Entomologist — living in Maryland. Times have changed, and we can no longer rely on human volunteers to determine an organism's potential hazard to man. We must rely instead on animal data. The laboratory data support the conclusion that entomopathogens are not infectious to humans, despite their persisting for various lengths of time in mammalian tissue. Based on our laboratory data, immunodeficient humans may not be at greater risk of infection from these agents than immune intact humans. Neither are entomopathogens significant toxicants for man, although entomopathogenic bacteria have proved toxic to rats and mice after massive quantities of viable bacteria were administered by ingestion and invasive routes. Using conventional routes of exposure, the entomopathogens tested were avirulent and nonpathogenic. The data support the belief that they can be safely used in environments in which human exposure is likely to occur.

ACKNOWLEDGMENT

This research was supported by WHO/World Bank/UNDP Special Programme for Research and Training in Tropical Diseases.

ADDENDUM

The issue of a standard definition of the term *infection* has been sidestepped in regulatory agency guidelines, which fail to provide an operational definition. This failure, it seems to us, leaves a serious issue — decisions on the safety of microbial insecticides — in a state of limbo. Historically, there have been two schools of thought concerning the meaning of infection, with each school's definition having a different implication for the relationship between microorganism and host. One school defines infection as the simple presence of microorganisms within living tissues, noting that their presence does not inevitably lead to damage. In this view, all animals are colonized shortly after birth by myriad microorganisms, which are restricted by host defenses to areas where they can be tolerated, such as the gastrointestinal tract or the upper respiratory tract.[31-33] The equilibrium between the host and its flora is maintained by a variety of tissue structures and defense mechanisms. Infectious disease results when microorganisms penetrate these defenses and disrupt the status quo. The term pathogenicity denotes the intrinsic capability of a microorganism to penetrate host defenses, the term virulence referring to the speed by which this is accomplished.[31]

The second school of thought inextricably links infection to disease by defining the former as occurring when living agents "enter an animal body and set up a disturbance"[34] either through multiplication in tissue, production of toxins, or both.[35-36] From the viewpoint of safety testing, the second definition of infection, which links the presence of a microorganism to tissue damage, is the more useful one. Safety tests by design introduce microbial insecticides into the host via a diversity of routes. Thus, transient disturbances in the normal flora should be expected. Also, recovery of some portion of the inoculum from host tissue may occur over a variable length of time, depending on the route of administration and the magnitude of the dose. Under these circumstances, concluding that a microbial insecticide is infectious merely because it is recovered, is not particularly enlightening.

We suggest that infection in mammalian safety studies be considered established when there is evidence of multiplication of the microbial control agent in mammalian tissue, coupled with tissue damage. Evidence of multiplication includes a measurable increase in the total amount of that agent recovered above the amount administered, recovery of its vegetative stages when the inoculum consisted solely of spores or conidia and/or failure of the inoculum to clear over time. Infection cannot be determined solely on the basis of lesions, since the injection of foreign material can elicit an inflammatory response.[37,38] For example, microbial control agents produced in insects may contain insect fragments as contaminants in the test preparation, which can then produce inflammation independent of the agent. To summarize: in this chapter, our working definition of infection involves multiplication of the microbial control agent combined with tissue damage. The other situations described, such as simple recovery of such agents, do not constitute infections.

REFERENCES

1. **Burges, H. D.,** Safety, safety testing and quality control of microbial pesticides, in *Microbial Control of Pests and Plant Diseases* 1970—1980, Burges, H. D., Ed., Academic Press, New York, 1981, 737.
2. **Schatzow, S.,** The role of the Environmental Protection Agency, in *Banbury Report #22: Genetically Altered Viruses and the Environment,* Fields, B., Martin, M. A., and Kamely, D., Eds., Cold Spring Harbor Laboratory, Cold Spring Harbor, NY, 1985, 49.
3. **Heimpel, A. M.,** Safety of insect pathogens for man and vertebrates, in *Microbial Control of Insects and Mites,* Burges, H. D. and Hussey, N. W., Eds., Academic Press, New York, 1971, 469.
4. **Ignoffo, C. M.,** Effects of entomopathogens on vertebrates, *Ann. N.Y. Acad. Sci.,* 217, 141, 1973.
5. **Fisher, R. and Rosner, L.,** Toxicology of the microbial insecticide Thuricide, *Agric. Food Chem.,* 7, 686, 1959.
6. **Shadduck, J. A.,** Some observations on the safety evaluation of nonviral microbial pesticides, *Bull. W.H.O.,* 61, 117, 1983.
7. **Thomas, W. E. and Ellar, D.,** *Bacillus thuringiensis* var. *israelensis* crystal delta endotoxin: effect on insect and mammalian cells *in vitro* and *in vivo, J. Cell Sci.,* 60, 181, 1983.
8. **Armstrong, J. L., Rohrmann, G. F., and Beaudreau, G. S.,** Delta endotoxin of *Bacillus thuringiensis* subsp. *israelensis, J. Bacteriol.,* 161, 39, 1985.
9. **Gill, S. S., Singh, G. J. P., and Hornung, J. M.,** Cell membrane interaction of *Bacillus thuringiensis* subsp. *israelensis,* cytolytic toxins, *Infect. Immun.,* 55, 1300, 1987.
10. **Shadduck, J. A., Singer, S., and Lause, S.,** Lack of mammalian pathogenicity of entomocidal isolates of *Bacillus sphaericus, Environ. Entomol.,* 9, 403, 1980.
11. **Siegel, J. P. and Shadduck, J. A.,** Mammalian safety of *Bacillus sphaericus,* in *Bacterial Larvicides for Control of Mosquitoes and Blackflies,* deBarjac, H. and Sutherland, O. J., Eds., Rutgers University Press, New Brunswick, NJ, in press.
12. **Anon.,** Mammalian safety of microbial agents for vector control: a WHO memorandum, *Bull. W.H.O.,* 59, 857, 1981.
13. **Anon.,** Pesticide Assessment Guidelines: Subdivision M, Biorational Pesticides, PB83-153965, U.S. Environmental Protection Agency, Washington, D.C., 1982.
14. *Dorland's Illustrated Medical Dictionary,* 24th ed., W. B. Saunders, Philadelphia, 1965.
15. **Samples, J. R. and Buettner, H.,** Corneal ulcer caused by a biologic insecticide *(Bacillus thuringiensis),* *Am. J. Ophthalmol.,* 95, 258, 1983.

16. **Siegel, J. P. and Shadduck, J. A.,** Safety of the entomopathogen *Bacillus thuringiensis* var. *israelensis* for mammals, *J. Econ. Entomol.,* 80, 717, 1987.

17. **Hadley, W. M., Burchiel, S. W., McDowell, T. D., Thilsted, J. P., Hibbs, C. M., Whorton, J. A., Day, P. W., Friedman, M. B., and Stoll, R. E.,** Five-month oral (diet) toxicity/infectivity study of *Bacillus thuringiensis* insecticides in sheep, *Fundam. Appl. Toxicol.,* 8, 236, 1987.

18. **Belschner, H. G.,** *Sheep Management and Diseases,* Angus and Robertson, London, 1959, 415.

19. **Newsom, I. E.,** *Sheep diseases,* Williams & Wilkins, Baltimore, 1952, 3.

20. **Jones, T. C. and Hunt, R. D., Eds.,** *Veterinary Pathology,* 5th ed., Lea & Febiger, Philadelphia, 1983, 581.

21. **Siegel, J. P. and Shadduck, J. A.,** Mammalian safety of *Bacillus thuringiensis* var. *israelensis,* in *Bacterial Larvicides for Control of Mosquitoes and Blackflies,* deBarjac, H. and Sutherland, O. J., Eds., Rutgers University Press, New Brunswick, NJ, in press.

22. **deBarjac, H., Larget, I., Benichou, L., Cosmao-DuManoir, V., Viviani, G., Ripoteau, H., and Papion, S.,** Test d'innocuité sur mammifères avec du sérotype H_{14} de *Bacillus thuringiensis* var. *israelensis,* WHO/VBC/80.761, mimeographed document, WHO, Geneva, 1980.

23. **Isaacson, P., Jacobs, P. H., Mackenzie, M. R., and Mathews, A. W.,** Pseudotumor of the lung caused by infection with *Bacillus sphaericus, J. Clin. Pathol.,* 29, 806, 1976.

24. **deBarjac, H., Larget, I., Cosmao, V., Benichou, L., and Viviani, G.,** Innocuité de *Bacillus sphaericus,* souch 1593, pour les mammifères. WHO/VBC/79.731, mimeographed document, WHO, Geneva, 1979.

25. **Shadduck, J. A., Roberts, D. W., and Lause, S.,** Mammalian safety tests of *Metarhizium anisopliae:* preliminary results, *Environ. Entomol.,* 11, 189, 1982.

26. **Siegel, J. P. and Shadduck, J. A.,** Safety of the entomopathogenic fungus *Lagenidium giganteum* (Oomycetes: Lagenidiales) to mammals, *J. Econ. Entomol.,* 80, 994, 1987.

27. **Undeen, A. H. and Alger, N. E.,** *Nosema algerae:* infection of the white mouse by a mosquito parasite, *Exp. Parasitol.,* 40, 86, 1976.

28. **Undeen, A. H.,** Growth of *Nosema algerae* in pig kidney cell cultures, *J. Protozool.,* 22, 107, 1975.

29. **Lamanna, C. and Jones, L.,** Lethality for mice of vegetative and spore forms of *Bacillus cereus* and *Bacillus cereus*-like insect pathogens injected intraperitoneally and subcutaneously, *J. Bacteriol.,* 85, 532, 1963.

30. **Davidson, E. W.,** Effects of *Bacillus sphaericus* 1593 and 2362 spore/crystal toxin on cultured mosquito cells, *J. Invertebr. Pathol.,* 47, 21, 1986.

31. **Davis, B. D., Dulbecco, R., Eisen, H. N., Ginsberg, H. S., and Wood, W. B., Jr., Eds.,** *Microbiology,* 2nd ed., Harper & Row, New York, 1973.

32. **Einstein, L. and Swartz, M. N.,** Pathogenic properties of invading microorganisms, in *Pathologic Physiology — Mechanisms of Disease,* 5th ed., Sodeman, W. A., Jr. and Sodeman, W. A., Eds., W. B. Saunders, Philadelphia, 1974, 454.

33. **Kissane, J. M.,** Bacterial diseases, in *Pathology,* Vol. 1, Anderson, W. A. D. and Kissane, J. M., Eds., C. V. Mosby, St. Louis, 1977, 369.

34. **Bruner, W. B. and Gillespie, J. H., Eds.,** *Hagan's Infectious Diseases of Domestic Animals,* 6th ed., Cornell University Press, Ithaca, NY, 1973.

35. **Apperly, F. L., Ed.,** *Patterns of Disease on a Basis of Physiologic Pathology,* Lippincott, Philadelphia, PA, 1951.

36. **Boyd, W. and Sheldon, H., Eds.,** *An Introduction to the Study of Disease,* 7th ed., Lea & Febiger, Philadelphia, 1977.

37. **Siegel, J. P., Shadduck, J. A., and Szabo, J.,** Safety of the entomopathogen *Bacillus thuringiensis* var. *israelensis* for mammals, *J. Econ. Entomol.,* 80, 717, 1987.

38. **Siegel, J. P. and Shadduck, J. A.,** Safety of the entomopathogenic fungus *Lagenidium giganteum* (Oomycetes: Lagenidiales) to mammals, *J. Econ. Entomol.,* 80, 994, 1987.

Chapter 9

SAFETY OF MICROBIAL INSECTICIDES TO VERTEBRATES — DOMESTIC ANIMALS AND WILDLIFE

J. E. Saik, L. A. Lacey, and C. M. Lacey

TABLE OF CONTENTS

I. INTRODUCTION

A. OVERVIEW

Arthropod pests and vectors continue to pose a major threat to agricultural productivity and human and animal health. Control of these vectors has centered predominantly on development and use of chemical pesticides. Originally good results were obtained with large-scale chemical pest control; however, use over the past 50 years has revealed its various shortcomings.[30,74] Side effects in nontarget organisms, resistance of target insects to pesticides, pollution of the environment, disruption of balanced insect communities, and the potential for teratogenicity and carcinogenicity has stimulated interest in a means of control other than chemical pesticides.

Accelerated interest in the use of microbial insecticides began sometime after 1880 but was limited by the fastidious growth requirements of insect pathogens.[55] By the 1940s several useful agents were discovered and used. At first very little was required to prove that an organism was effective and safe. Some demonstration of safety to vertebrates was among the requirements.[55] Dutky isolated and described the milky disease organism, *Bacillus popillae*, in the late 1930s and was able to obtain a U.S. patent for distribution of his organism in nature after satisfying only limited tests involving feeding of spores to starlings and chickens without adverse effects.

It became obvious that there was a need for systematic safety testing of pesticides to predict their hazard to mammals, and, by extension, to humans.[55] The U.S. was the first country to create a law dealing with registration of materials used to treat food crops. This was the Federal Insecticide, Fungicide and Rodenticide Act of 1947.

Within the next 10 to 15 years other countries using pesticides developed regulations similar to those instituted in the U.S. Sweden, West Germany, Holland, New Zealand, Canada, France, the U.K., Japan, Belgium, Czechoslovakia, Iran, Poland, Italy, Spain, the U.S.S.R., and Switzerland all have pesticide acts regulated by governmental bodies.[55]

As safety testing was developed for pesticides, it became obvious that biological pesticides posed unique testing requirements not necessary for chemical pesticides. Of primary concern was the ability of these agents to replicate with possible establishment in the environment and the infection of nontarget organisms (NTOs). Other concerns included production of toxic metabolites, genetic mutation into more pathogenic organisms, induction of allergic reactions, and contact irritancy.[102,117]

B. SAFETY TEST DESIGN

Although chemical and biological pesticides share the end result of arthropod control, their basic mechanisms of control can be quite different. Earlier tests designed to evaluate chemical pesticides only partially answered safety questions about biological pesticides. Therefore, in 1981, the World Health Organization (WHO) produced a memorandum addressing safety testing and regulation of biological pesticides for mammals.[119] The main purpose of the tests designed by WHO were to perform a risk/benefit analysis for each of the different uses envisaged for the various biological agents. They developed a three-tier mammalian safety testing scheme (for greater detail see companion Chapters 1 and 8); recommendations for performing regular serological surveys on all humans exposed to the pathogen during development and application; and suggestions for when to begin limited field trials and advancement to large-scale field trials.

WHO listed specific routes of exposure, types of tests to be performed, and the type and number of mammalian species to be used in each test. Tier I tests should include oral, intraperitoneal (i.p.), respiratory, eye, and dermal exposure along with allergenicity, hypersensitivity, and mutagenicity tests. Because fungal agents are more likely to cause allergic reactions, additional indicators, such as Arthus, or delayed hypersensitivity reactions should be included in fungal safety testing. Emphasis with the protozoa and viruses should be on infectivity since little toxicity is associated with these organisms. Infectivity testing in mammalian tissue cultures with these agents should be included in Tier I.[119] Testing of microsporidian protozoa should include intracranial exposure due to their known pathogenicity for the nervous system.[119] In addition, protozoa which do not replicate at mammalian body temperatures should be injected into tissue with lower temperatures, such as tails or footpads, replacing i.p. exposure.

Various groups have designed protocols for testing agents which demonstrate toxicity to vertebrates[60,61] (see companion Chapter 8). The toxins produced by these organisms should be isolated and characterized and their modes of action defined. In addition, metabolism of the toxin by vertebrates should be evaluated and acute and chronic toxicity testing be set up using testing protocols for chemical insecticides as guidelines.[117] Evaluation of long-term effects on the host such as carcinogenicity and teratogenicity should be included.[102]

In summary, the design of safety studies of entomopathogens in vertebrates should include testing of infectivity, toxicity, allergenicity, and irritancy.[117] Infection of vertebrates is based on identification and characterization of the agent and its host range. Studies on temperature growth dependence, survival in tissue cultures, and survival, distribution, and replication potentials in recipient mammals through different routes of exposure (including immunosuppressed animals) should also be included. Toxins that are produced by a candidate organism should be evaluated based on principles of safety evaluation of chemical pesticides. Allergenicity testing should include sensitivity testing in experimental animals and observations of humans handling the pathogen for skin sensitivity and respiratory symptoms. Irritancy testing both of the primary product and formulated product can be performed through skin and eye irritation tests in lab animals and observation of humans handling the product during development.[117]

II. PATHOGEN GROUPS

A. BACTERIA

Both sporeforming and nonsporeforming bacteria have been considered for use as entomopathogens.

The sporeforming bacteria have received the most widespread attention as microbial insecticides. Of this group, varieties of *Bacillus thuringiensis* and *B. sphaericus* have been the most widely studied and tested.

1. *Bacillus thuringiensis*

B. thuringiensis has been used extensively in agriculture, forestry, and public health programs for the control of lepidopterous pests and certain nematocerous Diptera. Varieties of *B. thuringiensis* are commercially produced and registered in the U.S., Western Europe, and numerous nations elsewhere using biological pesticides.

Safety testing and use of this organism dates back to the 1950s with development of Thuricide.[55] Initial studies included human volunteer tests with both oral and inhalation exposure and inoculation of mice with several varieties of *B. thuringiensis* in an attempt to identify strains that were pathogenic for mice.[45,55,56,109] Further tests included serial passage through mice, observation for persistence in the blood of mice, i.p. injection of guinea pigs, inhalation toxicity for mice, allergenicity to guinea pigs, and oral toxicity to rats.[74] No toxicity or pathogenicity was demonstrated with this organism in any of these tests. Safety tests carried out by other groups showed no ill effects on chicks, laying hens, young and mature pigs, fish, wild pheasant, and partridge.[13,60,116]

Hadley et al.[52] instituted a 5-month oral toxicity/infectivity study of *B. thuringiensis* in sheep. This study was initiated because Mordan and Herein (unpublished results) reported that *B. thuringiensis* insecticides fed to sheep might produce pneumonia, myocarditis, and hepatic lesions. Hadley et al., however, found no toxic effects or infectivity of *B. thuringiensis* spp. *kurstaki* for sheep. Concern over the possible ability of *B. thuringiensis* to mutate to *B. anthracis* (cause of anthrax in animals and man) in the gastrointestinal tract initiated another study.[107] Adding to this concern was the isolation of a phage from a *B. thuringiensis* culture that is capable of infecting cultures of *B. anthracis*. Concern over possible mutation and proliferation in mammals prompted a study involving two separate exposure techniques. Antibiotic-treated mice were fed *B. thuringiensis* and *Pseudomonas aeruginosa,* and excised mouse intestine was tied off and injected with bacterial culture. In neither case was there an indication of mutation, and after 24 h, the population of *B. thuringiensis* was negligible, while *P. aeruginosa* was drastically reduced.[107]

Within the literature there have been rare reports of mammalian pathogenicity by *B. thuringiensis*. Accidental splashing of Dipel® (an Abbott Laboratories formulation of *B. thuringiensis*) in a farmer's eye resulted in a corneal ulcer which yielded *B. thuringiensis* ssp. *kurstaki* on culture[98,99] (see Section II. B, Chapter 8) There also exists a report of fatal bovine mastitis from *B. thuringiensis*.[51] Several authors have reported noninfectious bacteremia with presence of bacteria in vertebrate tissues after inoculation with heavy doses of *B. thuringiensis*.[60]

In the 1960s, several reports appeared stating that certain strains of *B. thuringiensis* produced a heat stable exotoxin.[9,43,54,89] It was soon discovered that four separate toxic entities could be isolated or detected in cultures of crystalliferous bacteria. These comprised α-exotoxin,[53,54,75,113] β-exotoxin,[12,54,56] δ-endotoxin, and γ-exotoxin.[54,56] The δ-endotoxin and β-exotoxin quickly assumed importance in further development of varieties of *B. thuringiensis*. Studies indicated that a wider spectrum of insects was affected by varieties producing exotoxins, whereas endotoxin producers were remarkably specific for their respective target organisms.

a. δ-Endotoxin

The pathogenic effects in Lepidoptera and Diptera of the various varieties of commercially produced *B. thuringiensis* are due for the most part to the endotoxins produced during sporulation. These are contained predominantly in parasporal inclusions and must be ingested in order to be active.[39,54,61,76] A number of intrinsic and extrinsic factors govern the susceptibility of a given species to these toxins. Feeding strategies and rate, pH, and enzymatic environment of the gut of the target organism and binding site of the toxin are some of the more important intrinsic factors that influence susceptibility and level of larvicidal activity.

Extensive safety tests have been performed on the endotoxin producer *B. thuringiensis* spp. *israelensis,* also known as *B. thuringiensis* (H-14). Early tests demonstrated it was safe to man and other vertebrates.[96] In later studies this organism was used as a model in developing maximum challenge tests for mammalian safety testing.[4,96] These tests included injecting high dosages of the organism intracranially and intraocularily. Conventional routes of exposure such as oral, parenteral, respiratory, and dermal were also used. These, together with allergenicity tests, use of immunosuppressed animals, and mutagenicity screens showed no evidence that *B. thuringiensis* ssp. *israelensis* poses any hazard to mammals.[96] Further studies including those in mice, rats, guinea pigs, and rabbits confirmed that mammals are highly tolerant of the agent, and there is rapid elimination as well as lack of multiplication of the agent.[91,96] Other groups indicated the safety of this organism for amphibians.[79,90] Siegel et al.[105] have performed a series of mammalian safety tests on *B. thuringiensis* ssp. *israelensis* which also demonstrate safety of this organism to mammals (see companion Chapter 8). Registration requirements and results of mammalian safety testing in Japan are presented in Chapter 4.

In contrast to these studies, injurious effects have been elicited in safety studies by directly administering the dissolved buffered endotoxin of this organism. It has been shown to be cytotoxic to five mammalian cultured cell lines and causes hemolysis of mammalian erythrocytes.[3,19,48,110] Armstrong et al.[3]; also demonstrated the ability of solubilized endotoxin to lyse red blood cells and in addition showed it to be lethal to mice when injected i.p. Another study showed i.p. injection of mice resulted in effects resembling that of a neuromuscular toxin.[19] This same group found that injection of the alkaline-dissolved *B. thuringiensis* spp. *kurstaki* toxin into mice did not produce neuromuscular symptoms. Kalmakoff and Pillai isolated two proteins (Protein A and B) from *B. thuringiensis* spp. *israelensis* crystals. They found Protein A to be hemolytic to human and rabbit erythrocytes, and demonstrated neurotoxic activity using the sixth abdominal ganglion of the American cockroach.[21,22]

Thomas and Ellar[110] conducted a study to demonstrate the difference in effects on mammals between administration of native δ-endotoxin crystal and the solubilized endotoxin of *B. thuringiensis* ssp. *israelensis*. They demonstrated no detectable toxicity of the native endotoxin when given *per os,* subcutaneous (s.c.), or intravenous (i.v.) to BALB/c mice. By contrast, the solubilized δ-endotoxin caused rapid cytological and cytopathic changes in mouse fibroblasts, primary pig lymphocytes and three mouse epithelial carcinoma cell types. In addition, there was hemolysis of rat, mouse, sheep, horse, and human red blood cells and i.v. administration to BALB/c mice caused rapid paralysis and death. Administration of solubilized crystal δ-endotoxin was not toxic when given *per os.* Tests with *B. thuringiensis* spp. *kurstaki* in native crystalline form is nontoxic to mice and noncytopathic *in vitro*, however, ingestion by susceptible insects was rapidly lethal, indicating protoxin activation in the alkaline insect gut.[110]

b. β-Exotoxin

Certain strains of *B. thuringiensis* also produce the heat stable β-exotoxin,[15,27,87] an adenine nucleotide.[43,108] Its broader spectrum of activity against a number of invertebrates has enabled its use for control of muscoid flies,[11,47] lice,[57] and a variety of other insect pests. Because of its stability, it has been used as a feed-through larvicide for control of flies in bovine and chicken manure.[7,50,80] Unfortunately, its toxic effects are not specific for insects. Some of the most deleterious effects of entomopathogens or their metabolites on vertebrates are reported for this toxin.

Early safety tests on β-exotoxin produced death in mice after i.p. injection.[9,43,100] No histopathological lesions were detected in these studies in brain, liver, lymphatic system and bone marrow. Other investigations,[43] however, revealed hepatic necrosis and lesions in the kidneys and spleen. Sublethal doses showed no cumulative effects.[9] Characterization of the toxin by Sebesta et al.[100] [also cited in Reference 4], revealed that it caused marked inhibition of RNA synthesis *in vitro* in mice. It was suggested that the physiological activity of β-exotoxin was due to competition between the exotoxin and ATP and that it exhibited slight mutagenic activity.[108]

Other toxicity studies revealed sex-related mortality differences in mice injected with the exotoxin. Tests using 60 mice of each sex estimated the LD_{50} and LD_{90} for males to be 184.8 and 290.6 µg/g body weight, respectively. These values for females were 135.6 and 226.9 µg/g body weight, respectively.[10,56a]

Feeding studies with β-exotoxin have shown mixed results. Feed additives containing enough β-exotoxin to pass through the feces in sufficient quantities to kill *Musca domestica*, the horn fly *Haemotobia irritans*,[43,57] and the face fly *M. autumnalis*[58] were given. No mammalian toxicity was noted from administration of this material to steers, poultry, Japanese quail, and mice.[10,11,80] In contrast, more recent studies[7,43] revealed lesions after oral administration of β-exotoxin to chickens. These included gizzard erosion, enteritis, proventriculitis, anemia, petechial hemorrhages in the proventriculus, and regressed ovaries. In addition, Galichet[46] observed anorexia and weight loss in treated pigs, and Ode and Matthysse[86] showed a refusal by cows of β-exotoxin-containing feed.

2. *B. sphaericus*

Several isolates of *B. sphaericus* have demonstrated larvicidal activity for mosquitoes.[24,25,76] Those exhibiting the greatest efficacy are in the 5a,5b and 25 serotypes. Although *B. sphaericus* is not yet produced commercially, it is used experimentally in a variety of mosquito habitats worldwide.

In 1980 an extensive safety study was initiated to test three strains of this organism (SSII-1, 1404-9, 1593-4).[103] Species tested included mice, rats, and rabbits. They received s.c., i.p., intracerebral, and intraocular injections. Results of these tests were encouraging. The s.c. injections produced only one local abscess which did not contain *B. sphaericus*. Intraperitoneal and intracerebral injections in rabbits and intracerebral injection of mice produced no significant histological lesions. Rats receiving an intracerebral injection of large numbers of organisms developed mild meningitis with perivascular cuffing and yielded organisms on culture. Intraocular injection of rabbits produced a moderate opthalmitis to panophthalmitis which was more severe in animals receiving viable organisms, although autoclaved material also produced lesions. Studies designed to test replication of the organism in rat brains noted that number of spores gradually declined until by day 14, the brains yielded negative cultures for *B. sphaericus*. Acute and chronic toxicity studies on rats, mice, and guinea pigs undergoing oral, s.c., i.p., intracerebral, inhalation, and dermal exposure were unaffected, both during the tests and on postmortem examination.[28]

More recent studies[27] using mice given injections of viable spores of serotypes 5a,5b, and strain 2362 by s.c., i.p., and i.v. routes with additional exposure orally, percutaneously, and by inhalation revealed no pathologic lesions on postmortem examination. Other groups have reported the safety of this organism for crayfish (Arthropoda/Decapoda), larvivorous fish, and tadpoles,[18] although Matavan and Velpandi report toxicity to *Rana bufo* tadpoles at high concentrations.[79]

There have been rare reports of noninsecticidal *B. sphaericus* infecting humans. Allen and Wilkinson[2] report a case of meningitis and generalized Schwartzman reaction in a 64-year-old man.[103] *B. sphaericus* was cultured from blood, spinal fluid, urine, and subarachnoid ventricular fluid. I.v. and i.p. injections of rabbits with this isolate failed to produce infection or lesions. Farrar[42] reviewed 12 infections of humans with ''nonpathogenic'' bacteria of the genus *Bacillus*. Most of these affected the central nervous system with a direct portal of entry to the meninges or a persistent bacteremia preceding meningeal infection. One of Farrar's own cases within the group was due to an infection by *B. sphaericus*. Injection of mice i.v., i.p., and intracerebrally with this isolate did not produce infection. Isaacson et al.[63] reported isolation of *B. sphaericus* from a pseudotumor of the lung, and Elter[40] reported six cases of food poisoning from sausages containing *B. sphaericus*.[103]

3. Nonsporeforming Bacteria

There are only two nonsporeforming bacteria that have been considered as potential microbial pesticides.[55] Unfortunately, both of these agents are known pathogens for animals and humans.

a. Pseudomonas aeruginosa

This organism has been associated with sporadic infection of plants, animals and man, as well as insects.[65] Infections in animals and man are usually associated with secondary invasion by this organism following a debilitating condition, although it may occasionally be a primary pathogen.[65,70,71] There are reports of infection of nearly all domestic and laboratory animals, and severe epizootics have occurred.[37,55,65]

b. Serratia marcescens

Serratia, like *Pseudomonas*, is predominantly an opportunistic organism. It is particularly common in reptiles although there have been numerous reports of its infection in mammals including man.[55,64]

B. VIRUSES

Viruses pathogenic to insects are generally divided for discussion into (1) those that are occluded in paracrystalline protein bodies from 0.5 to 20 μm in diameter, and (2) those that are not occluded. The occlusion bodies are thought to protect the virions from adverse environmental conditions.[5] Viruses containing occluded virions include Baculoviridae (although one member of this group is nonoccluded), Poxviridae, and Reoviridae. The Iridoviridae, Rhabdoviridae, Picornaviradae, and Parvovirus groups are not occluded.[5]

1. Baculoviridae

The Baculoviridae is divided into three subgroups: (1) nuclear polyhedrosis viruses (NPV); (2) granulosis viruses (GV); and (3) a small group of nonoccluded rod-shaped viruses.[5,44] This family of viruses has seen widespread development, testing, and use both in biological pest control,[30,55,60] and more recently as recombinant viruses to produce such diverse products as human interleukin 2,[106] human c-myc protein,[82] and mouse interleukin-3.[81]

a. Nuclear Polyhedrosis Viruses

The NPVs have been the most extensively safety tested of all the entomopathogenic viruses.[5] They appear to be particularly suited to biological pest control because of their host specificity (primarily Lepidoptera); variety of species-specific viruses in this group; high degree of virulence; and the fact that the virus is relatively well protected from the elements by its occlusion protein.[74]

In the 1960s the *Heliothis zea* NPV underwent testing as extensive as that required for chemicals by the Environmental Protection Agency (EPA) in the U.S. Long-term carcinogenicity and teratogenicity studies on man and other primates were included. The NPVs were unable to replicate or cause pathogenic effects on microorganisms, noninsect invertebrate cell lines, vertebrate cell lines, vertebrates, plants, and nonarthropod invertebrates.[14,55,60] Tests were generally conducted at doses 10 to 100 times the average field rate per acre (0.405 hectare) converted to a ratio of weight of test animal to weight of a 70-kg human. Routes of exposure included oral, inhalation, dermal application, intradermal, intramuscular, intracerebral, i.v., i.p., and s.c. Vertebrates exposed (besides humans and other primates) included mouse, rat, guinea pig, rabbit, dog, quail, chicken, house sparrow (*Passer domesticus*), mallard duck (*Anas platyrhynchos*), pheasant, deer, immunosuppressed rats, and several species of fish.

Wells and Heimpel[118] reported occluded virus infections in bacteria. Their results suggested that under certain conditions a *Baculovirus* may infect and replicate in cells of other organisms. These tests suggested a need for further monitoring. Tests by Doller[29,30] showed a weak virus-specific antibody production after i.v. injection of virions into mice. Larvae fed viscera from these infected animals, however, demonstrated no virus activity.[30] Oral feeding of NPV[30,34] to swine showed no evidence of inapparent virus replication. Tests with vertebrate cell lines conducted by McIntosh and Shamy[30,78] showed in two of five experiments that replication was observed following inoculation of Chinese hamster cells (CHO), however, CHO is routinely grown at 35°C (whereas insect cell lines grow at lower temperatures: 28 to 30°C). When CHO cultures were incubated at the lower temperature of 28°C, no cell replication took place. Tjia et al.[112] were also unable to get an NPV to multiply in various mammalian cell lines.

Further studies in the 1980s by Döller and Gröner,[32,33] Döller and Enzmann,[31] and Reimann and Miltenburger[93-95] have found no virus replication in chickens, rainbow trout, and carp, and no increased rates of sister chromatid exchanges or chromosomal aberrations in Chinese hamsters or mice exposed orally or injected i.p. Serology of humans, horses, cattle, pigs, and sheep performed by several groups[30] have demonstrated an antibody response to Baculovirus proteins. It could not be determined whether this was from exposure to the viral antigen or a cross-reacting antigen which is widely distributed in the environment.

Carbonell et al.[16] conducted a unique safety test in which they designed a recombinant Baculovirus and tested its ability to enter and express viral DNA in mammalian cells that are considered refractory to baculoviral replication. Their results demonstrated that baculoviral DNA was unable to enter efficiently into the nuclei of mammalian cells. Their conclusions expressed the belief that these viruses demonstrated an inherent level of safety with regard to mammalian species.

b. Granulosis Viruses

The granulosis viruses are cosmopolitan in distribution and tend to be genus specific.[5] In contrast to NPV, the body of safety testing has been considerably smaller for the GV.[30]

Bailey et al.[6] performed tests in which they examined small mammals exposed to field spraying of a GV. Detectable levels of antibody were found in these animals.[30] Döller and Huber[35] collected fecal samples daily for 3 weeks after exposing mice to virus. Biologically active virus was detected along with antibody production, however, within 80 d postfeeding, no virus-specific antibodies were detected by radioimmunoassay.[30] Tests for vertical transmission of the virus included oral infection of mice before and after fertilization. Sera obtained from the young were found free of virus antibodies.[30] Both Bailey et al.[6] and Döller and Huber[35] concluded from their tests that virus transmission had not taken place in investigated animals after exposure to single doses of GV, and production of antibody levels only followed exposure to multiple doses of virus.

Other safety studies performed with the GV again indicate a lack of infectivity. Mice and guinea pigs exposed *per os,* by inhalation, injection, and through contact with skin and eye demonstrated no adverse effects.[14] When injected, the *Estigmene acraea*-GV did not impair the health of mice and guinea pigs.[14,55] Further immunization experiments by various routes using capsules or isolated virions in mice, guinea pigs, and rabbits again showed no ill effects.[14,60] No growth or cytopathic effects of two GVs was demonstrated for avian embryos.[60]

Results of tests designed to detect sister chromatid exchanges and chromosomal aberration rates were similar to those obtained for the NPVs. In addition, serology from humans, horses, cattle, sheep, and swine demonstrated positive reactions for GV proteins as they had for NPV proteins.[30]

In conclusion, Baculovirus replication could not be detected for mammals, birds, and fish although biologically active virus was detected for short periods from fecal samples of several species. This indicates that birds might be able to spread the virus in nature and over crops.[30] Studies designed to detect chromosome aberration rates and sister chromatid exchanges were negative. Baculovirus antibodies were detected in human and animal sera and positive reactions appeared dependent only on the relative concentration of immunoglobulins.[30] Based on these results, some authors have stated it is difficult to decide whether there is actual potential risk from these viruses.[30]

2. Poxviridae

This family includes a wide range of genera including those affecting mammals: *Orthopoxvirus, Parapoxvirus, Avipoxvirus, Leporipoxvirus, Capripoxvirus,* and *Suipoxvirus*; and those affecting insects: *Entomopoxvirus* (EPV), with subgenus A (Coleoptera EPV), subgenus B (Lepidoptera and Orthoptera EPV), and subgenus C (Diptera EPV).[5,66]

There is no serological cross reactivity of subgenera A and B with vaccinia virus (human smallpox vaccine virus), and nongenetic reactivation has not been demonstrated. Coleoptera EPV has been subjected to small-scale outdoor trials in France with marginal success. Certain species of Lepidoptera are susceptible to subgenus B entomopoxvirus. Field trials were conducted in Canada in the early 1970s with moderate success. Limited studies have been conducted on the Diptera EPV; field trials have not been reported.[5]

Safety tests on entomopoxviruses have included intracerebral and i.p. injection of suckling mice; oral exposure of rats and mice; i.p. and intranasal exposure of 10-week-old mice; and exposure of caged wild mammals and laboratory mice to high concentrations of field-applied virus.[60] Results of these tests indicated no ill effects or pathological lesions in these animals. *In vitro* tests including avian embryos and vertebrate cell lines demonstrated a lack of virus multiplication or cytopathic effects.[60]

3. Iridoviridae

Iridoviruses, also known as iridescent viruses, infect 50 insect species and cause such diverse diseases in vertebrates as African swine fever, lymphocystis disease of fish, gecko virus, and frog virus.[3,5,66] Safety tests with vertebrates have indicated some problems with iridoviruses. Intraperitoneal injection of large doses of two arthropod iridoviruses showed lethal toxicity to the Oriental frog, *Rana limnocharis,* without viral multiplication.[87] Further tests using mice resulted in immediate death after i.p. injection of untreated virus.[88] Injection of heat-inactivated virus produced no adverse effects. Tinsley and Harrap[111] reported that sera from lab workers reacted with iridovirus, types 1 and 4.[5]

Attempts to grow insect iridoviruses in vertebrate tissue cultures have given negative results.[5,68] These viruses are not under development at present for control of agricultural or forest pests.[5]

4. Parvoviridae

This family contains three genera. They include *Parvovirus, Densovirus* (also known as densonucleosis virus), and the Adeno-associated virus group. Approximately 20 viruses from the Parvoviridae (parvoviruses and Adeno-associated virus genera) infect mammals, while four viruses (*Densovirus* genus) infect insects.[5] The densonucleosis viruses are of particular interest because of their similarity to the vertebrate-infecting Parvoviridae. The mammalian parvoviruses infect a variety of mammals including man and produce such diverse diseases as feline panleukopenia, mink enteritis, canine parvovirus enteritis, Kilham rat virus, and infectious gastroenteritis of humans.[8,66]

Safety tests on *Aedes aegypti* densonucleosis virus showed no pathogenic effect on mice, rats, or chick embryos.[5] The densonucleosis virus of *Galleria mellonella,* however, transforms and produces viral antigens in vertebrate cells[5,72,73] as cited in Reference 5. Giran[49] (cited in Reference 55) performed safety tests on this virus in both adult and newborn mice and rabbits, using i.p., s.c., and intracerebral exposure and was unable to cause infections or pathological lesions.[55] Tinsley and Harrap[111] collected sera from laboratory workers and found 8 of 23 reacted with one or more antigens of a densonucleosis virus.

The virulence of these viruses has produced interest in using them for insect control, but possible problems of vertebrate infection is a major concern.[5]

5. Reoviridae

The Reoviridae include the following genera: *Reovirus* and *Rotavirus* which infect vertebrates;[20] *Orbivirus* ssp. which infects both arthropods and vertebrates; and cytoplasmic polyhedrosis viruses, which infect arthropods only. In vertebrates, the reoviruses cause respiratory infections in humans; tracheobronchitis in dogs, hepatoencephalitis in mice; and arthritis in chickens.[66] Rotaviruses cause enteritis in humans and animals and oribiviruses cause a number of diseases including bluetongue (sheep), Colorado tick fever (human), epidemic hemorrhagic disease (deer), and african horse sickness.[20,66]

Of the two genera that infect insects, the cytoplasmic polyhedrosis viruses have stimulated the most interest due to their arthropod specificity and severe debilitative and lethal effects on their hosts. The orbiviruses multiply in arthropods without any deleterious effects.[5]

Ignoffo[60] reports exposure of two cytoplasmic polyhedrosis viruses to many types of vertebrates by numerous routes and found no ill effects. There was no multiplication or cytopathic effects of this virus in mammalian, fish or avian cell lines.[60] Aizawa[1] (also cited in Reference 5) reports exposure of mice, rabbits, and hamsters to occlusion bodies both orally and by injection without adverse effects or multiplication of the virus.

Arata et al.[5] suggest that more extensive safety testing of this group of viruses should be undertaken since the frequency of genetic recombination in vertebrate-infective Reoviridae indicates this group of viruses may not be genetically stable.

6. Rhabdoviridae

The genera of this family are *Vesiculovirus, Lyssavirus, Sigmavirus,* and a group of plant rhabdoviruses. *Vesiculovirus* causes disease in both vertebrates and insects. Examples of mammalian disease include vesicular stomatitis (horse, pig, cow), ephemeral fever (cow), and viral hemorrhagic septicemia (trout). The genus *Lyssavirus* also infects both vertebrates and insects. In mammals, it causes rabies, Mokola virus, Lagos bat virus, Nigerian shrew virus, and Katonkan virus.[20,23,66] Sigma virus is a congenitally transmitted virus of *Drosophila*. The plant rhabdoviruses multiply in several species of insects and plants.

These viruses have not been safety tested and are considered insufficiently deleterious to insects to be seriously considered as microbial control agents.[5]

7. Picornaviridae

Picornaviridae include the genera *Enterovirus, Rhinovirus,* and *Apthovirus.*[20] The enteroviruses cause polio in humans, swine, and mice; avian encephalomyelitis; Coxsackie virus (humans); and bee paralysis, cricket paralysis, and Nodamura virus in insects. *Rhinovirus* and *apthorvirus* include the mammalian diseases hepatitis A, foot and mouth disease (cow, pig), and the human "cold" viruses.[66]

The Nodamura virus multiplies in mosquitoes without causing symptoms and kills bees, caterpillars, and mice.[5] In the U.K., Ig M antibodies have been detected in mammals to the gonometa paralysis virus[5,7] (also cited in Reference 5). Due to the vertebrate pathogenecity, none of these viruses are currently under development.[74]

C. FUNGI

Approximately 500 species of fungi are now known to infect insects.[18] Some of these are infectious to a wide range of NTOs, whereas others appear to be more host-specific. Most can be cultured on artificial media and for some, the medium has been examined for toxicity to insects and the principal active compound is known.[97] Fungi belonging to the genera *Beauveria, Metarhizium, Culicinomyces, Entomophthora, Nomuraea, Aspergillus, Paecilomyces, Tolypocladium, Leptolegnia, Coelomomyces, Lagenidium,* and others, have been either used or considered for use as microbial insecticides.[18,76,97] The genera which have undergone some safety testing will be discussed.

1. *Beauveria*

These fungi are often isolated from diseased insects and are frequently used in microbial control tests.[97] They produce toxic compounds which rapidly debilitate their host after invasion.[97]

Safety tests have been conducted predominantly on *Beauveria bassiana*. Subcutaneous and i.v. injections of rats produced no lesions and lung tissue of mice exposed to *B. bassiana* spores for 90 d were negative when inoculated into nutrient agar. In contrast, inhalation exposure of mice, rats and guinea pigs to spores caused symptomology.[60] In another series of studies, oral high dose exposure of rats for 21 d to spores and mycelium produced three deaths in ten animals.[60] It was felt by groups reporting adverse reactions that they were due to the particulate nature of the preparations or other causes unrelated to *B. bassiana*.[60] Other tests on the safety of *B. bassiana* include oral dosing of frogs with spores. No pathologic lesions or infection of viscera were produced by this agent.[36] Additional reports indicate *Beauveria*'s inability to grow at the mammalian body temperature of 37°C.[60]

There have been reports of pathogenic effects on animals and humans by *Beauveria*. Müller-Kögler[83] (as cited in Reference 97) reported moderate to severe allergic reactions to spore preparations from scientists working with the organism;[55] however, Dresner and Schaerffenberg reported no deleterious effects in humans repeatedly handling cultures of *B. bassiana*.[60] Multiple reports of *B. bassiana* causing fungal keratitis prompted Ishibashi et al.[62] to conduct a series of tests comparing the pathogenicity of corneal lesions from *B. bassiana* and *Candida albicans*. Histopathology revealed only a weak pathogenicity by *B. bassiana* for the cornea and no invasion of the anterior chamber.[62] To add to these records, Ignoffo[60] reports indirect evidence of possible toxicity, pathogenicity, and allergenicity in humans; and isolation of species of *Beauveria* from vertebrate tissue. *Beauveria bassiana* has been reported to cause pulmonary mycosis in the giant land tortoise.[60] *B. shiotae* was isolated from a human bile duct abscess.[55] Due to the multiplicity of conflicting reports, Ignoffo states additional testing is needed.[60]

2. *Metarhizium*

Pathologic effects of this organism include disruption of mitochondria and rough endoplasmic reticulum.[97] Safety tests have included oral dosing of frogs with *M. anisopliae*;[36] and oral, s.c., inhalation, and i.v. 90-d exposure of rats with no adverse effects.[60] Inhalation, i.p., s.c., and intraocular exposure of mice, rats, and rabbits produced granulomas at injection sites, and recovery of fungi from the spleen for 18 d following injections but without production of lesions. Exposure to dry spores (dust) yielded recovery of *Metarhizium* spores for 2 weeks following exposure but produced no lesions. No adverse effects resulted from oral or intraocular exposure and there was no evidence of replication in mammalian tissues.[101] Dresner and Schaerffenberg reported no deleterious effects in humans handling cultures of *M. anisopliae*.[60]

3. *Culicinomyces*

The host range of the genus *Culicinomyces* is probably restricted to mosquitoes and other aquatic dipterous larvae.[38] Because of the organism's broad mosquito host range, it is a promising candidate for production as a microbial control agent.[76]

Safety tests have included oral dosing of mice, rats, guinea pigs, sheep, and cattle without adverse effects.[101] Another group[38] orally dosed rats, mice, guinea pigs, sheep, cattle, and two species of wild duck with fungal suspensions. These suspensions were pathogenic for mosquito larvae but produced no observable effects on the health of the test animals. Hematological and biochemical values were monitored, with no changes noted. Necropsy demonstrated no signs of sporulation or tissue invasion. Serology was performed on the ruminants, ducks, nine humans (two lab technicians working with the agent), and 30 sheep from New South Wales. The ruminants and sheep produced titers to *Culicinomyces* conidia of 20 to 80. Ducks and humans had titers less than 10. The titers in the treated animals were no higher than the controls. These results suggest that the organism, or one antigenically related to it, is common in the environment. The failure of threshold antibody levels to rise after dosing indicate lack of infection.[38]

Mulley et al. developed further safety tests for this fungus. This group's concern was that the fungus could invade skin wounds or induce allergic responses in animals chronically exposed to it.[85] Intradermal and s.c. injections of the fungus were administered to normal and immunosuppressed mice and a horse. They also orally and s.c. exposed two species of native Australian lizard, freshwater tortoise (one species), and one species of toad. Results in the mammals included transient local inflammatory reactions at the injection sites, but inability to culture the organisms. Granulomas were produced in lizards, tortoises, and toads at the injection sites, and fungus was isolated. There was no indication of systemic spread, however. The conclusion drawn by this group from these tests is that *Culicinomyces* is an unlikely pathogen for mammals. Its inability to grow above 30°C and lack of case reports of wild or domestic animal infection in the literature supports their claims.[85]

4. *Lagenidium giganteum*

Both nonmammalian and mammalian safety tests have recently been completed on this organism.[69,104] Exposure of fish, mallard ducks, and bobwhite quail (*Colinus virginianus*) produced no ill effects.[69] Various mammals were exposed by intratracheal instillation, oral administration, i.p. injection, and dermal and ocular application, again, without adverse effects.[104] It was concluded that this organism should be safe due to its temperature tolerance limit which precludes growth and reproduction above 32°C. There was no evidence of toxin production.[69]

5. Other Fungi

Entomophthora primarily infects Lepidoptera and Diptera. Safety tests have been few and not extensive. Frogs were orally dosed with *E. virulenta* while Japanese quail (*Coturnix coturnix*) were dosed with a suspension of *E. ignobilis*.[36] An additional organism in this genus, *Conidiobolus coronatus*, has been considered for development, however, numerous reports of infections in both man and horses indicate lack of safety of this organism.[41,59,60]

Safety tests with *Paecilomyces* have included oral dosing of frogs and mice with *P. farinosus (Spicaria farinosus)* and *P. fumosoroseus*.[36] No pathologic lesions or infection of viscera were produced by this agent. Müller-Kögler[83] reports no adverse effects in rabbits after injection.[55] Aizawa reports injecting rabbits i.v. with *p. farinosus* 22 times over 6 months without adverse effects[1] (as cited in Reference 60).

Dosing of frogs with *Nomuraea rileyi* produced no adverse effects or lesions.[36]

Aspergillus has some species that are insect pathogens but also has members which are pathogenic to vertebrates.[55] *Aspergillus flavus* and *A. ochraceus* have been considered for use as entomopathogens.[97]

Aspergillus infections of birds, other animals and humans have been widely reported.[67] The organism is a particularly aggressive secondary invader following debilitating conditions and prolonged antibiotic therapy but may be a primary pathogen in some instances.[67] *Aspergillus* is the second most common opportunistic mycosis among humans with malignant disease.[17] *A. ochraceus* produces an ochratoxin which is nephrotoxic to many mammals while *A. flavus* produces aflatoxins which produce a toxic hepatitis and hepatic neoplasia in mammals.[67] It has been suggested that due to this organism's pathogenicity and toxicity to mammals, it should be avoided as a biocontrol agent.[18]

The genus *Coelomomyces* has potential as a microbial control agent of mosquitoes.[76] Exposure of mice and birds produced no abnormalities.[101]

Interest in *Hirsutella thompsonii* as a potential microbial control agent has initiated limited safety testing. Rats were exposed orally for 21 d to spores and mycelium without evidence of toxicity or pathologic lesions.[60]

D. PROTOZOA

Among the most frequently occurring protozoans in insects are the Microspora.[55] They have been evaluated as biological control agents for mosquitoes, locusts, and Lepidoptera. Unfortunately, other Microspora also infect fish, other animals, and humans, although most of the species found in vertebrates appear to be taxonomically distinct.[18,55,67] *Nosema*, which has generated interest as a microbial control agent, has surfaced in the literature periodically in infections of vertebrates.[55] This organism is also related to *Encephalitozoon cuniculi* which commonly causes disease in rabbits and several other mammals.[67] Among the various microsporidians considered for use, *N. algerae* and *N. locustae* have stimulated the most interest in safety testing.

N. algerae was injected i.p. into normal and immunosuppressed mice, as well as into the footpad, and tail skin of mice, intracerebrally into weanling rabbits, and administered orally to rats. Necropsy and histopathology on these test animals revealed no lesions.[101,102] Other tests have demonstrated transient local infections at s.c. injection sites and antibody response in mice[101,102,115,116] and growth in pig kidney cell cultures at 35°C.[114] No microscopic or clinical alterations were demonstrated in athymic (nude) mice exposed to *N. algerae*[101,102]

N. locustae has been used by aerial application for control of rangeland grasshoppers in Montana with no adverse health effects being reported.[18] Safety tests of this organism included exposure of rabbits by simultaneous intracerebral, intraocular, and i.p. exposure, along with simultaneous intracerebral and i.p. exposure of mice. No significant lesions were produced.[101,102]

E. NEMATODES

There are 19 families of nematodes containing members that are facultative or obligate parasites of insects.[91] Of these, the Mermithidae have attracted substantial attention as potential biological control agents. Mermithids are specific to one or a few species of insects and potential hazards of these parasites to mammals appears small.[91] Limited safety testing has included a report of oral, intranasal, i.p., and dermal exposure of *Romanomermis culicivorax* to suckling and adult mice and adult rats, including immunosuppressed animals.[91] No clinical illness or histopathologic lesions were detected. Similar findings have been reported for *R. iyengari* in India and *R. jingdeensis* in the People's Republic of China where suckling mice were injected dermally; adult rats were exposed *per os*; and three species of fish were exposed.[91]

There have been occasional reports of mermithids found as accidental parasites of man.[91] Poinar[92] reviewed the literature and concluded that the data were insufficient to determine if infection had indeed occurred.[91]

III. CONCLUSIONS

Testing of entomopathogens for safety to animals and humans is of the utmost importance. No matter how promising a microbial control agent is, if it infects or causes toxicity to exposed vertebrates, it is more deleterious than beneficial.

The guidelines suggested by Ignoffo[60] for planning and interpreting studies of safety of entomopathogens to vertebrates are as applicable today as when he proposed them in 1973. These include:

1. The presumption of safety or lack of safety to vertebrates and other organisms that is based upon studies of closely related species may be a guideline for initial studies but must be confirmed by direct experiments that are designed to evaluate a specific entomopathogen.
2. The safety of entomopathogens to vertebrates is relative. Interpretation of results should consider the dose administered, how it was administered, and how safety or lack of safety was evaluated.

3. Absolute safety cannot be guaranteed in all living systems for all time. Toxicity or pathogenicity can generally be demonstrated if no limitation is imposed on dosage or type of vertebrate system.
4. Reports of toxicity pathogenicity of a specific entomopathogen should be carefully and prudently evaluated against reports of absence of toxicity pathogenicity.

REFERENCES

1. **Aizawa, K., Ed.,** Recent development in the production and utilization of microbial insecticides in Japan, *Proc. 1st Int. Coll. Invertebr. Pathol.,* Queens University, Kingston Ontario, 1976, 59—63.
2. **Allen, B. T. and Wilkinson, H. A.,** A case of meningitis and generalized Schwartzman reaction caused by *Bacillus sphaericus, Johns Hopkins Med. J.,* 125, 8, 1969.
3. **Armstrong, J. L., Rohrmann, G. F., and Beaudreau, G. S.,** Delta-endotoxin of *Bacillus thuringiensis* subsp. *israelensis, J. Bacteriol.,* 161, 39, 1985.
4. **Angus, T. A.,** *Bacillus thuringiensis* as a microbial insecticide, in *Naturally Occurring Insecticides,* Jacobson, M., and Crosby, D. G., Eds., Marcel Dekker, New York, 1971, 463.
5. **Arata, A. A., Roberts, D. W., Shadduck, J. A., and Shope, R. E.,** Public health considerations for the use of viruses to control vectors of human diseases, in *Viruses and Environment,* Kurstak, E. and Maramorosch, K., Eds., Academic Press, New York, 1978, 593.
6. **Bailey, M. J., Field, A. M., and Hunter, F. R.,** Environmental impact of spraying apple orchards with the granulosis virus of the codling moth (*Cydia pomonella*). I. Field studies, in *Proc. 3rd Int. Coll. Invertebr. Pathol.,* University of Sussex, 1982, 182.
7. **Barker, R. J. and Anderson, W. F.,** Evaluation of Beta-exotoxin of *Bacillus thuringiensis* Berliner for control of flies in chicken manure, *J. Med. Entomol.,* 12, 103, 1975.
8. **Barker, I. K. and Van Dreumel, A. A.,** The alimentary system, in *Pathology of Domestic Animals,* Vol. 2, 3rd ed., Jubb, K. V. F., Kennedy, P. C., and Palmer, N., Eds., Academic Press, New York, 1985, 1.
9. **Bond, R. P. M., Boyce, C. B. C., Rogoff, M. H., and Shieh, T. R.,** The thermostable exotoxin of *Bacillus thuringiensis,* in *Microbial Control of Insects and Mites,* Burges, H. D. and Hussey, N. W., Eds., Academic Press, New York, 1971, 275.
10. **Borgatti, A. and Guyer, G.,** The effectiveness of commercial formulations of *B. thuringiensis* Berliner on house fly larvae, *J. Insect Pathol.,* 5, 377, 1963.
11. **Briggs, J. D.,** Reduction of adult house fly emergence by the effects of *Bacillus* spp. on the development of immature forms, *J. Insect. Pathol.,* 2, 418, 1960.
12. **Burgerjon, A.,** Le titrage biologique des cristaux de *Bacillus thuringiensis* Berliner par réduction de consommation au Laboratoire de La Minière, *Entomophaga,* 10, 21, 1965.
13. **Burges, H. D.,** Safety, safety testing and quality control of microbial pesticides, in *Microbial Control of Pests and Plant Diseases 1970—1980,* Burges, H. D., Ed., Academic Press, New York, 1981, 737.
14. **Burges, H. D., Croizer, G., and Huber, J.,** A review of safety tests on baculoviruses, *Entomophaga,* 25, 329, 1980.
15. **Cantwell, G. E., Heimpel, A. M., and Thompson, M. J.,** The production of an exotoxin by various crystal-forming bacteria related to *Bacillus thuringiensis* var. *thuringiensis* Berliner, *J. Insect Pathol.,* 6, 466, 1964.
16. **Carbonell, L. F., Klowden, M. J., and Miller, K. K.,** Baculovirus-mediated expression of bacterial genes in dipteran and mammalian cells, *J. Virol.,* 56, 153, 1985.
17. **Chandler, F. W. and Watts, J. C.,** Mycotic, actinomycotic, and algal infections, in *Anderson's Pathology,* Vol. 1, 8th ed., Kissane, J. M., Ed., C.V. Mosby, St. Louis, 1985, 371.
18. **Chapman, H. C., Davidson, E. W., Laird, M., Roberts, D. W., and Undeen, A. H.,** Safety of microbial control agents to nontarget invertebrates, *Environ. Conserv.,* 6, 278, 1979.
19. **Cheung, P. Y. K., Roe, R. M., Hammock, B. D., Judson, C. L., and Montague, M. A.,** The apparent *in vivo* neuromuscular effects of the delta endotoxin of *Bacillus thuringiensis* var. *israelensis* in mice and insects of four orders, *Pestic. Biochem. Physiol.,* 23, 85, 1985.
20. **Cheville, N. F.,** Pathogenic intracellular microorganisms, in *Cell Pathology,* 2nd Ed., Cheville, N. F., Ed., Iowa State University Press, Ames, 1983, 451.
21. **Chilcott, C. N., Kalmakoff, J., and Pillai, J. S.,** Neurotoxic and haemolytic activity of a protein isolated from *Bacillus thuringiensis* var. *israelensis* crystals, *FEMS Microbiol. Lett.* 25, 259, 1984.

22. **Chilcott, C. N., Kalmakoff, J., and Pillai, J. S.,** Cytotoxicity of two proteins isolated from *Bacillus thuringiensis* var. *israelensis* crystals to insect and mammalian cell lines, *FEMS Microbiol. Lett.,* 26, 83, 1985.

23. **Costa, J. and Rabson, A. S.,** Viral diseases, in *Anderson's Pathology,* Vol. 1, 8th ed., Kissane, J. M., Ed., C.V. Mosby, St. Louis, 1985, 345.

24. **Davidson, E. W.,** Bacteria for the control of arthropod vectors of human and animal disease, in *Microbial and Viral Pesticides,* Kurstak, E., Ed., Marcel Dekker, New York, 1982, 289.

25. **Davidson, E. W.,** *Bacillus sphaericus* as a microbial control agent for mosquito larvae, in *Mosquito Control Methodologies,* Vol. 2, Laird, M. and Miles, J., Eds., Academic Press, New York, 1985, 213.

26. **deBarjac, H., Burgerjon, A., and Bonnefoi, A.,** The production of heat-stable toxin by nine serotypes of *Bacillus thuringiensis, J. Invertebr. Pathol.,* 8, 537, 1966.

27. **deBarjac, H., Dumanoir, V. C., Hamon, S., and Thiery, I.,** Safety tests on mice with *Bacillus sphaericus* serotype H-5a,5b, strain 2362, W.H.O. mimeographed document, WHO/VBC/87.948, 1987.

28. **deBarjac, H., Larget, I., Cosmao, V., Benechan, L., and Viviani, G.,** Innocuité de *Bacillus sphaericus* souche 1593 pour les mammiferes, W.H.O. mimeographed document, WHO/VBC/79.731, 1979.

29. **Döller, G.,** Solid phase radioimmunoassay for the detection of polyhedrin antibodies, in *Safety Aspects of Baculoviruses as Biological Insecticides,* Miltenburger, H. G., Ed., Sym. Proc. Fed. Min. Res. Technol., Bonn, 1978, 203.

30. **Döller, G.,** The safety of insect viruses as biological control agents, in *Viral Insecticides for Biological Control,* Maramorosch, K. and Sherman, K. E., Eds., Academic Press, New York, 1985, 399.

31. **Döller, G. and Enzmann, H.-J.,** Induction of baculovirus specific antibodies in rainbow trout and carp, *Bull. Eur. Assoc. Fish Pathol.,* 2, 53, 1982.

32. **Döller, G. and Gröner, A.,** Sicherheitsstudie zur Prufung einer Virusvermehrung des Kernpolyedervirus aus *Mamestra brassicae* in Vertebraten, *Z. Angew. Entomol.,* 92, 99, 1981.

33. **Döller, G. and Gröner, A.,** Safety studies for the control of baculovirus replication in vertebrates, in *Proc. 3rd Int. Coll. Invertebr. Pathol.,* University of Sussex, 1982, 198.

34. **Döller, G., Gröner, A., and Straub, O. C.,** Safety evaluation of nuclear polyhedrosis virus replication in pigs, *Appl. Environ. Microbiol.,* 45, 1229, 1983.

35. **Döller, G. and Huber, J.,** Sicherheitsstudie sur Prufung einer Vermehrung des Granulosevirus aus *Laspeyresia pomonella* in Saugern, *Z. Angew. Entomol.,* 95, 64, 1983.

36. **Donovan-Peluso, M., Wasti, S. S., and Hartmann, G. C.,** Safety of entomogenous fungi to vertebrate hosts, *Appl. Entomol. Zool.,* 15, 498, 1980.

37. **Dungworth, D. L.,** The respiratory system, in *Pathology of Domestic Animals,* Vol. 2, 3rd ed., Jubb, K. V. F., Kennedy, P. C., and Palmer, N., Eds., Academic Press, New York, 1985, 413.

38. **Egerton, J. R., Hartley, W. J., Mulley, R. C., and Sweeney, A. W.,** Susceptibility of laboratory and farm animals and two species of duck to the mosquito fungus *Culicinomyces* sp., *Mosq. News,* 38, 260, 1978.

39. **Ellar, D. J., Thomas, W., Knowles, B. H., Ward, S., Todd, J., Drobniewski, F., Lewis, J., Sawyer, T., Last, D., and Nichols, C.,** Biochemistry, genetics, and mode of action of *Bacillus thuringiensis* delta-endotoxins, in *Molecular Biology of Microbial Differentation,* Hock, J. A. and Setlow, P., Eds., ASM Publications, Washington, D. C., 1985, 230.

40. **Elter, V. B.,** Beitrag zum Problem der Lebensmitelvergiftungen durch aerobe Sporenbildner, *Z. Gesamte Hyg. Ihre Grenzgeb.,* 12, 65, 1966.

41. **Emmons, C. W., Binford, C. H., and Utz, J. P., Eds.,** *Medical Mycology,* 2nd ed., Lea & Febiger, Philadelphia, 1970, 508.

42. **Farrar, W. E.,** Serious infections due to "non-pathogenic" organisms of the genus *Bacillus, Am. J. Med.,* 34, 134, 1963.

43. **Faust, R. M.,** The *Bacillus thuringiensis* beta-exotoxin: current status, *Bull. Entomol. Soc. Am.,* 19, 153, 1973.

44. **Fenner, F., Ed.,** *Classification and Nomenclature of Viruses,* S. Karger, Basel, 1976.

45. **Fisher, R. and Rosner, L.,** Toxicology of the microbial insecticide Thuricide, *J. Agric. Food Chem.,* 7, 686, 1959.

46. **Galichet, P. F.,** Administration aux animaux domestiques d'une toxine thermostable secretée par *Bacillus thuringiensis* Berliner, en vue d'êmpecher la multiplication de *Musca domestica* Linnaeus dans les fèces, *Ann. Zootechnol. Paris,* 15, 135, 1966.

47. **Galichet, P. F.,** Sensitivity of the soluble heat-stable toxin of *Bacillus thuringiensis* of strains of *Musca domestica* tolerant to chemical insecticides, *J. Invertebr. Pathol.,* 9, 261, 1967.

48. **Gill, S. S., Hornung, J. M., Ibarra, J. E., Singh, G. J. P., and Federici, B. A.,** Cytolytic activity and immunological similarity of the *Bacillus thuringiensis* subsp. *israelensis* and *Bacillus thuringiensis* subsp. *morrisoni* isolate PG-14 toxins, *Appl. Environ. Microbiol.,* 53, 1251, 1987.

49. **Giran, F.,** Action de la «Densonucleosé» de létidottères sur les mammifères, *Entomophaga,* 11, 405, 1966.

50. **Gingrich, R. E.,** *Bacillus thuringiensis* as a feed additive to control dipterous pests of cattle, *J. Econ. Entomol.,* 58, 363, 1965.

51. **Gordon, R. E.,** Some taxonomic observations on the genus *Bacillus,* in *Biological Regulation of Vectors: The Saprophytic and Aerobic Bacteria and Fungi,* NIH-77-1180, U.S. Department of Health, Education and Welfare, Washington, D.C., 1977, 67.

52. **Hadley, W. M., Burchiel, S. W., McDowell, T. D., Thilsted, J. P., Hibbs, C. M., Whorton, J. A., Day, P. W., Friedman, M. B., and Stoll, R. E.,** Five-month oral (diet) toxicity/infectivity study of *Bacillus thuringiensis* insecticides in sheep, *Fund. Appl. Toxicol.,* 8, 236, 1987.

53. **Heimpel, A. M.,** Investigations of the mode of action of strains of *Bacillus cereus* Frankland and Frankland pathogenic for the larch sawfly, *Pristiphora erichsonii* (Htg.), *Can. J. Zool.,* 33, 311, 1955.

54. **Heimpel, A. M.,** A critical review of *Bacillus thuringiensis* var. *thuringiensis* Berliner and other crystalliferous bacteria, *Ann. Rev. Entomol.,* 12, 287, 1967.

55. **Heimpel, A. M.,** Safety of insect pathogens for man and vertebrates, in *Microbial Control of Insects and Mites,* Burges, H. D. and Hussey, N. W., Eds., Academic Press, New York, 1971, 469.

56. **Heimpel, A. M. and Angus, T. A.,** Diseases caused by certain spore forming bacteria, in *Insect Pathology: An Advanced Treatise,* Vol. 2, Steinhaus, E. A., Ed., Academic Press, New York, 1963, 21.

56a. **Hauffler, M. and Kunz, S.,** Laboratory evaluation of an exotoxin from *Bacillus thuringiensis* subsp. *morrisoni* to hornfly larvae (Diptera: Muscidae) and mice, *J. Econ. Entomol.,* 8, 613, 1985.

57. **Hoffman, R. A. and Gingrich, R. E.,** Dust containing *Bacillus thuringiensis* for control of chicken body shaft, and wing lice, *J. Econ. Entomol.,* 61, 85, 1968.

58. **Hower, A. A. and Cheng, T. H.,** Inhibitive effect of *Bacillus thuringiensis* on the development of the face fly in cow manure, *J. Econ. Entomol.,* 61, 26, 1968.

59. **Hutchins, D. R. and Johnston, K. G.,** Phycomycosis in the horse, *Aust. Vet. J.,* 48, 269, 1972.

60. **Ignoffo, C. M.,** Effects of entomopathogens on vertebrates, *Ann. N.Y. Acad. Sci.,* 217, 141, 1973.

61. **Ignoffo, C. M.,** Vertebrates and entomopathogens, *Ann. N.Y. Acad. Sci.,* 217, 165, 1973.

62. **Ishibashi, Y., Kaufam, H. E., Ichinoe, M., and Kagawa, S.,** The pathogenicity of *Beauveria bassiana* in the rabbit cornea, *Mykosen,* 30, 115, 1987.

63. **Isaacson, P., Jacobs, P. H., Mackenzie, A. M. R., and Mathews, A. W.,** Pseudotumor of the lung caused by infection with *Bacillus sphaericus, J. Clin. Pathol.,* 29, 806, 1976.

64. **Jacobson, E. R.,** Biology and diseases of reptiles, in Fox, J. G., Cohen, B. J., and Loew, F. M., Eds., *Laboratory Animal Medicine,* Academic Press, New York, 1984, 449.

65. **Jones, T. C. and Hunt, R. D.,** Diseases due to simple bacteria, in *Veterinary Pathology,* 5th ed., Lea & Febiger, Philadelphia, 1983, 574.

66. **Jones, T. C. and Hunt, R. D.,** Diseases caused by viruses, in *Veterinary Pathology,* 5th ed., Lea & Febiger, Philadelphia, 1983, 286.

67. **Jones, T. C. and Hunt, R. D.,** Diseases caused by higher bacteria and fungi, in *Veterinary Pathology,* 5th ed., Lea & Febiger, Philadelphia, 1983, 638.

68. **Kelly, D. C. and Robertson, J. S.,** Icosahedral cytoplasmic deoxyriboviruses, *J. Gen. Virol.,* 20, 17, 1973.

69. **Kerwin, J. L., Dirtz, D. A., and Washino, R. K.,** Non-mammalian safety tests for *Lagenidum giganteum* (Oomycetes: Lagenidiales), *J. Econ. Entomol.,* 81, 158, 1988.

70. **Kissane, J. M.,** Bacterial diseases, in *Anderson's Pathology,* 8th ed., Kissane, J. M., Ed., C. V. Mosby, St. Louis, 1985, 278.

71. **Kuhn, C., III and Askin, F. B.,** Lung and mediastinum, in *Anderson's Pathology,* 8th ed., Kissane, J. M., Ed., C. V. Mosby, St. Louis, 1985, 833.

72. **Kurstak, E.,** Répercussions éventuelle de la lutte microbiologique sur les vertebrés. Cas des insecticides biologiques à base, *Ann. Parasitol. Hum. Comp.,* 46, 277, 1971.

73. **Kurstak, E.,** *Adv. Virus Res.,* 17, 207, 1972.

74. **Kurstak, E., Tijssen, P., and Maramorosch, K.,** Safety considerations and development problems make an ecological approach of biocontrol by viral insecticides imperative, in *Viruses and Environment,* Kurstak, E. and Maramorosch, K., Eds., Academic Press, New York, 1978, 571.

75. **Kushner, D. J. and Heimpel, A. M.,** Lecithinase production by strains of *Bacillus cereus* Fr. and Fr. pathogenic for the larch sawfly *Pristiphora erichsonii* (Htg.), *Can. J. Microbiol.,* 3, 547, 1957.

76. **Lacey, L. A. and Undeen, A. H.,** Microbial control of black flies and mosquitoes, *Annu. Rev. Entomol.,* 31, 265, 1986.

77. **Longworth, J. F., Robertson, J. S., Tinsley, T. W., Rowlands, D. J., and Brown, F.,** Reactions between an insect picornavirus and naturally occurring Ig M antibodies in several mammalian species, *Nature (London),* 242, 314, 1973.

78. **McIntosh, A. H. and Shamy, R.,** Biological studies of a baculovirus in mammalian cell line, *Intervirology,* 13, 331, 1980.

79. **Mathavan, S. and Velpandi, A.,** Toxicity of *Bacillus sphaericus* strains to selected target and non-target aquatic organisms, *Indian J. Med. Res.,* 80, 653, 1984.

80. **Millar, E.,** *Bacillus thuringiensis* in the control of flies breeding in the droppings of caged hens, *N. Z. J. Agric. Res.,* 8, 721, 1965.

81. **Miyajima, A., Schreurs, J., Otsu, K., Kondo, A., Arai, K., and Maeda, S.,** Use of the silkworm, *Bombyx mori,* and an insect baculovirus vector for high-level expression and secretion of biologically active mouse interleukin-e, *Gene,* 58, 273, 1987.

82. **Miyamoto, C., Smith, G. E., Farrell-Towt, J., Chizzonite, R., Summers, M. D., and Ju, G.,** Production of human c-myc protein in insect cells infected with a baculovirus expression vector, *Mol. Cell Biol.,* 5, 2860, 1985.

83. **Müller-Kögler, E.,** *Pilzkrankheiten bei Insekten,* Paul Parey, Berlin, 1965, 130.

84. **Müller-Kögler, E.,** Nebenwirkungen Insektenpathogener Pilze auf Mensch und Wirbeltiere: Aktuelle Fragen, *Entomophaga,* 12, 429, 1967.

85. **Mulley, R. C., Egerton, J. R., Sweeney, A. W., and Hartley, W. J.,** Further tests in mammals, reptiles, and an amphibian to delineate the host range of the mosquito fungus *Culicinomyces* sp., *Mosq. News,* 41, 528, 1981.

86. **Ode, P. E. and Matthysse, J. G.,** Feed additive larviciding to control face fly, *J. Econ. Entomol.,* 57, 637, 1964.

87. **Ohba, M. and Aizawa, K.,** Lethal toxicity of arthropod iridoviruses to an amphibian, *Rana limnocharis, Arch. Virol.,* 68, 153, 1981.

88. **Ohba, M. and Aizawa, K.,** Mammalian toxicity of an insect iridovirus, *Acta Virol.,* 26, 165, 1982.

89. **Ohba, M., Tantichodok, A., and Aizawa, K.,** Production of heat-stable exotoxin by *Bacillus thuringiensis* and related bacteria, *J. Invertbr. Pathol.,* 38, 26, 1981.

90. **Paulov, S.,** Interactions of *Bacillus thuringiensis* var. *israelensis* with developmental stages of amphibians *(Rana temporaria* L.), *Biologia,* 40, 133, 1985.

91. **Petersen, J. J.,** Nematodes as biological control agents. I. Mermithidae, *Adv. Parasitol.,* 24, 307, 1985.

92. **Poiner, G. O., Jr.,** *Nematodes for Biological Control of Insects,* CRC Press, Boca Raton, FL, 1979.

93. **Reimann, R. and Miltenburger, H. G.,** Cytogenetic investigations in mammalian cells *in vivo* and *in vitro* after treatment with insect pathogenic virus (Baculoviridae), in *Proc. 3rd Int. Coll. Invertebr. Pathol.,* University of Sussex, 1982, 236.

94. **Reimann, R. and Miltenburger, H. G.,** Cytogenetic studies in mammalian cells after treatment with insect pathogenic viruses (Baculoviridae). I. *In vivo* studies with rodents, *Entomophaga,* 27, 25, 1982.

95. **Reimann, R. and Miltenburger, H. G.,** Cytogenetic studies in mammalian cells after treatment with insect pathogenic viruses (Baculoviridae). II. *In vitro* studies with mammalian cell lines, *Entomophaga,* 28, 33, 1983.

96. **Rishikesh, N., Burges, H. D., and Vandekar, M.,** Operational use of *Bacillus thuringiensis* serotype H-14 and environmental safety, mimeographed document WHO/VBC/83.371, World Health Organization, 1983.

97. **Roberts, D. W.,** Toxins of entomopathogenic fungi, in *Microbial Control of Pests and Plant Diseases 1970—1980,* Burges, H. D., Ed., Academic Press, New York, 1981, 441.

98. **Samples, J. R. and Buettner, H.,** Corneal ulcer caused by a biologic insecticide *(Bacillus thuringiensis), Am. J. Ophthalmol.,* 95, 258, 1983.

99. **Samples, J. R. and Buettner, H.,** Ocular infection caused by a biological insecticide, *J. Infect. Dis.* 148, 614, 1983.

100. **Sebesta, K., Horska, K., and Ankova, J.,** Inhibition of the novo RNA synthesis by the insecticidal exotoxin of *Bacillus thuringiensis* var. *gelechiae, Collect. Czech. Chem. Commun.,* 34, 891, 1969.

101. **Shadduck, J. A.,** The safety of entomopathogens for mammals: present evaluation methods and approaches and suggestions for the future, mimeographed document, TDR/BCV/SWG.79/WP.04/ Add.1, World Health Organization, 1979.

102. **Shadduck, J. A.,** Some considerations on the safety evaluation of nonviral microbial pesticides, *Bull. W.H.O.,* 61, 117, 1983.

103. **Shadduck, J. A., Singer, S., and Lause, S.,** Lack of mammalian pathogenicity of entomocidal isolates of *Bacillus sphaericus, Environ. Entomol.,* 9, 403, 1980.

104. **Siegel, J. P. and Shadduck, J. A.,** Safety of the entomopathogenic fungus *Lagenidium giganteum* (Oomycetes: Lagenidiales) to mammals, *J. Econ. Entomol.,* 80, 994, 1987.

105. **Siegel, J. P., Shadduck, J. A., and Szabo, J.,** Safety of the entomopathogen *Bacillus thuringiensis* var. *israelensis* for mammals, *J. Econ. Entomol.,* 80, 717, 1987.

106. **Smith, G. E., Ju, G., Erickson, B. L., Moschera, J., Lahm, H. W., Chizzonite, R., and Summers, M. D.,** Modification and secretion of human interleukin 2 produced in insect cells by a baculovirus expression vector, *Proc. Natl. Acad. Sci. U.S.A.,* 82, 8404, 1985.

107. **Som, N. C., Ghosh, B. B., and Majumdar, M. K.,** Effects of *Bacillus thuringiensis* and insect pathogen, *Pseudomonas aeruginosa,* on mammalian gastrointestinal tract, *Indian J. Exp. Biol.,* 24, 102, 1986.

108. **Sorsa, M., Carlberg, G., Gripenberg, U., Linnainmaa, K., Meretoja, T., and Troil, H. V.,** Mutagenic activity of *Bacillus thuringiensis* exotoxin, the potential biological insecticide, *Hereditas,* 84, 253, 1976.

109. **Steinhaus, E. A.,** On the improbability of *Bacillus thuringiensis* Berliner mutating to forms pathogenic for vertebrates, *J. Econ. Entomol.,* 52, 506, 1959.

110. **Thomas, W. E. and Ellar, D. J.,** *Bacillus thuringiensis* var. *israelensis* crystal delta-endotoxin: effects on insect and mammalian cells *in vitro* and *in vivo, J. Cell Sci.,* 60, 181, 1983.

111. **Tinsley, T. and Harrap, K.,** *Baculoviruses for Insect Pest Control: Safety Considerations,* Summers, M., Engler, R., Falcon, L. A., and Vail, P., Eds., American Society for Microbiology, Washington, D. C., 1975.

112. **Tjia, S. T., Lubbert, H., Kruczek, J., Meyer Z. Altenschildesche, G., and Doerfler, W.,** Studies on the persistence of *Autographa californica* nuclear polyhedrosis virus and its genome in mammalian cells, Proc. 5th Int. Congr. Virol., Strasbourg, 1982, 289.

113. **Toumanoff, C.,** Description de quelques souches entomophytes de *Bacillus cereus* Frank. and Frank. avec remarques sur leur action et celle d'autres bacilles sur le jaune d'oeuf, *Ann. Inst. Pasteur,* 85, 90, 1953.

114. **Undeen, A. H.,** Growth of *Nosema algerae* in pig kidney cell cultures, *J. Protozool.,* 22, 107, 1975.

115. **Undeen, A. H. and Alger, N. E.,** *Nosema algerae:* infection of the white mouse by a mosquito parasite, *Exp. Parasitol.,* 40, 86, 1976.

116. **Undeen, A. H. and Alger, N. E.,** Agglutination and immunofluorescent tests for infection of mammals by *Nosema algerae* (Cnidospora: Microsporida), *Sci. Biol. J.,* p. 259, 1977.

117. **Vandekar, M.,** The safety of entomopathogens for mammals: present evaluation methods and approaches and suggestions for the future, Progress Report to WHO, mimeographed document, TDR/BCV/SWG.79/WP.04, World Health Organization, 1979.

118. **Wells, F. E. and Heimpel, A. M.,** Replication of insect viruses in insect hosts, *J. Invertebr. Pathol.,* 16, 301, 1970.

119. **WHO,** Mammalian safety of microbial agents for vector control: a WHO Memorandum, Bull. W.H.O., 59, 857, 1981.

IV. Safety of Microbial Insecticides to Nontarget Invertebrates

Chapter 10

SAFETY TO NONTARGET INVERTEBRATES OF BACULOVIRUSES

A. Gröner

TABLE OF CONTENTS

I. INTRODUCTION

Controlling pests in field crops and forests to keep or reduce the pest population below the economic threshold is a prerequisite for producing food and commodities for man and domestic animals. With increasing knowledge about contamination of the environment by excessive use of chemical pesticides, the search for alternative agents and methods for plant protection was intensified in recent years. The use of biological insecticides, either alone in crops with one (main) pest or in combination with other agents and methods in an integrated pest management system in crops with several pests, was studied mainly with respect to efficacy for the target pest. Research on deleterious effects on the ecosystem was conducted relatively rarely, except for toxicological and/or pathological effects on vertebrates (for more information see Chapter 9). This chapter deals with safety of baculoviruses with regard to invertebrates, because, after demonstration of the potency of baculoviruses in natural epizootics in insect pest populations, the use of baculoviruses is destined to be a major component in integrated pest management systems.

The family Baculoviridae, consisting of only one genus — *Baculovirus* — is characterized by virions with double-stranded circular DNA with about 100 to >150 kilobase pairs. Around an empty core the DNA-protein complex forms a superhelix which is surrounded by a protein layer, the so-called intimate membrane; these structures collectively represent the nucleocapsid. Finally, the nucleocapsid(s) is/are enveloped by a developmental membrane. The genus *Baculovirus* is divided into three morphologically based subgroups. The members of subgroup A, the nuclear polyhedrosis viruses (NPVs), have many virions — containing one or many nucleocapsids per virion — in each occlusion body or polyhedron. These occlusion bodies form in the nuclei of NPV-infected cells. In subgroup B, the granulosis viruses (GV), only one nucleocapsid is integrated per virion, the latter being incorporated singly in an occlusion body (capsule). The virus replication and occlusion body formation of GV occurs in the cytoplasm of infected cells after disintegration of the nuclear envelope, suggesting a mixing of nuclear and cytoplasmic material. Subgroup C (nonoccluded viruses) consists of singly enveloped nucleocapsids which are not occluded.

Baculoviruses have been found only in invertebrates; no member of this family is known to infect vertebrates or higher plants. Baculoviruses have been recorded from the following Orders of insects: Coleoptera, Diptera, Hymenoptera, Lepidoptera, Neuroptera, and Trichoptera. Baculoviruses (or *Baculovirus*-like particles) have been isolated also from various Crustacea, mites, and an entomopathogenic fungus.[1]

Any baculovirus-caused epizootic, whether occurring naturally or artificially induced after viral application, has an impact on the environment. Before the practical application of baculoviruses as insecticides, their impact on the target pests, nontarget organisms (NTOs), and beneficial insects, has to be considered. Because the host range of baculoviruses has been documented extensively,[2,3] primarily to determine new hosts capable of being controlled by a virus application, this chapter will concentrate upon the interaction of baculoviruses and beneficial insects.

II. SPECIFICITY OF BACULOVIRUSES FOR INSECT PESTS (AS ALTERNATE HOSTS AND "INDIFFERENT SPECIES")

Cross-infectivity studies using the oral route of infection and high doses of virus inoculum show that few baculoviruses are species specific. Indeed, most have a narrow host range, but never exceeding the order and usually not the family of the host from which the virus was originally isolated. Commonly, the host range is restricted to the genus of the competent host.[2,3] In general, the host range of an NPV from Lepidoptera is wider than that from a GV or an NPV from sawflies, from which some appear to be species specific. The nonoccluded *Baculovirus* from *Oryctes rhinoceros* (Coleoptera: Scarabaeidae) infects certain other dynastine beetles,[4] while that from *Panonychus citri* (Acari: Tetranychidae) is infectious (to a certain degree) only for *Tetranychus urticae*.[5] A *Baculovirus* from *Penaeus duorarum* has been isolated from several other shrimp species.[6,7]

However, the interpretation of results of cross-infectivity studies may be difficult, if not impossible, in the case of test insects already (although inapparently) infected with a virus. It is known that viruses "noninfectious" to a particular host may nevertheless be potent stressors for the activation therein of an occult virus.[8-11] Therefore, a characterized virus source with known biochemical properties has to serve as virus inoculum and has to be identified as the lethal virus in order to gain reliable results; plaque-purified virus is essential in a stringent cross-infectivity study. Many results of cross-infectivity studies are thus questionable, due to the fact that progeny virus was not identified and compared with the virus inoculum.

Nevertheless, striking differences in susceptibility can be demonstrated, regardless of the taxonomic status of the alternate host in comparison to the competent host.[12] The low susceptibility of *Heliothis subflexa* to the *H. zea*-NPV, relative to *H. virescens,* is due to the interaction of virions and midgut epithelial cells, and appears to be controlled by a single gene.[13,14] The mechanism of the host range has yet to be determined, although some requirements for a successful infection are known:[15]

1. The virus has to be ingested for a naturally occurring infection.
2. The occlusion bodies of NPV and GV have to be dissolved in the gut of the potential host, liberating virions.
3. Before inactivation by the gut juice, free virions have to pass through the peritrophic membrane.
4. Virions have to be adsorbed to midgut epithelial cells at specific receptors on the microvilli.
5. Virus uptake, transportation of the nucleocapsid to the cell nucleus, uncoating of the viral DNA, DNA replication and translation, and assembly to mature virus particles have to occur.
6. In the case of Lepidoptera, virions have to be released through the basal plasma membrane of midgut cells to cause a generalized infection which finally kills the caterpillar.

Restriction of the host range of baculoviruses to the family or even the genus of the competent host from which the virus was isolated, and never going beyond the order of the competent host,[2,3] may depend on the fact that in all other organisms at least one of the above-mentioned requirements is not fulfilled.

III. SAFETY OF BACULOVIRUSES FOR BENEFICIAL INSECTS

Because in their role as microbial insecticides baculoviruses have to be registered as any other pesticide prior to marketing, the protocols for testing the side effect of baculoviruses on beneficial insects have been similar to those for conventional chemical pesticides. For example, most emphasis has been laid on the effect of baculoviruses on the honeybee (*Apis mellifera*) and such other useful insects as silkworms (*Bombyx mori* and others — see Table 2). In addition, standard protocols for testing the side effect of chemical pesticides on certain predators and parasites were used to evaluate the impact of a baculoviral application on these beneficial insects. Usually, these tests with beneficial insects were conducted in the laboratory and rarely in the field. Following standard protocols only those pesticides showing harmful effects in the stringent laboratory tests were also tested under field conditions.[16]

Honeybees and Silkworms — Accepting the restriction of infectivity of a certain

TABLE 1
Toxicity Studies of Baculoviruses in Honeybees (*Apis mellifera*)

Virus	Isolated from	Exposure	Effect[a]	Ref.
NPV	*Autographa californica*	*Per os* to adult bees either individually or to whole colony	—	44, 45
NPV	*Choristoneura fumiferana*	After aerial application monitoring the impact on colonies	—	46
NPV	*Heliothis zea*	*Per os* to adult bees either individually or to whole colony	—	45, 47
NPV	*Lymantria dispar*	*Per os* to bees in a colony	—	47
NPV	*Mamestra brassicae*	Contact, *per os* to adult bees	—	48
NPV	*Orgyia pseudotsugata*	*Per os* to bees in a colony	—	47
NPV	*Spodoptera frugiperda*	*Per os* to bees in a colony	—	47
NPV	*Thymelicus lineola*		—	49
NPV	*Trichoplusia ni*	*Per os* to bees in a colony	—	47
NPV	*Neodiprion lecontei*	After aerial application monitoring the impact on colonies	—	50
NPV	*Neodiprion sertifer*	*Per os* to bees in a colony	—	47
GV	*Cydia pomonella*	Contact, *per os* to adult bees either individually or to a whole colony	—	47, 48
GV	*Estigmene acrea*	*Per os* to bees in a colony	—	47

[a]   — no deleterious effect.

Baculovirus to the family or at least order of the original host, the lack of hazard to pollinators and silkworms posed by baculoviruses is not surprising. Table 1 lists the attempts to challenge bees with these entomopathogens; no deleterious effect has ever been found. Treatments of whole colonies have never revealed any abnormalities in egg production, brood rearing, worker and queen mortality, and general colony behavior. Table 2 shows the sensitivity of silkworms to baculoviruses from other hosts. With the exception of the *Autographa californica*-NPV, the *Baculovirus* with the broadest host range known so far, and the *Bombyx mori*-NPV, no other baculoviruses had shown a harmful effect on different silkworms.

A. PREDATORS

Laboratory studies with several predators of lepidopteran larvae (pentatomids, lacewings, ladybirds and scavenger beetles) have established that baculoviruses pose no adverse effect; neither when fed via NPV-infected larvae, nor in consequence of their being fed baculoviral occlusion bodies suspended in semisynthetic diets, nor by contact with NPV- and GV-preparations.[17-20]

Furthermore, it is demonstrable that predators are potential dispersal agents of baculoviruses. This is due to the fact that they often feed on virus-infected larvae as well as on larvae that have died from the effects of a *Baculovirus* and therefore yield infectious occlusion bodies. Studies on the dispersal of the baculoviruses by predators are listed in Table 3. Results from field tests suggest that the predator complex enhances the epidemic potential of baculoviruses by contaminating the foliage with occlusion bodies; either directly after individuals clean their mouthparts with the tarsi, or via feces. Because viral occlusion bodies are retained in the gut of heteropteran nymphs which preyed on virus-diseased hosts until after the final moult, the adults (being strong fliers) appear capable of introducing the viral inoculum to healthy pest populations.[21,22]

B. PARASITOIDS

There is a possible impact of an application of a virus preparation on adult parasitoids

TABLE 2
Sensitivity of Silkworms to Baculoviruses

Virus	Isolated from	Silkworm	Result[a]	Ref.
NPV	*Autographa californica*	*Anisota senatoria*[b]	+	51
NPV	*Bombyx mori*	*Samia cynthia*[b]	+	52
NPV	*Euproctis similis*	*Bombyx mori*[c]	−	53
		Samia cynthia[b]	−	53
NPV	*Heliothis zea*	*B. mori*[c]	−[d]	54
NPV	*Hyphantria cunea*	*B. mori*[c]	−	55
NPV	*Mamestra brassicae*	*Antheraea pernyi*[b]	−	3
		B. mori[c]	−	55
NPV	*Pseudaletia separata*	*A. mylitta*[b]	−	56—58
		Samia ricini[b]		
		B. mori[c]		
NPV	*Spodoptera litura*	*B. mori*[c]	−	59
GV	*Artona funeralis*	*B. mori*[c]	−	60
GV	*Cydia pomonella*	*A. pernyi*[b]	−	3
GV	*Hyphantria cunea*	*B. mori*[c]	−	61
GV	*Pieris rapae*	*Bombyx* spp.[c]	−	62

[a]   + lethal infection; − no deleterious effect.
[b]   (Lep.: Saturniidae).
[c]   (Lep.: Bombycidae).
[d]   Virus preparation with a high contamination by bacteria were harmful to silkworms.

as well as the more important consequences of a parasitized pest larva becoming infected and an infected host becoming parasitized.

Few results of studies of viral impact on adult parasites are available. However, according to studies of host range,[2,3] no deleterious effect would be expected. *Trichogramma cacoeciae* imagines were exposed to or had ingested different dosages of polyhedra of the *Mamestra brassicae*-NPV with no alteration of their parasitism rates.[18] Similar results were reported for *Cotesia melanoscelus,* where mated females were fed *Lymantria dispar*-NPV with no effect on longevity of the wasps, their parasitism rate, and sex ratio of emerging next-generation wasps.[23] Imagines of *Chelonus blackburni, Meteorus leviventris, Brachymeria intermedia, Campoletis sonorensis,* and *Voria ruralis* were exposed to *Heliothis zea*-NPV with no adverse effects.[17]

Most feeding tests were conducted with the larval endoparasitoids where actual amounts of the virus consumed could not be exactly quantified. Histological studies demonstrated that parasitoid larvae were never infected by a *Baculovirus;*[24-28] however, the parasitoid larvae ingested infected host cells and free occlusion bodies.[25,26,28,29] Such ingested polyhedra were voided in the parasite's meconium during pupation or voided from adults soon after emergence and before parasitization.[26,28]

Tests were run either by infecting a preparasitized host larva with a *Baculovirus* or by parasitization of a preinfected host. Deleterious effects to the parasitoid larvae occurred if the hosts were killed or physiologically altered by the virus prior to pupation of the parasite. The toxic factor(s) in the hemolymph of larvae respectively infected with the synergistic strain of the *Mythimna unipuncta*-GV, or the hypertrophic strain of the *M. unipuncta* NPV, are the most pronounced virus-dependent alterations in infected hosts. With the typical GV and NPV strains, however, the toxic factor(s) are not induced, and therefore no deleterious effects on *Glyptapanteles militaris* are demonstrable.[30,31]

To summarize, baculoviruses have a narrow host range and no evidence of direct del-

TABLE 3
Predators Disseminating Baculoviruses

Predator	Virus	Prey	Test site	Result	Ref.
Orthoptera (Acridiidae) *Acrotylus patruelis*	NPV	*Spodoptera exempta*	Field/lab	Virus activity in feces and gut content	63
(Oecanthidae) *Oecanthus* sp.	NPV	*Anticarsia gemmatalis*	Field	Virus activity in feces (predator homogenates)	22
(Tettigoniidae)	NPV	*A. gemmatalis*	Field	Virus activity in feces (predator homogenates)	22
Dermaptera	NPV	*A. gemmatalis*	Field	Virus activity in feces (predator homogenates)	22
Heteroptera (Lygaeidae) *Geocoris* spp.	NPV	*A. gemmatalis*	Field	Virus activity in feces (predator homogenates)	22
(Miridae) *Span(a)gonicus* sp.	NPV	*A. gemmatalis*	Field	Virus activity in feces (predator homogenates)	22
(Nabidae) *Nabis tasmanicus*	NPV	*Heliothis punctigera*	Field	Virus activity in feces	21
Nabis spp.	NPV	*A. gemmatalis*	Field	Virus activity in feces (predator homogenates)	22
(Pentatomidae) *Euryrhynchus floridanus*	NPV	*Urbanus proteus*	Field	Feeding on larval cadavers	33
Oechalia schellenbergii	NPV	*H. punctigera*	Field/lab	Virus activity in feces	21
Podisus sp.	NPV	*A. gemmatalis*	Field	Virus activity in feces (predator homogenates)	22
(Reduviidae) *Rhinocorus annulatus*	NPV	*Neodiprion sertifer*	Lab	Virus activity in feces	64
Coleoptera (Anthicidae) *Notoxus* sp.	NPV	*A. gemmatalis*	Field	Virus activity in feces (predator homogenates)	22
(Coccinellidae)	NPV	*A. gemmatalis*	Field	Virus activity in feces (predator homogenates)	22
(Carabidae)	NPV	*A. gemmatalis*	Field	Virus activity in feces (predator homogenates)	22
Calosoma sayi	NPV	*S. frugiperda*	Lab	Virus activity in feces	20
C. sycophanta	NPV	*Lymantria dispar*	Field/lab	Virus activity in feces	65
Hymenoptera (Vespidae) *Polistes* sp.	NPV	*U. proteus*	Field	Feeding on larval cadavers	33

eterious effects to parasites has been documented. All lethal and sublethal effects are indirect, being caused by the host's unsuitability due to virus infection. The exception is the atypical GV (synergistic)- and NPV (hypertrophic)-strain of *M. unipuncta,* which induces toxic factors in caterpillars. All lethal effects were confined to immature parasitoids developing in virus-infected hosts (Table 4) and decreased with increasing intervals between parasitoid oviposition and virus infection.

The ability to discriminate virus-infected hosts from healthy hosts would have a selective advantage for the parasites. Host discrimination was reported for *Cotesia melanoscelus,* parasitizing *Lymantria dispar.*[32] The parasite-host contact did not differ significantly between infected and noninfected *L. dispar* larvae, but the ovipositional attempts in noninfected larvae were 68.7%, which was significantly greater than in virus-infected larvae (32.1%).

The tachinid, *Chrysotachina alcedo,* emerged only from noninfected *Urbanus proteus* larvae, in contrast to the sarcophagid, *Sarcodexia innota,* as field studies showed.[33] Most *S. innota* emerged as larvae from diseased *U. proteus* larvae, pupated outside the host larvae, but failed to undergo successful adult eclosion. Many of the *S. innota* pupae were abnormally small indicating that virus-infected *U. proteus* larvae were unsuitable hosts. However, the same sarcophagid was observed emerging from NPV-infected *Epargyreus clarus* larvae.[33]

Hyposoter exiguae females do not discriminate between virus-infected and noninfected hosts. Parasitoid larvae within hosts exposed to virus prior to parasitization died when their hosts died of virus infection; but those larvae in hosts exposed to virus after parasitization completed their development before their host died. Parasites spent significantly less time in their hosts, if the latter were exposed to the virus soon after parasitization.[27] In addition, the sensitivity of a parasitized *Trichoplusia ni* larva to NPV is distinctly lower than that of a nonparasitized larva.[34] In contrast to the results with *H. exiguae* and *T. ni,* in *L. dispar* larva there is a synergistic effect of the two mortality factors NPV and the tachinid, *Blepharipa pratensis.* The combination of NPV and parasitoid increases the mortality, but at the expense of the parasitoid.[35]

The transmission of a baculoviral infection by parasites from an infected population of pest insects to a healthy one can occur mechanically. This happens via virus-contaminated parasites which have contacted recently virus-killed larvae, or which have become contaminated by virus-containing meconium through transfer of the entomopathogen to foliage later consumed by the host larvae,[27,36,37] or to a fresh host by a virus-contaminated ovipositor.[26,27]

C. AQUATIC NONTARGET ORGANISMS (NTOs)

Side effects on such nontarget invertebrates resulting from the application of baculo-viruses have been the subject of studies using species from different orders listed in Table 5. As in the other studies on the environmental impact of these entomopathogens, no adverse effects could be demonstrated.

IV. EFFECT OF BACULOVIRAL FIELD APPLICATIONS

The results of laboratory experimentation indicate that no deleterious effects of an application of a baculoviral preparation are to be expected on beneficial insects such as *Trichogramma.* Viral contamination of host eggs had no repellent effect on the parasite, and the virus did not adversely affect the development of the parasite in the egg.[18,38]

In long-term field trials in the Federal Republic of Germany, attention was paid to the effect of the codling moth granulosis virus (CpGV) on the fauna of apple trees.[39] By decimating the host population, the CpGV treatments have a notable effect on the population of codling moth parasites, in contrast to the parasite complex of the leafroller species. Because the leafrollers were not infected by CpGV, their population level remained unaltered; so, therefore, did that of their parasites, and presumably also, those of other pests (e.g.,

TABLE 4
Development of Parasites in *Baculovirus*-infected Hosts

Parasite	Host	Virus	Result	Ref.
Hymenoptera				
(Braconidae)				
Cotesia glomerata	*Pieris rapae*	GV	Parasites survive only if larvae were infected 4 days after parasitization; survival rate of the parasite is negative correlated to virus dosage	66
C. marginiventris	*Spodoptera mauritius*	NPV	Death of the host causes death of the (immature) parasite; parasites pupate in living infected hosts	24
C. marginiventris	*Mythimna unipuncta*	GV-s[a]	No mortality	67
C. marginiventris	*Spodoptera exigua*	GV-s	Significantly higher mortality rate; death short after emergence from host, before completion of cocoon spinning	67
C. marginiventris	*M. unipuncta*	NPV-h[b]	Normal development of the parasite	67
Glyptopantales militaris	*M. unipuncta*	NPV	Normal development of the parasite	31
G. militaris	*M. unipuncta*	NPV-h	Mortality of parasites — caused by a toxic factor in the hemolymph of infected larvae	31 67
G. militaris	*M. unipuncta*	GV-s	Mortality of parasites	67
Chelonus insularis	*M. unipuncta*	GV-s	Death of the host causes death of the (immature) parasite	67
C. insularis	*S. exigua*	GV-s	No mortality, but significant longer time for development	67
C. insularis	*M. unipuncta*	NPV-h	Death of the host causes death of the (immature) parasite — parasites were alive in moribund hosts	67
(Encyrtidae)				
Copidosoma truncatellum	*Trichoplusia ni*	NPV	Death of the host causes death of the (immature) parasite	28
(Ichneumonidae)				
Campoletis sonorensis	*Heliothis virescens*	NPV	Death of the host causes death of the (immature) parasite (if infection prior to parasitization); parasites survive, if host is infected 48 h or more after parasitization	26
C. sonorensis	*M. unipuncta*	GV-s	No mortality, but significant longer time for development	67
C. sonorensis	*M. unipuncta*	NPV-h	No mortality and alteration of development	67
Hyposoter exiguae	*T. ni*	NPV	Death of the host causes death of the (immature) parasite (if infection prior to parasitization); parasites survive if host is infected after parasitization; time for development of parasites in infected hosts significantly reduced	27
H. exiguae	*M. unipuncta*	GV-s	No mortality, but significantly longer time for development	67
H. exiguae	*S. exigua*	GV-s	No mortality, but significantly longer time for development	67

TABLE 4 (continued)
Development of Parasites in *Baculovirus*-infected Hosts

Parasite	Host	Virus	Result	Ref.
H. exiguae (Pteromalidae)	*M. unipuncta*	NPV-h	No mortality and alteration of	67
Pteromalus puparum (Calliphoridae)	*P. rapae*	GV	Death of the host causes death of the (immature) parasite; surviving parasites smaller and shorter-living	25
Sarcodexia innota	*Urbanus proteus*	NPV	Development of the parasite in diseased larvae; however, failure to emerge from puparia	33
S. innota (Tachinidae)	*Epargyreus clarus*	NPV	Emerging from diseased larvae	33
Blepharipa pratensis	*Lymantria dispar*	NPV	Death to the host causes death of the (immature) parasite; survival rate of the parasite is negative correlated to virus dosage	35, 68
Compsilura concinnata	*M. unipuncta*	GV-s	No effect on development, but fewer parasites emerged from infected hosts	67
C. concinnata	*M. unipuncta*	NPV-h	No effect on development, but fewer parasites emerged from infected hosts	67
Parasarchopaga misera	*Spodoptera litura*	NPV	No mortality and alteration of development	69
Voria ruralis	*T. ni*	NPV	Death of the host causes death of the (immature) parasite	28

[a]   Synergistic strain of the *Mythimna unipuncta* GV.[30]
[b]   Hypertrophic strain of the *Mythimna unipuncta* NPV.[31]

TABLE 5
Effect of Baculoviruses on Aquatic Invertebrates

Virus	Isolated from	Exposed species	Result[a]	Ref.
NPV	*Autographa californica*	Penaeid shrimps	−	70
NPV	*Heliothis zea*	Crayfish, brown shrimp, grass shrimp, Daphnia, oyster	−	71
NPV	*Lymantria dispar*	Daphnia, Chironomus, Notonecta, waterboatmen	−	23
NPV	*Neodiprion lecontei*	Daphnia	−	72
GV	*Pieris rapae*	Shrimps	−	62
Baculovirus from penaeid shrimps		Shrimps	+	73,74

[a]   − no adverse effect; + lethal infection.

red mites and aphids) in the virus-treated plots. Due to the narrow host range of this *Baculovirus* (and by skipping chemical insecticides), the population of European red mites and woolly apple aphids remained below the economic threshold in the virus-treated plots.[39]

Tests in Canadian apple orchards showed little effect of CpGV on populations on insects predaceous on the codling moth (thrips, clerids, pentatomids, and mirids). Numbers of the European red mite averaged around 1% of the number in chemically treated plots, providing further evidence of greater predator activity in trees treated with CpGV.[40]

Application of the *Autographa californica*- and *Heliothis zea*-NPV to cotton had no deleterious effect to the predators observed.[41,42] Application of *H. zea*-NPV to sorghum for *H. armigera* control has a significant influence on the level of parasitism by *Microplitis* sp. In parasitized *H. armigera* larvae, the NPV infection was almost completely suppressed, whereas the suppression of parasitism occurred in NPV-infected hosts.[43]

V. CONCLUSIONS

Since baculoviruses are naturally occurring, beneficial insects have always had contact with these natural regulatory agents. Deleterious effects of baculoviruses to pollinators, predators, and adult parasitoids have never been reported from nature. Atypical development of entomophagous larvae in virus-infected host larvae proved to be entirely due to the unsuitability of the host for the parasitoids in question. Host discrimination on the basis of viral infection has been documented, implying that some parasitoid species do not "waste" eggs on a host which is soon to die. Parasitoids which develop exclusively in eggs or pupae will be unaffected by or after a virus application because these stages are (nearly) insensitive to viral infection. The decrease in numbers of beneficial insects after pest control based on baculoviruses is due to the decreased number of hosts. In crops with a complex of pests the selective baculoviral application will allow the survival of all other insects and mites except the target pest. Therefore, alternative hosts for the predators and parasitoids are still available. In contrast to experience with most chemical insecticides, directly adverse effects of baculoviruses on beneficial insects have never been reported from field tests.

REFERENCES

1. **Tinsley, T. W. and Kelly, D. C.,** Taxonomy and nomenclature of insect pathogenic viruses, in *Viral Insecticides for Biological Control,* Maramorosch, K. and Sherman, K. E., Eds., Academic Press, New York, 1985, 3.
2. **Ignoffo, C. M.,** Specificity of insect viruses, *Bull. Entomol. Soc. Am.,* 14, 265, 1968.
3. **Gröner, A.,** Specificity and safety of baculoviruses, in *The Biology of Baculoviruses, Vol. 1,* Granados, R. R. and Federici, B. A., Eds., CRC Press, Boca Raton, FL, 1986, 177.
4. **Lomer, C. J.,** Infection of *Strategus aloeus* (L.) (Coleoptera: Scarabaeidae) and other Dynastinae with *Baculovirus oryctes, Bull. Entomol. Res.,* 77, 45, 1987.
5. **Beaver, J. B. and Reed, D. K.,** Susceptibility of seven tetranychids to the nonoccluded virus of the citrus red mite and the correlation of the carmine spider mite as a vector, *J. Invertebr. Pathol.,* 20, 279, 1972.
6. **Lightner, D. V.,** Diseases of cultured penaeid shrimp, in *CRC Handbook of Mariculture, Vol. 1,* McVey, J. P., Ed., CRC Press, Boca Raton, FL, 1983, 289.
7. **Brock, J. A., Nakagawa, L. K., van Campen, H., Hayashi, T., and Teruta, S.,** A record of *Baculovirus penaei* from *Penaeus marginatus* Randall in Hawaii, *J. Fish Dis.,* 9, 353, 1986.
8. **Longworth, J. F. and Cunningham, J. C.,** The activation of occult nuclear polyhedrosis virus by foreign nuclear polyhedra, *J. Invertebr. Pathol.,* 10, 361, 1968.
9. **Maleki-Milani, H.,** Influence de passages répétés du virus de la polyédrose nucléaire de *Autographa californica* chez *Spodoptera littoralis* (Lep.: Noctuidae), *Entomophaga,* 23, 217, 1978.
10. **Jurkovicova, M.,** Activation of latent virus infections in larvae of *Adoxophyes orana* (Lepidoptera: Tortricidae) and *Barathra brassicae* (Lepidoptera: Noctuidae) by foreign polyhedra, *J. Invertebr. Pathol.,* 34, 213, 1979.
11. **McKinley, D. L., Brown, D. A., Payne, C. C., and Harrap, K. A.,** Cross-infectivity and activation studies with four baculoviruses, *Entomophaga,* 26, 79, 1981.
12. **Gröner, A.,** Studies on the specificity of the nuclear polyhedrosis virus of *Mamestra brassicae* (L.) (Lep.: Noctuidae), in *Safety Aspects of Baculoviruses as Biological Insecticides,* Miltenburger, H. G., Ed., Symp. Proc., Bundesminsterium für Forschung und Technologie, Projektträger Biotechnologie in der Kernforschungsanlage Jülich GmbH, 1978, 265.
13. **Ignoffo, C. M., McIntosh, A. H., Huettel, M. D., and Garcia, C.,** Relative susceptibility of *Heliothis subflexa* (Guenée) (Lepidoptera: Noctuidae) and other species of *Heliothis* to nonoccluded *Baculovirus heliothis. Ann. Entomol. Soc.,* 78, 740, 1985.

145

14. **Ignoffo, C. M., Huettel, M. D., McIntosh, A. H., Garcia, C., and Wilkening, P.,** Genetics of resistance of *Heliothis subflexa* (Lepidoptera: Noctuidae) to *Baculovirus heliothis, Ann. Entomol. Soc.,* 78, 468, 1985.

15. **Granados, R. R. and Williams, K. A.,** In vivo infection and replication of baculoviruses, in *The Biology of Baculoviruses, Vol. 1,* Granados, R. R. and Federici, B. A., Eds., CRC Press, Boca Raton, FL, 1986, 89.

16. **Hassan, S. A.,** Nebenwirkungen von Pflanzenschutzmitteln auf Nützlinge, *Nachrichtenbl. Dtsch. Pflanzenschutzdienstes (Braunschweig),* 36, 6, 1984.

17. **Wilkingson, J. D., Biever, K. D., and Ignoffo, C. M.,** Contact toxicity of some chemical and biological pesticides to several insect parasitoids and predators, *Entomophaga,* 20, 113, 1975.

18. **Hassan, S. A. and Gröner, A.,** Die Wirkung von Kernpolyedern *(Baculovirus* spec.) aus *Mamestra brassicae* auf *Trichogramma cacoeciae* (Hym.: Trichogrammatidae) und *Chrysopa carnea* (Neur.: Chrysopidae), *Entomophaga,* 22, 281, 1977.

19. **Abbas, M. S. T. and Boucias, D. G.,** Interaction between nuclear polyhedrosis virus-infected *Anticarsia gemmatalis* (Lepidoptera: Noctuidae) larvae and predator *Podisus maculoventris* (Say) (Hemiptera: Pentatomidae), *Environ. Entomol.,* 13, 599, 1984.

20. **Young, O. P. and Hamm, J. J.,** The effect of the consumption of NPV-infected dead fall armyworm larvae on the longevity of two species of scavenger beetles, *J. Entomol. Sci.,* 20, 90, 1985.

21. **Cooper, D. J.,** The role of predatory Hemiptera in disseminating a nuclear polyhedrosis virus of *Heliothis punctigera, J. Aust. Entomol. Soc.,* 20, 145, 1981.

22. **Boucias, D. G., Abbas, M. S. T., Rathbone, L., and Hostetter, N.,** Predators as potential dispersal agents of the nuclear polyhedrosis virus of *Anticarsia gemmatalis* (Lep.: Noctuidae) in soybean, *Entomophaga,* 32, 97, 1987.

23. **Lewis, F. B. and Podgwaite, J. D.,** Safety evaluations, in *The Gypsy Moth: Research towards Integrated Pest Management,* Doane, C. C. and McManus, M. L., Eds., U.S. Department of Agriculture, 1981, 475.

24. **Laigo, F. M. and Tamashiro, M.,** Virus and insect parasite interaction in lawn armyworm, *Spodoptera mauritia acronyctoides* (Guenée), *Proc. Hawaii. Entomol. Soc.,* 19, 233, 1966.

25. **Laigo, F. M. and Paschke, J. D.,** *Pteromalus puparum* L. parasites reared from granulosis and microsporidiosis infected *Pieris rapae* L. chrysalids, *Philipp. Acric.,* 52, 430, 1968.

26. **Irabagon, T. A. and Brooks, W. M.,** Interaction of *Campoletis sonorensis* and a nuclear polyhedrosis virus in larvae of *Heliothis virescens, J. Econ. Entomol.,* 67, 229, 1974.

27. **Beegle, C. C. and Oatman, E. R.,** Effect of a nuclear polyhedrosis virus on the relationship between *Trichoplusia ni* (Lepidoptera: Noctuidae) and the parasite, *Hyposoter exiguae* (Hymenoptera: Ichneumonidae), *J. Invertebr. Pathol.,* 25, 59, 1975.

28. **Vail, P. V.,** Cabbage looper nuclear polyhedrosis virus-parasitoid interaction, *Environ. Entomol.,* 10, 517, 1981.

29. **Smith, R. P. and Kurczewski, F. E.,** The gypsy moth, *Lymantria dispar* (L.) (Lepidoptera, Lymantriidae), its parasitoid (Hymenoptera, Braconidae) and the nuclear polyhedrosis virus: an ultrastructural study, *Pol. Pismo Entomol.,* 50, 189, 1980.

30. **Kaya, H. K.,** Toxic factor produced by a granulosis virus in armyworm larva: Effect on *Apanteles militaris, Science,* 168, 251, 1970.

31. **Kaya, H. K. and Tanada, Y.,** Hemolymph factor in armyworm larvae infected with a nuclear-polyhedrosis virus toxic to *Apanteles militaris, J. Invertebr. Pathol.,* 21, 211, 1973.

32. **Versoi, P. L. and Yendol, W. G.,** Discrimination by the parasite, *Apanteles melanoscelus,* between healthy and virus-infected gypsy moth larvae, *Environ. Entomol.,* 11, 42, 1982.

33. **Temerak, S. A., Boucias, D. G., and Whitcomb, W. H.,** A singly-embedded nuclear polyhedrosis virus and entomophagous insects with populations of the bean leafroller *Urbanus proteus* L. (Lepid.: Hesperiidae), *Z. Angew. Entomol.,* 97, 187, 1984.

34. **Beegle, C. C. and Oatman, E. R.,** Differential susceptibility of parasitized and nonparasitized *Trichoplusia ni* larvae to a nuclear polyhedrosis virus, *J. Invertebr. Pathol.,* 24, 188, 1974.

35. **Godwin, P. A. and Shields, K. S.,** Effects of *Blepharipa pratensis* (Dip.: Tachinidae) on the pathogenicity of nucleopolyhedrosis virus in stage V of *Lymantria dispar* (Lep.: Lymantriidae), *Entomophaga,* 29, 381, 1984.

36. **Stairs, G. R.,** Quantitative aspects of virus dispersion and the development of epizootics in insect populations, in *Proc. Joint U.S. — Japan Seminar on Microbial Control of Insect Pests, Fukuoka,* 1967, 19.

37. **Mohamed, M. A., Coppel, H. C., Hall, D. J., and Podgwaite, J. D.,** Field release of virus-sprayed adult parasitoids of the European pine sawfly (Hymenoptera: Diprionidae) in Wisconsin, *Great Lakes Entomol.,* 14, 177, 1981.

38. **Vybornov, R. V., Moiseev, V. A., Kogan, V. Sh.,** [On the question of the effect of viruses of the spotted cutworm *Amathes c-nigrum* L. on trichogramma (*Trichogramma euproctidis* Gir., *T. evanescens* Westw.)], *Izv. Akad. Nauk Tadzh. SSR Otd. Biol. Nauk,* 1, 77, 1982 (Russian).

39. **Dickler, E.,** Einfluss von Behandlungen mit Apfelwickler-Granulosevirus (CpGV) und breitenwirksamen chemischen Insektiziden auf Parasiten des Apfelwicklers und Parasiten von Schalenwickler-Arten, *WPRS Bull.*, 9, 90, 1986.

40. **Jaques, R. P., Laing, J. E., MacLellan, C. R., Proverbs, M. D., Sanford, K. H., and Trottier, R.,** Apple orchard tests of the efficacy of the granulosis virus of the codling moth, *Laspeyresia pomonella* (Lep.: Olethreutidae), *Entomophaga,* 26, 111, 1981.

41. **Vail, P. V., Soo Hoo, C. F., Seay, R., and Ost, R.,** Experiments on the integrated control of a cotton and lettuce pest, *J. Econ. Entomol.,* 69, 787, 1976.

42. **Gröner, A.,** unpublished data.

43. **Teakle, R. E., Jensen, J. M., and Mulder, J. C.,** Susceptibility of *Heliothis armigera* (Lepidoptera: Noctuidae) on sorghum to nuclear polyhedrosis virus, *J. Econ. Entomol.,* 78, 1378, 1985.

44. **Morton, H. L., Moffett, J. O., and Stewart, F. D.,** Effect of alfalfa looper nuclear polyhedrosis virus on honey bees, *J. Invertebr. Pathol.,* 26, 139, 1975.

45. **Cantwell, G. E. and Lehnert, T.,** Lack of effect of certain microbial insecticides on the honey bee, *J. Invertebr. Pathol.,* 33, 381, 1979.

46. **Buckner, C. H., McLeod, B. B., and Kingsbury, P. D.,** The effect of an experimental application of nuclear polyhedrosis virus upon selected forest fauna, *Report CC-X-101, Canadian Forest Service, Sault Ste. Marie, Ont.,* 1975.

47. **Knox, D. A.,** Tests of certain insect viruses on colonies of honeybees, *J. Invertebr. Pathol.,* 16, 152, 1970.

48. **Gröner, A., Huber, J., Krieg, A., and Pinsdorf, W.,** Bienenprüfung von zwei Baculovirus-Präparaten, *Nachrichtenbl. Dtsch. Pflanzenschutzdienst (Berlin),* 30, 39, 1978.

49. **Smirnoff, W. A., McNeil, J. N., and Lamothe, P.,** Safety tests for the baculovirus of *Thymelicus lineola* (Lepidoptera: Hesperiidae), *Can. Entomol.,* 111, 459, 1979.

50. **Kingsbury, P., McLeod, B., and Mortensen, K.,** Impact of applications of the nuclear polyhedrosis virus of the red-headed pine sawfly, *Neodiprion lecontei* (Fitch), on non-target organisms in 1977, *Report FPM-X-11, Canadian Forest Service, Sault Ste. Marie, Ont.,* 1978.

51. **Kaya, H. K.,** Transmission of a nuclear polyhedrosis virus isolated from *Autographa californica* to *Alsophila pometaria, Hyphantria cunea,* and other forest defoliators, *J. Econ. Entomol.,* 70, 9, 1977.

52. **Aratake, Y. and Kayamura, T.,** Pathogenicity of a nuclear-polyhedrosis virus of the silkworm, *Bombyx mori,* for a number of lepidopterous insects, *Jpn. J. Appl. Entomol. Zool.,* 17, 121, 1973.

53. **Chu, K. K., Hsieh, Y. T., Chang, H. C., Yao, Y. E., and Fang, C. C.,** On a nuclear polyhedrosis of mulberry tussock moth, *Euproctis similis* Fuessly (Lepidoptera: Lymantriidae) and field tests for the moth control, *Acta Microbiol. Sin.,* 15, 93, 1975.

54. **Padhi, S. B. and Maramorosch, K.,** *Heliothis zea* baculovirus and *Bombyx mori:* Safety considerations, *Appl. Entomol. Zool.,* 18, 136, 1983.

55. **Aruga, H., Yoshitake, N., Watanabe, H., and Hukuhara, T.,** Studies on the nuclear polyhedrosis and their induction in some Lepidoptera, *Jpn. J. Appl. Entomol. Zool.,* 4, 51, 1960.

56. **Dhaduti, S. G. and Mathad, S. B.,** Effect of NPV of the armyworm *Mythimna (Pseudaletia) separata* on the silkworm *Bombyx mori, Experientia,* 35, 81, 1979.

57. **Dhaduti, S. G. and Mathad, S. B.,** Effect of nuclear polyhedrosis virus of the armyworm *Mythimna (Pseudaletia) separata* on the tasar silkworm *Antheraea mylitta, Curr. Sci.,* 48, 750, 1979.

58. **Dhaduti, S. G. and Mathad, S. B.,** Effect of NPV of the armyworm *Mythimna (Pseudaletia) separata* on the silkworm *Philosamia ricini* (Hutt.), *Entomon,* 6, 115, 1981.

59. **Hwang, G. H. and Tsuey, D.,** Studies on the nuclear polyhedrosis virus disease of the cotton leafworm, *Prodenia litura* F., *Acta Entomol. Sin.,* 18, 17, 1975.

60. **Wu, J. F., Tai, G. G., Shi, M. B., Huang, Z. H., and Xian, B. C.,** [A preliminary study on granulosis of the zygaenid, *Artona funeralis* Bulter], *J. Bamboo Res.,* 2, 102, 1983 (Chinese).

61. **Tomita, K. O. and Ebihara, T.,** Cross-transmission of the granulosis virus of the *Hyphantria cunea* Drury (Lepidoptera: Arctiidae), to other lepidopterous insect species, *Jpn. J. Appl. Entomol. Zool.,* 26, 224, 1982.

62. **Technical staff,** Insect Virus Laboratory, Department of Virus, [Safety tests of a GV insecticide against cabbage butterfly *Pieris rapae* larvae], *Wuhandaxue Xuebao,* 2, 77, 1981 (Chinese).

63. **Odindo, M. O.,** Epizootiological observations on a nuclear polyhedrosis of the African armyworm *Spodoptera exempta* (Walk.), *Insect Sci. Appl.,* 4, 291, 1983.

64. **Franz, J., Krieg, A., and Langenbuch, R.,** Untersuchungen über den Einfluß der Passage durch den Darm von Raubinsekten und Vögeln auf die Infektiosität insektenpathogener Viren, *Z. Pfl. Krankh. Pfl. Schutz,* 62, 721, 1955.

65. **Capinera, J. L. and Barbosa, P.,** Transmission of nuclear-polyhedrosis virus to gypsy moth larvae by *Calosoma sycophanta, Ann. Ent. Soc. Amer.,* 68, 593, 1975.

66. **Levin, D. B., Laing, J. E., and Jacques, R. P.,** Interactions between *Apanteles glomeratus* (L.) (Hymenoptera: Braconidae) and granulosis virus in *Pieris rapae* (L.) (Lepidoptera: Pieridae), *Environ. Entomol.,* 10, 65, 1981.

67. **Hotchkin, P. G. and Kaya, H. K.,** Interactions between two baculoviruses and several insect parasites, *Can. Entomol.,* 115, 841, 1983.

68. **Godwin, P. A. and Shield, K. S.,** Some interactions of *Serratia marcescens,* nucleopolyhedrosis virus and *Blepharipa pratensis* (Dip.: Tachinidae) in *Lymantria dispar* (Lep.: Lymantriidae), *Entomophaga,* 27, 189, 1982.

69. **Battu, G. S. and Dilawari, V. K.,** Preliminary investigations on the safety evaluation of *Spodoptera litura* (Fabricius) nuclear polyhedrosis virus (SLNPV) against a parasitoid, *Parasarcophaga misera* (Walker), *Entomol. Newsl.,* 8, 6, 1978.

70. **Lightner, D. V., Proctor, R. R., Sparks, A. K., Adams, J. R., and Heimpel, A. M.,** Testing penaeid shrimp for susceptibility to an insect nuclear polyhedrosis virus, *Environ. Entomol.,* 2, 611, 1973.

71. **Ignoffo, C. M.,** Evaluation of *in vivo* specificity of insect viruses, in *Baculoviruses for Insect Pest Control: Safety Considerations,* Summers, M., Engler, R., Falcon, L. A., and Vail, P. V., Eds., American Society for Microbiology, Washington, D.C., 1975, 52.

72. **Hicks, B. D., Geraci, J. R., Cunningham, J. C., and Arif, B. M.,** Effects of red-headed pine sawfly, *Neodiprion lecontei,* nuclear polyhedrosis virus on rainbow trout, *Salmo gairdneri,* and *Daphnia pulex, J. Environ. Sci. Health,* B16, 493, 1981.

73. **Couch, J. A.,** An enzootic nuclear polyhedrosis virus of penaeid shrimp: ultrastructure, prevalence and enhancement, *J. Invertebr. Pathol.,* 24, 311, 1974.

74. **Lightner, D. V., Redman, R. M., and Bell, T. A.,** Infectious hypodermal and hematopoietic necrosis (IHHN), a newly recognized virus disease of penaeid shrimp, *J. Invertebr. Pathol.,* 42, 62, 1983.

Chapter 11

SAFETY TO NONTARGET INVERTEBRATES OF LEPIDOPTERAN STRAINS OF *BACILLUS THURINGIENSIS* AND THEIR β-EXOTOXINS

B. E. Melin and E. M. Cozzi

TABLE OF CONTENTS

I. INTRODUCTION

From the earliest usage of commercial preparations of *Bacillus thuringiensis* more than 25 years ago, safety has been one of the greatest advantages in using microbial insecticides.[1] These microbial products, in contrast to chemical pesticides, have been shown to be highly specific in their insecticidal action. They have demonstrated little or no observable toxicity to nontarget organisms (NTOs) in both controlled laboratory testing and actual field usage. Recently, however, as a result of increasing concern for the environment and a better understanding of the toxins associated with *B. thuringiensis*, more attention has been focused on the effects of these toxins on NTOs, particularly invertebrates. This chapter reviews these studies.

The first commercial products based on this microorganism were introduced in the U.S. in 1958, for use against lepidopteran larvae. The variety used in the early products was *B. thuringiensis* ssp. *thuringiensis*, which is known to produce both δ-endotoxin and β-exotoxin. In 1973,[2,3] commercial products containing β-exotoxin were prohibited in the U.S. due to concern over the safety of this metabolite, and all *B. thuringiensis* products made since then have been free of β-exotoxin.[4]

Table 1 lists 23 varieties for which lepidopteran activity has been demonstrated in laboratory and/or field tests.[2,3,5-7] Most of the existing NTO safety data have been generated on existing commercial products or as supportive information to gain U.S. Environmental Protection Agency (EPA) product registration. Because of this, most of our relevant knowledge concerns only three of the recognized subspecies, *thuringiensis*, *kurstaki*, and *galleriae* and very little has been reported concerning the remaining 19. In the following discussion, many commercial formulations are mentioned which have differing potencies and use different subspecies of *B. thuringiensis*. These data are summarized in Table 2.

II. LEPIDOPTERAN-ACTIVE STRAINS

A. EFFECTS UPON FRESHWATER ORGANISMS

It has been shown that spores of *B. thuringiensis* ssp. *kurstaki* may persist for some time in freshwater systems. Menon and DeMestral[8] found that 50% of the spores remained viable after 50 d. It is likely that usage of this species near a freshwater habitat would result in the presence of small quantities of spores and δ-endotoxin in the water.

In two field studies, *B. thuringiensis* ssp. *kurstaki* formulations were applied to ponds at recommended field application rates for agricultural crops. Anderson evaluated the effects of Dipel® 4L on zooplankton, particularly *Daphnia* sp., in small ponds over a 3-month period. Zooplankton, including rotifers, copepods, cladocerans, and phantom midges (*Chaoborus* sp.) were monitored throughout the study, without the finding of significant affects as compared with the pretreatment situation, for any of the taxa studied.[9] Laszlo[10] conducted a similar investigation in ponds where Dipel WP was applied at field rates, and found no toxic effects to microcrustaceans (Copepoda, Ostracoda), mites (Hydracarina), and insects (Diptera, Heteroptera, Ephemeroptera, Odonata, and Coleoptera).

Laboratory tests have been conducted on stream-collected insects to determine the potential effects on NTOs associated with a large-scale forestry spray. Eidt[11] collected larvae of species of Trichoptera, Plecoptera, Ephemeroptera, Megaloptera, and Diptera and exposed them to concentrations of Thuricide 32LV, equivalent to the worst-case field situation, and 10 and 100 times this amount.

Of the insects tested, only *Simulium vittatum* was found to be susceptible to the product. In related studies, Lacey et al.[12,13] found this species to be more susceptible to *B. thuringiensis* ssp. *kurstaki* and more sensitive than were other Simuliidae tested. Trichopteran species inhabiting a stream, within the boundaries of an aerial *B. thuringiensis* forestry spraying

TABLE 1
Subspecies of *Bacillus thuringiensis* Reported to Have Lepidopteran Activity

Subspecies	H-serotype	δ-endotoxin	β-exotoxin
thuringiensis	1	+	+ (or 0)
finitimus	2	0	0
alesti	3a	+	0
kurstaki	3a,3b	+	+ (or 0)
dendrolimus	4a,4b	+	0
sotto	4a,4b	+	0
kenyae	4a,4c	+	+ (or 0)
galleriae	5a,5b	+	0
entomocidus	6	+	0
aizawai	7	+	0
morrisoni	8a,8b	+	+ (or 0)
ostriniae	8a,8c	+	0
tolworthi	9	+	+ (or 0)
darmstadiensis	10	+	+ (or 0)
toumanoffi	11a,11b	+	0
kyushuensis	11a,11c	+	0
pakistani	13	+	0
israelensis	14	+	0
dakota	15	+	?
wuhanensis	—	+	+
kumamotoensis	18	+	+ (or 0)
japonensis	23	+	0

TABLE 2
Commercial Formulations of *B. thuringiensis* for Control of Lepidopteran Larvae (Referenced in the Studies Discussed)

Product	*B. thuringiensis* subspecies	Type	Potency[a]	Company
Bactospeine	*thuringiensis*	Powder	1,000 IU/mg	Solvay
	kurstaki	Liquid	8,800 IU/mg	
Biotrol XK		Powder	7,500 IU/mg	Nutrilite
Bitoxibacillin	*thuringiensis*	Powder	4XE10 sp+cry/g (0.6—0.8% exotoxin)	(Produced in U.S.S.R.)
Certan	*galleriae*	Liquid		Sandoz
Dipel WP	*kurstaki*	Powder	16,000 IU/mg	Abbott
Dipel 4L	*kurstaki*	Liquid	8,800 IU/mg	
Dipel 6L	*kurstaki*	Liquid	13,200 IU/mg	
Dipel 8L	*kurstaki*	Liquid	17,600 IU/mg	
Enterobacterin	*galleriae*	Powder	1X10E9 sp/g	(Produced in U.S.S.R.)
Thuricide 30B	*thuringiensis*	Liquid		Sandoz
Thuricide 50B	*thuringiensis*	Liquid		
Thuricide HP	*kurstaki*	Powder	16,000 IU/mg	
Thuricide 16B	*kurstaki*	Liquid	3,430 IU/mg	
Thuricide HPC	*kurstaki*	Liquid	4,000 IU/mg	
Thuricide 32B/LV	*kurstaki*	Liquid	8,000 IU/mg	
Thuricide 48LV	*kurstaki*	Liquid	12,000 IU/mg	

[a] Potency information derived from product labels and literature cited in this review.

program, were found to have concentrated spores and crystals in their midguts following the application, but no toxic effects were noted.[14]

B. EFFECTS UPON MARINE ORGANISMS

Spores of *B. thuringiensis* ssp. *kurstaki* do not survive as well in sea water as in freshwater, and in general, sea water is considered to be bactericidal to nonmarine bacteria. Menon and DeMestral found that 90% of the spores of this subspecies died after 30 d exposure to sea water.[8] Even though long-term *B. thuringiensis* survival is not likely in a marine habitat, it is nevertheless possible for organisms in these habitats to ingest spores and/or crystals if treatments are made near coastal areas.

When *B. thuringiensis* cells and spores were injected into adult oysters (*Crassostrea virginica*), these rapidly cleared the microorganisms from their systems. At 10 d post-treatment, the test animals were indistinguishable from the untreated controls. Vegetative cells were nearly 100% phagocytized 24 h postinjection.[15] Larval oysters (*C. gigas*) held in aquaria containing wide-ranging concentrations of Dipel 8L, developed normally.[16]

Several marine species were tested in the lab by Alzieu et al.[17] Mussels (*Mytilus edulis*), oysters (*Crassostrea gigas*), common periwinkle (*Littorina littorea*), shrimps (*Crangon crangon*), and the brine shrimp (*Artemia salina*) were exposed in aquaria to Dipel concentrations of 10 to 400 mg/l for 96 h. In field terms, 20 mg/l represents a concentration of 1000 times expected levels of *B. thuringiensis* in an oyster bed following aerial application at recommended rates. *A. salina* was the only species found to be susceptible to the *B. thuringiensis* in this study, having an LC_{50} of 65 mg/l. Inert formulation particulates may have contributed to the death rate in this case since mortalities were substantially reduced following removal of the particulates from the test suspension.

C. EFFECTS UPON SOIL ORGANISMS

Spores of *B. thuringiensis* ssp. *kurstaki* have been reported to persist in soils for up to a year or more.[18] If conditions are favorable, vegetative growth may occur and numbers of the cells in the soil may increase significantly.[19] Spore levels, though, have been shown to decrease rapidly in the first few months following a *B. thuringiensis* application, remaining stable for several months thereafter. The crystalline δ-endotoxin, however, is believed to be degraded by soil microorganisms in 1 to 3 months.[20,21]

In an early study, the common earthworm, *Lumbricus terrestris,* was exposed to soil containing 1 to 10% (w/w) Thuricide 30B or 50B for 2 months, resulting in 100% mortality.[22] Mortality may have been due to the excessively high application rates, septicemia, or to β-exotoxin possibly present in the formulations. Benz and Altwegg[23] applied Dipel WP to soil at 1, 10, and 100 times recommended field rates (60, 600, and 6000 mg/m²) and Bactospeine at 30 g/m² to small field plots. During a 2-month test period, no adverse effect on the earthworm population was seen in the treated plots, and no dead or diseased worms were found in the treatment area.[23]

Enterobacterin was evaluated in soil-filled pots in the laboratory at rates of 0.2 and 2.0%. No decrease in earthworm survival was noted after 110 d, however, earthworm activity was judged to be reduced in the treated pots. Levels of microarthropods (Acarina and Collembola) were two to ten times lower following treatment.[24]

D. EFFECTS UPON TERRESTRIAL ORGANISMS

On foliage, spores and crystals of *B. thuringiensis* persist for relatively short periods of time.[25] Salama et al.[26] found that spores had half-lives between 75 and 256 h on cotton leaves, not due to high temperatures, but rather to the effect of ultraviolet radiation. Crystalline δ-endotoxin of *B. thuringiensis* ssp. *kurstaki* has been shown to lose its lepidopteran activity after 40 h of exposure to simulated sunlight.[27] This is in contrast to its much longer persistence in freshwater environments.[8]

1. Orthoptera

It is a matter of concern that a beneficial insect predator may become intoxicated or infected when feeding upon a pest species that has ingested *B. thuringiensis* spores and/or crystals. Yousten[28] fed lethal quantities of *B. thuringiensis* ssp. *kurstaki* to larval cabbage loopers (*Trichoplusia ni*) and just prior to death, offered these larvae to young Chinese praying mantids (*Tenodera aridifolia* ssp. *sinensis*). Quantification of the dosage ingested by the mantids was not possible, but it was concluded that mantids were not susceptible to spore/crystal mixtures in an intact insect host.

2. Dermaptera

Workman[29] conducted lab tests on the striped earwig (*Labidura riparia*), an important insect predator of lepidopteran larvae. He exposed the earwigs to Dipel WP applied to the soil at rates equivalent to 10 times the normal field application rate and obtained no mortality in the treated group.

3. Heteroptera

In a laboratory study, Hamed[30] found that *Picromerus bidens* was not adversely affected after feeding upon prey larvae, *Yponomeuta evonymellus*, that had fed upon leaves treated with Dipel WP or Thuricide HP.

Many field studies with commercial *B. thuringiensis* formulations have demonstrated safety to predatory hemipterans. Spined stiltbugs (*Jalysus spinosus*) are important predators on lepidopteran eggs, particularly those of the tobacco budworm (*Heliothis virescens*). In field applications of Dipel WP on tobacco, Elsey[31] reported effective control of lepidopteran pest species with no detrimental effect on nymphs or adults of *J. spinosus* during the 2-month study. A similar study in soybeans using Dipel WP against green cloverworm (*Plathypena scabra*) and velvetbean caterpillar (*Anticarsia gemmitalis*) demonstrated no effect of *B. thuringiensis* ssp. *kurstaki* on Nabidae (damsel bugs, *Nabis* spp.) or Lygaeidae (bigeyed bugs, *Geocoris* spp.).[32] On cotton, Harding et al.[33] conducted a 2-year study to evaluate the effects of the same bacterial subspecies on natural enemies of the bollworm, *Heliothis zea*, on cotton. Following applications of the microbial preparation against this pest at rates of 6 to 24 lb/acre, they reported no detectable effects on Anthocoridae (minute pirate-bugs, *Orius* spp.), *Geocoris* spp., Nabidae, or Reduviidae (assassin bugs).[33] Wallner and Surgeoner[34] found no effect on the spined soldier bug, *Podisus maculiventris* (Pentatomidae), following forest sprays of Thuricide and Dipel WP on the variable oakleaf caterpillar, *Heterocampa manteo*.

4. Coleoptera

Salama and Zaki[35] reared *Spodoptera* larvae on a diet containing *B. thuringiensis* ssp. *entomocidus* and then fed these larvae to adult staphylinid beetles (*Paederus alfierii*). Predator longevity was not significantly affected, and no difference was seen in prey acceptance between untreated larvae and those exposed to the microbial agent.[35] In another study, Salama et al.[36] treated aphids with sprays of *B. thuringiensis* ssp. *entomocidus* and provided these treated insects to newly hatched coccinellid larvae (*Coccinella undecimpunctata*). The survival of larvae of predators was not affected by feeding on the treated prey. However, the duration of predator larval development was increased in the group treated with *B. thuringiensis* spp. *entomocidus* and there was a definite reduction in prey consumption.[36] From these data, beetles appear to be able to distinguish prey surface treated with the microbial agent and such prey may produce a feeding deterrence or repellency in their beetle predators. In contrast to this, the coccinellids are apparently unable to distinguish prey that have ingested *B. thuringiensis* spp. *entomocidus*.

In small plot and laboratory tests with one of the predatory Carabidae, *Bembidion lampros,* Obadofin and Finlayson[37] found that Dipel WP had a minimal effect on this beetle.

Contact activity of Thuricide HPC was evaluated by Wilkinson et al.[38] and adult *Hippodamia convergens* (Coccinellidae) showed no susceptibility after 5 d when treated at levels equivalent to field rates.

In the field, beetle populations have been monitored for sensitivity to *B. thuringiensis* sprays on various crops. For example, Johnson[39] evaluated several commercial formulations as both sprays and bait formulations on tobacco. Populations of two species, *H. convergens* and another coccinellid *Coleomegilla maculata,* were not affected by the microbial treatments during the 2-year study.[39] Asquith[40] found the black lady beetle, *Stethorus punctum,* to be unaffected by combination treatments of Dipel WP and Guthion on apple trees. Harding et al.[33] detected no reduction in population levels in the Dipel-treated plots of ladybird, rove, and checkered beetles in their field study. Wallner and Surgeoner[34] found no effects on coccinellids (*Cycloneda munda, Chilocorus bivulnerus, Adalia bipunctata*) following forest sprays of Thuricide and Dipel WP for variable oakleaf caterpillar. They also saw no effect on a predatory carabid, *Pinacodera platicollis.* Buckner et al.[42] monitored populations of ground beetles following aerial spraying of spruce with Thuricide 16B and Dipel WP and found no effect on these predators. Although it has been reported by Kazakova and Dzhunusov[43] that the commercial product Bitoxibacillin-202 may cause mortality to *Coccinella septempunctata,* these lethal effects are most likely due to the presence of the *B. thuringiensis* β-exotoxin in these preparations, and not the bacterium or the crystalline δ-endotoxin.

5. Neuroptera

Larvae of the lacewing, *Chrysopa carnea,* are important insect predators and several workers have tested this species for susceptibility to *B. thuringiensis* products. Wilkinson et al.[38] found that Thuricide HPC at recommended field rates had essentially no effect on either adults or larvae of this species when applied as a contact spray. Hassan[44] judged Dipel to be harmless to adult *C. carnea* when they were exposed to dried films at normal field rates concentrations.

Salama et al.[36] evaluated the effect of *B. thuringiensis* ssp. *entomocidus* on larval *C. carnea,* by presenting them with either sprayed aphids or treated *Spodoptera* larvae. When fed the latter, the duration of larval development was significantly extended and the prey consumption was significantly reduced. The same trend was seen with the sprayed aphid larvae, indicating a feeding deterrence or repellency.

6. Diptera

Hamed[30] found that two tachinid species, *Bessa fugax* and *Zenillia dolosa,* were not affected after being fed suspensions of Dipel containing 5×10^8 spores per ml. Horn[45] observed a reduction in numbers of syrphid larvae on collards sprayed with Dipel WP, possibly due to a repellent effect on adults.

Field observations of the effects of *B. thuringiensis* applications on various dipterans have not revealed any adverse effect on the percentage of parasitism. In their evaluation of aerial applications of Thuricide HPC for control of gypsy moth and elm spanworm, Dunbar et al.[46] found no such effect for two tachinids, *Blepharipa scutellata* and *Parasitigena agilis.* Fusco[47] actually reported increased parasitism by the tachinids, *Compsilura concinnata* and *Blepharipa pratensis.*

7. Hymenoptera — Honeybees, *Apis mellifera*

Early research by several workers[48-50] failed to reveal *B. thuringiensis* activity against adult or immature honeybees. Martouret and Euverte[51] fed worker bees cultures of *B. thuringiensis* ssp. *thuringiensis* incorporated into mixtures of sugar, honey, and clay. Complete mortality was seen at 7 d for the spore-crystal-exotoxin preparation and at 14 d for the spore-crystal complex alone. Krieg and Herfs,[52] while finding that vegetative *B. thuringiensis* cells did not harm bees, reported toxicity in preparations containing β-exotoxin.

Cantwell et al.[53] fed bees sugar solutions containing *B. thuringiensis* ssp. *thuringiensis* spores (0.67 and 1.67 × 10^9 per bee), supernatant (2.5 mg per bee), and crystals (0.5 to 16 × 10^6 per bee), and also crystals of ssp. *alesti* and *sotto* (both at 0.5 × 10^6 per bee). All three crystal types failed to harm the bees, but the β-exotoxin (supernatant) gave nearly 100% mortality at 7 d. Significant mortality was seen in the spore treatment at 8 d. This was probably a consequence of septicemia. The dosages in this text were many times higher than bees would be exposed to in the course of a lepidopteran control program.

Whole nonsporulated cultures of *B. thuringiensis* ssp. *thuringiensis* and ssp. *kurstaki* were fed to adult bees by Krieg.[54] He found mortality in the bees due to the β-exotoxin in the *B. thuringiensis* ssp. *thuringiensis* culture, and mortalities from both cultures to a thermolabile toxin, possibly α-exotoxin. Since this toxin is inactivated during sporulation, it is not a concern in sporulated commercial products.[54] When Krieg et al.[55] fed fully sporulated cultures of the typical subspecies and *B. thuringiensis* ssp. *kurstaki* to adult bees at concentrations of 1 × 10^8 spores + crystals per bee over a 7-d period, no harmful effects were noted for either preparation.

B. thuringiensis ssp. *galleriae* is a lepidopteran-active subspecies which has proved a commercial success in controlling wax moth (*Galleria* spp.) associated with bee hives.[56,57] This bacterium can be sprayed on the surface of the honeycomb or impregnated into the wax. With these methods of application, there is little chance for NTOs other than honeybees, to come into contact with this bacterium.

Cantwell and Shieh[58] fed a 1:20 dilution of *B. thuringiensis* ssp. *galleriae* in a 40% sucrose solution to newly emerged adult bees. At the end of 14 d, there were no differences in mortality between treated and untreated groups. The treated hives showed no adverse effect on the adult workers or colony life as determined by egg laying, brood production, brood capping, or honey production. Burges[59] speculated that spores of this subspecies may remain viable in honey virtually indefinitely. Due to their tolerance to this bacterium, honeybees would not appear threatened by such persistence. In Canada, Buckner et al.[42] observed no effect on honeybees following aerial spraying of spruce with Dipel WP and Thuricide 16B.

8. Hymenoptera — Parasitoids

Insect parasitoids, being very important regulators of insect pest populations, have been extensively tested for sensitivity or susceptibility to *B. thuringiensis*.

In feeding studies with suspensions of this microorganism, mortality has been noted in some species. Krieg et al.[55] fed washed spores and crystals of *B. thuringiensis* sspp. *thuringiensis* and *kurstaki* (5 × 10^7 spores + crystals per ml) to adult *Trichogramma cacoeciae* and observed no mortality or reduced capacity to parasitize after 7 d feeding. When Hassan and Krieg[60] fed suspensions of Bactospeine, Dipel, and Thuricide to adult *T. cacoeciae* at similar levels, they saw no harmful effects except for the Bactospeine treatment which resulted in reduced ability to parasitize. This may have been due to the β-exotoxin in the Bactospeine. Thoms and Watson[61] found lower survival in adult *Hyposoter exiguae* fed suspensions of Dipel (2 mg/ml). They concluded that the mortality was due to the spore-crystal complex. Muck et al.[62] conducted similar tests with Dipel WP on adults of both *Cotesia glomerata* and *Pimpla turionellae,* finding significant mortality in *C. glomerata* at rates of 10^8 and 10^9 spores per ml, but observing little effect at any rate on *P. turionellae.* They reported seeing midgut epithelial damage in *P. turionellae* at the highest rate, due to the δ-endotoxin. Dunbar and Johnson[63] collected adult *Cardiochiles nigriceps* in the field and fed them suspensions of Dipel WP and Biotrol (XK and WP). They reported shorter life spans in the group provided with *B. thuringiensis,* but could not be sure feeding actually took place. Avoidance and starvation may have been the cause of death.

Parasitoids may either avoid or be repelled by these preparations. Hassan[44] observed that while *T. cacoeciae* was not affected by exposure to dried surface films of *B. thuringiensis* sspp. *thuringiensis* and *kurstaki,* the adults were repelled by formulations of Bactospeine and Dipel in choice tests. Horn,[45] however, found no decrease in the percentage of parasitism of aphids by *Diaretiella rapae* on collards treated with Dipel WP.

Intoxication of host larvae by these microbial preparations may render them less attractive to adult parasitoids. Salama and Zaki[35] observed that *Zele chlorophthalma* had reduced levels of parasitism of *Spodoptera* larvae intoxicated with *B. thuringiensis* ssp. *entomocidus.* Temerak[64] injected *Sesamia cretica* larvae with the bacterium and found these larvae to be significantly less acceptable to *Bracon brevicornis.* Weseloh and Andreadis[65] found that female *C. melanoscelus* could not discern intoxicated host larvae 3 d after treatment, but were able to do so after 10 d.

In contrast to these findings, Dunbar et al.[46] were the first to report an increase in the percentage of parasitism of gypsy moth and elm spanworm larvae in forestry plots treated with Thuricide HPC. Wallner and Surgeoner[34] observed no effect on parasitoids following treatments with Dipel WP and Thuricide for control of the notodontid moth, *Heterocampa manteo.* Fusco[47] reported an increase in the percentage of parasitism of gypsy moth larvae by *Cotesia melanoscelus* and *Phobocampe unicincta* following aerial sprays of Dipel 4L. He determined that this was due to increased density of proper-sized host larvae and suggested synergism between *B. thuringiensis* and *C. melanoscelus.* Weseloh and Andreadis[65] confirmed synergism in lab tests with gypsy moth larvae fed Dipel 4L and exposed to *C. melanoscelus.* The percentage of parasitism was increased in bacteria-intoxicated larvae since these grew more slowly and were at the appropriate size for parasitization for a longer time. Wallner et al.[66] obtained similar results in the laboratory with *Rogas lymantriae* and gypsy moth hosts fed *B. thuringiensis* ssp. *kurstaki.* In the case of *Rogas,* however, the sex ratio of parasitoids was skewed to produce more males in the treated larvae, since the females lay more fertilized eggs in larger host larvae. In applying these results to the field, Weseloh et al.[67] demonstrated 6- to 12-fold increases in the percentage of parasitism of gypsy moth larvae by *C. melanoscelus* in forestry plots treated with *B. thuringiensis.* Hamel[68] found that parasitoids attacking early instar budworm larvae increased in number following aerial application of Dipel WP, while parasitoids of older budworm larvae were reduced in number. Large operational usage of *B. thuringiensis* ssp. *kurstaki* formulations have not resulted in any detrimental effects to parasitoid populations.[42,69,70]

Parasitoid larvae developing within a *B. thuringiensis*-intoxicated host larva have been shown to have longer development times in some cases. Salama and Zaki[35] reported increased development times for *Zele chlorophthalma* in *Spodoptera* treated with the microorganism. Slightly longer larval development periods have been observed for *C. melanoscelus* developing in Dipel-treated gypsy moth larvae.[65,71] These results are in agreement with Temerak[64] who found *B. thuringiensis*-intoxicated host larvae to be nutritionally unsuitable for either adult or larval parasitoids, adult *Bracon brevicornis* feeding on their body fluids exhibiting reduced longevity. Hamed[30] reported toxic effects in four parasitoid species, *Diadegma armillata, Pimpla turionellae, Ageniaspis fuscicollis,* and *Tetrastichus evonymellae,* which fed upon *B. thuringiensis*-infected host body fluids. In contrast, Hassan[44] showed that development of *Trichogramma cacoeciae* was not affected when host eggs were dipped into *B. thuringiensis* suspensions following parasitization. Kaya and Dunbar[72] obtained similar results with *Telonomus alsophilae* which emerged from egg masses dipped in a Dipel suspension.

Following their development in hosts exposed to *B. thuringiensis,* some parasitoid species have exhibited reduced adult emergence.[36,64] Salama et al.[35,36] observed lowered reproductive potential for both *Z. chlorophthalma* and *Microplitis demolitor*. Biache,[73] however, found no effect on either the percentage of emergence or fecundity of *P. turionellae*.

In host larvae which have ingested large dosages of *B. thuringiensis*, the consequence for developing parasitoids may be death, as Thoms and Watson[61] found to be the case — because of premature host death — in immature *Hyposoter exiguae*.

The above studies indicate that as the host condition worsens, there is a steady decline in its nutritional value, which can be harmful or fatal to developing parasitoids.

9. Acarina

Krieg[74] exposed two-spotted spider mites, *Tetranychus urticae*, to foliage treated with an early Thuricide formulation (1966 production). Significant mite mortalities resulted, in all probability because of the β-exotoxin in the formulation.

Under field conditions, mites have not been harmed by sprays of formulations based upon *B. thuringiensis* ssp. *kurstaki*. Horsburgh and Cobb[75] reported populations of *T. urticae* and *Panonychus ulmi* unaffected by biweekly sprays of Dipel WP. Weires and Smith[76] determined that sprays of Dipel on apples during the 4-month season had no effect on *T. urticae* and *P. ulmi* or on two predatory species, *Amblyseius fallacis* and *Zetzellia mali*.

III. β-EXOTOXIN OF *BACILLUS THURINGIENSIS*

Thuringiensin, or β-exotoxin, is a water-soluble heat-stable secondary metabolite of certain strains of *B. thuringiensis*. Exotoxin is a 701-kDa nucleotide containing adenine, glucose, and allaric acid.[77,78] Although exotoxin production has been historically associated with certain subspecies of *B. thuringiensis*, especially sspp. *thuringiensis*, *galleriae*, and *darmstadiensis*,[79-82] its production in these and other strains depends on the growth medium.[6,83] In the U.S., commercial products of the microorganism must be free of exotoxin as demonstrated by a housefly bioassay. Thuringiensin itself, however, is currently under development as a commercial insecticide.

A. MODE OF ACTION

The mode of toxic action of thuringiensin in living systems is by inhibition of RNA polymerase enzymes acting competitively with ATP.[84,87] In mammals, thuringiensin has been shown to preferentially block the synthesis of ribosomal RNA.[88] In the cabbage looper, *Trichoplusia ni*, thuringiensin injection inhibited radiolabel incorporation into protein, RNA and DNA.[89]

Since RNA synthesis is a vital process in all life, exotoxin exerts its toxicity in almost all forms of life tested.[2,90,91] In mammals the primary site of action is probably the liver.[92] Death is always delayed, even after direct injection of massive doses.[93] Lower lethal doses show longer delay periods.[94]

In insects, exotoxin has been shown to produce mortality,[95-97] reduce longevity, inhibit reproduction, and act as a teratogen and a feeding deterrent.[98-102] The toxic effect of thuringiensin is particularly effective in holometabolous larvae at the time of pupation.[95,96] The material is also toxic, however, to hemimetabolous insects, especially at the time of molting.[97] Time of death is usually delayed, which may be a function of the length of life stage. Mortality to adult insects occurs only at higher dose levels. Treated bollworms and armyworms showed reduced fecundity and longevity, although longevity of the armyworm was reduced only at higher doses.[98,99,103] Exotoxin also has an effect on the egg, inhibiting larval development of hatching and producing mortality at first molt apparently by direct penetration into the egg.[99,104] The sublethal teratogenic effect of thuringiensin has been observed in several orders of holometabolous insects, Diptera,[79,95] Lepidoptera,[100,101,105] and Coleoptera.[102]

B. FACTORS AFFECTING TOXICITY

In general, thuringiensin has been shown to be much less toxic orally than by direct routes of administration in both mammals and insects.[84,93,106-108] This phenonemon is probably due to the presence of large amounts of detoxifying phosphatases in the intestinal cells of mammals and insects.[77,84,109] The difference in toxicity between direct and oral administration is in the order of 60 to 100 for mammals,[93] and may be as high as 250 in insects.[79]

Differences in rate of elimination may also contribute, as less sensitive species such as the mouse appear to clear and detoxify thuringiensin more readily than the sensitive wax moth, *Galleria mellonella*.[110,111] Species differences in the sensitivity of RNA polymerase preparations to thuringiensin have been demonstrated between the rat and the flesh fly *Sarcophaga bullata*.[112]

Similarly, differences in enzyme sensitivity have been suggested as the cause of the difference between the adult and larval susceptibility outlined above.[113] It is postulated that thuringiensin is particularly toxic to holometabolous insects during molting, when thuringiensin-sensitive ecdysone-stimulated RNA polymerase activity is high.

C. TOXICITY TO NONTARGET INVERTEBRATES

Little work has been done on the toxicity of thuringiensin to invertebrate NTOs. Most available data address the toxicity to beneficial insect species related to target pests during commercial development trials. The susceptibility of arthropods to thuringiensin has been reviewed.[2] Most of the species included were insect pests. From this compilation it can be concluded that thuringiensin does not exhibit a high degree of selectivity among the insects, since most species tested in the laboratory were sensitive to the product. A quantitation of sensitivity is difficult since most experiments used thuringiensin-containing broths without quantitation of the dose of β-exotoxin. Since the medium effects the ability of a given strain of *B. thuringiensis* to produce β-exotoxin as well as the amount of β-exotoxin produced, direct comparisons of susceptibility are difficult.[83] Further, preparations previously denoted as "pure" may have contained 10 to 20% exotoxin as determined by more recent analytical techniques,[114] and preparations may also have contained more than one toxin. There is evidence that more than one heat-stable toxin is produced by three different subspecies of *Bacillus thuringiensis* based on differential heat sensitivities of toxicity[82] and isolation of several distinct toxic HPLC fractions.[115]

1. Freshwater Zooplankton

In a standard static acute bioassay, thuringiensin formulation was tested for its effect on *Daphnia magna*.[116] The 24- and 48-h LC_{50} values were greater than 15 and 6.3 mg/l, respectively. No abnormal effects were seen in the *Daphnia* at 24 h, which probably reflects thuringiensin's delayed mode of action. Since thuringiensin is stable to hydrolysis greater toxicity may have been evidenced at later time points.[117] These LC_{50} values indicate that at 24 h thuringiensin is not highly toxic to *Daphnia magna*. Since rates applied in the field for economic control are generally low (40 to 60 g/A), and thuringiensin does not have significant potential for environmental mobility due to strong soil binding, no effect on the *Daphnia magna* population would be expected by field application of thuringiensin.[118]

2. Benthic Aquatic Organisms

Benz remarks that thuringiensin is nontoxic to *Tubifex* spp. in preliminary experiments,[119] although the route of administration is not given. Unfortunately, no other data are available. The high soil binding of thuringiensin and the product's low application rates for commercial uses render hazard to such organisms in detritus and soil unlikely.[118]

3. Terrestrial Organisms

a. Thysanoptera

In a trial by Pan-Agricultural Labs,[120] no effect was observed on *Scolothrips sexmaculatus*, the predator of the two-spotted spider mite, for 28 d after treatment with 20 g/A. The population was significantly reduced at 14 d post-treatment at 10 g/A of thuringiensin (one of two formulations), but appeared to recover by 21 d. The author did not consider this a significant effect.

b. Heteroptera

No effect was seen by Schuster on *Geocoris* spp. in cotton fields treated at 45 g/A.[121] No effect was noted on *Nabis* spp. in field trials at rates of up to 60 g/A.[121,123] Although hemipterans are not generally regarded as susceptible to thuringiensin, these levels proved highly toxic to *Lygus* spp.

Sanford[124] showed that application of thuringiensin to apples did not affect hatching of the mirid predator, *Atractomus mali*, 30 d following treatment with 16 g/100 l to runoff. Field trials on the predacious minute pirate bug, *Orius tristicolor*, have shown it to be unaffected by application of thuringiensin at up to 80 g/A.[121-123] Two studies have investigated the possible effect of thuringiensin on the two-spotted stink bug, *Perillus bioculatus*, as it fed on treated Colorado potato beetle larvae. Burgerjon and Biache[126] fed third instar *Perillus* larvae on the cadavers of Colorado potato beetles intoxicated with thuringiensin with no affect on the survival and maturation of the heteropterans. No effect was seen on fourth-instar larvae feeding on moribund Colorado potato beetles or third-instar larvae feeding exposed to both Colorado potato beetle larvae and foliage containing exotoxin residue. Cantwell[127] also failed to show an effect on neonate *Perillus* feeding on Colorado potato beetle larvae which had fed on thuringiensin-treated tomato leaves or when the predator fed on thuringiensin-treated eggs. In the same trial the effect of thuringiensin on another predator of the Colorado potato beetle, *Podisus maculiventris*, was investigated. The feeding of neonate predators on intoxicated larvae or treated eggs had no effect on survival for 10 d.

c. Coleoptera

Hippodamia spp. populations were not affected by 40 g/A thuringiensin when applied to alfalfa, a rate which provides economic control of insect pests.[122] In the same study, however, the nitulid beetle, *Melegethes nigrescens*, was quite sensitive to treatment, at rates as low as 10 g/A.

d. Hymenoptera

The majority of studies on the toxicity of thuringiensin to NTOs include tests on *Apis mellifera*. Such studies have indicated lethality to adult honeybees. However, when thuringiensin is presented in the food source,[52,128,131] mortality among both adults and larvae is usually delayed, a high death rate only being observed after 3 to 5 d, respectively, at the doses providing 90 to 100% mortality. Exotoxin was also lethal to honeybee larvae but did not appear to affect adult emergence,[128] since all surviving larvae emerged normally. No teratogenic effects on honeybees have been reported. A colony of 10,000 bees treated with 2000 mg *B. thuringiensis* supernatant suffered almost complete mortality.[129] The material affected larvae, pupae, and adults. Vandenberg and Shimanuki[131] found that a dose of 0.002 mg per bee was effective in reducing longevity of adult bees fed thuringiensin in spore-free preparations. Dilutions of exotoxin preparations were toxic in the diet or by contact when sprinkled on caged honeybees. The contact activity was only seen at the most concentrated dilutions (100 times more concentrated than levels eliciting oral toxicity by lifetime feeding, but similar to levels affecting the LT_{50} by a single oral dose). Atkins[132] found that levels of 0.054 g per bee were not lethal to caged adult honeybees for 72 h when sprayed onto them as a kaolin (clay from decomposed feldspar) dust.[132] Mayer[133] demonstrated that caged adult honeybees were unaffected when allowed to forage for 24 h on alfalfa foliage collected 2 to 8 h after application of 40 g/A thuringiensin. Future such experiments should include longer observation periods, so as to take cognizance of delayed effects.

Field studies have evidenced that environmental dilution and feeding habits of honeybees afford protection from commercial applications of thuringiensin. In one trial on blooming alfalfa,[134] treatment with 40 g/A thuringiensin did not produce mortality in excess of the untreated check as measured by counts in dead bee traps for three days following treatment. The foraging of adult bees was also assessed by observation and was unaffected during the 2 d following treatment. The effect on larval honeybees in the colonies was also measured. Thuringiensin treatment did not affect brood survival or development at 6 d posttreatment, although it is not clear whether these numbers were subjected to statistical analysis and the number of pupae developing appeared low. In a study in pollen shedding sweet corn there was a large increase in the number of dead honeybees in traps the first day following treatment with 40 g/A thuringiensin; however, day two and three counts were normal.[135] The significance of the first-day mortality is questionable. Unfortunately, no untreated checks were available for comparison. There was no effect in this trial on the number of honeybees observed foraging or the number returning pollen to the hive for 2 d after application. Brood survival, although not evaluated against a control, was apparently not affected by treatment, since 90% survived 6 d after the application.

Two of the above studies also included the alfalfa leafcutting bee, *Megachile rotundata*. When caged adults of this species were allowed to forage on alfalfa foliage treated with thuringiensin at 40 g/A, no mortality was seen after 24 h. However, this observation period was probably inadequate for due recognition of mortality.[133]

In a field trial, mortality was assessed by observing the number of *M. rotundata* found in a dead bee trap for 2 d following application of thuringiensin to alfalfa. Increased mortality was not observed. No reduction in sightings of foraging bees was noted for 2 d after application. Again, longer observation periods would have been useful.[134]

Studies on hymenopteran parasitoids have, in general, shown that thuringiensin poses no hazard to these beneficial insects. *Edovum puttleri*, which parasitizes the Colorado potato beetle (*Leptinotarsa decemlineata*), was not affected by feeding on a 7.5 mg/ml dilution of thuringiensin in honey water as caged adults for 14 d.[127] Cantwell[136] also noted that the population of *E. puttleri* was not affected by treatment of thuringiensin at 30 g/A in the field although data were not available in the report. Similarly, *Pediobius foveolatus*, the (originally Indian) larval parasite of the Mexican bean beetle, was unaffected by an identical solution (7.5 mg/ml in honey water) when it fed for 11 d.[127] No effect on the parasitoid population was noted during treatment of Mexican bean beetle on snap beans at 30 g/A.[136,137] These two species are clearly less sensitive to thuringiensin toxicity than the honeybee. Hymenopterous parasitoids were not affected in one field trial by treatment with thuringiensin at up to 80 g/A in alfalfa.[124] Collins[122] noted that thuringiensin appeared to affect the population of ichneumon wasps in alfalfa at 40 g/A.

Two studies[138,139] investigated the susceptibility to thuringiensin of *Trioxys pallidus*, an aphidid (braconid) parasitoid brought from France to California to help control the walnut aphid, *Chromaphis juglandicola*, an Old World import to the U.S. Thuringiensin, at 22 ppm did not produce mortality in the adult parasitoid by contact or reduce parasitism when exposed by ingestion. Topical treatment of 22 ppm thuringiensin to late third instar pupae also had no effect on emergence.

e. Acarina

The toxicity of thuringiensin to two species of mites was investigated by Patterson et al.[140] Female adults and mixed sex nymphs and eggs of the predacious spider mites, *Neo-*

seiulus fallacis and *Typhlodromus occidentalis,* were treated by dipping into thuringiensin solutions at 10 to 1000 ppm. Mortality evaluated 24 h after treatment yielded $LD_{50}s$ of 150 and 35 ppm for *N. fallacis* adults and nymphs, respectively, and 300 and 35 ppm for *T. occidentalis* adults and nymphs. The last-mentioned were clearly more susceptible to thuringiensin toxicity, although less susceptible than the spider mite prey. The observations that mortality was greatest during the molting process, and the susceptibility of the nymphs agree with the generally recognized mode of action of thuringiensin.

f. Araneida

In a field trial on alfalfa, spider populations were reduced at 7 d following treatment with 20 to 80 g/A thuringiensin.[123]

IV. CONCLUSIONS

From the existing data, it is clear that lepidopteran-active varieties of *Bacillus thuringiensis* do exhibit toxic effects on some invertebrate NTOs. These effects may be due to (1) formulation components, (2) spore-crystal complex, and (3) β-exotoxin.

A. FORMULATIONS

Ingredients in commercial *B. thuringiensis* products are proprietary in nature and cannot be discussed in detail here. It may nevertheless be stated that these materials show nontarget toxicity, based upon lab and field testing conducted in the last 20 years, which is extremely low. It is on the basis of this safety assurance that the U.S. EPA has granted product registrations.

Limited published data exist for evaluation of formulation effects on NTOs. One study by Haverty[141] on *Chrysopa* and *Hippodamia* with Dipel® 4L carrier indicated significant mortalities at levels eight times greater than labeled field application rates. The *Artemia* mortality discussed previously was observed at concentrations in excess of 3000 times those expected following standard field applications. There does not seem to be any correlation between the nontarget effects and specific formulations reviewed in this paper.

Formulation components may have definite secondary effects on some insect predators and parasitoids in rendering prey or hosts treated with *B. thuringiensis* less attractive than untreated ones. While this would tend to improve pest control by diverting predators and parasitoids to seek prey which had been missed in a field spray, it could also result in movement of these beneficials out of the treatment area.

B. SPORE-CRYSTAL COMPLEX

Feeding deterrence or prey avoidance effects have also been reported for the unformulated spore-crystal complex in predators and parasitoids. In the case of some parasitoids, the avoidance of *B. thuringiensis*-treated hosts may be more related to the ability to detect an impaired physiological state rather than to the microbial agent itself.

Some predatory species of Neuroptera and Coleoptera have exhibited prolonged larval development following ingestion of *B. thuringiensis*-treated prey. Also, adult hymenopteran parasitoids have had their life span shortened after feeding on the body fluids of *B. thuringiensis*-intoxicated host larvae. Furthermore, the latter have sometimes yielded reduced emergence of adult parasitoids with lowered reproductive potential. One report has indicated gut damage in a hymenopteran due to the δ-endotoxin. While it is possible that the latter could be weakly toxic to the larvae, this is unlikely to be a widespread event since toxin activation is dependent upon specific pH conditions and membrane binding sites. At high spore concentrations, bacterial septicemia is likely to occur. A more probable explanation for these effects is that the *B. thuringiensis*-intoxicated hosts are nutritionally less suitable than healthy ones. Little has been done to establish the cause of death in many of the studies. More thorough microscopic and microbiological studies are needed to determine whether septicemia, intoxication, nutrition, or some other factor is responsible (see also Chapter 5).

A rather obvious direct cause of death to NTOs occurs in cases where lepidopteran larvae ingest lethal doses of *B. thuringiensis* and larval parasitoids are thus unable to complete their growth prior to host death. The consequences to the hymenopterans are as surely fatal as when host larvae are eaten by predators.

Some species of tachinid flies and hymenopteran parasitoids have exhibited increased rates of parasitism following field applications of *B. thuringiensis* as a result of the greater numbers of young larvae available as hosts. Other parasitoid species have shown skewed sex ratios following such field treatments as a result of their host larvae remaining in the early instars for extended periods of time.

Overall, the effects of the *B. thuringiensis* spore-crystal complex has been shown to have minimal effects on NTOs when applied at recommended label rates. Many of these effects are secondary in nature, resulting from the declining health of the host or prey larva.

C. β-EXOTOXIN (THURINGIENSIN)

Since thuringiensin has a mode of action which is toxic to all animals, high doses are likely to adversely affect all exposed fauna. Thuringiensin has little contact activity in insects. Exposure of predators and parasites to field-applied thuringiensin would be through the host or by contact. Ingestion of toxic doses via predation and parasitism would be difficult due to dilution in the host and rapid breakdown of thuringiensin within it. The water-soluble thuringiensin has little potential to accumulate in the environment. In addition, dephosphorylation in the host would lead to inactivity. In general, toxicity of thuringiensin via predation and parasitism has not been observed in the laboratory or in the field.

Species in the class Arachnida may to be particularly susceptible to the activity of thuringiensin. Thus field applications of thuringiensin may pose more hazard to beneficial mites and spiders than to insect NTOs. Most nontarget insect species so far investigated have not been affected by application of thuringiensin at commercial rates. Species selectivity may provide adequate safety to some of the species coincidentally with effective control of target pests.

Although thuringiensin is toxic to the honeybee by contact or ingestion, populations of *Apis mellifera* are unlikely to be affected in the field because of the low application rate, dilution, and insect feeding habits. Future field trials should include extended observation periods to accomodate the delayed effect of thuringiensin in this species at low doses.

REFERENCES

1. **Falcon, L. A.,** Microbial control as a tool in integrated control programs, in *Biological Control,* Huffaker, C. B., Ed., Plenum Press, New York, 1971, 346.
2. **Krieg, A. and Langenbruch, G. A.,** Susceptibility of arthropod species to *Bacillus thuringiensis,* in *Microbial Control of Pests and Plant Diseases 1970—1980,* Appendix 1, Burges, H. D., Ed., Academic Press, New York, 1981, 837.
3. **Faust, R. M.,** Toxins of *Bacillus thuringiensis*: mode of action, in *Biological Regulation of Vectors,* DHEW Publication No. 77-1180 (NIH), Briggs, J. D., Ed., U.S. Department of Health, Education, and Welfare, 1975, 31.
4. Federal Register, Vol. 38, #180.1011, July 17, 1973.
5. **Ohba, M. and Aizawa, K.,** *Bacillus thuringiensis* subsp. *japonensis* (flagellar serotype 23): a new subspecies of *Bacillus thuringiensis* with a novel flagellar antigen, *J. Invertebr. Pathol.,* 48, 129, 1986.
6. **Ohba, M., Tantichodok, A., and Aizawa, K.,** Production of heat-stable exotoxin by *Bacillus thuringiensis* and related bacteria, *J. Invertebr. Pathol.,* 38, 26, 1981.

7. **Ohba, M., Ono, K., Aizawa, K., and Iwanami, S.,** Two new subspecies of *Bacillus thuringiensis* isolated in Japan: *Bacillus thuringiensis* subsp. *kumamotoensis* (serotype 18) and *Bacillus thuringiensis* subsp. *tochigiensis* (serotype 19), *J. Invertebr. Pathol.,* 38, 184, 1981.

8. **Menon, A. S. and DeMestral, J.,** Survival of *Bacillus thuringiensis* var. *kurstaki* in waters, *Water Air Soil Pollut.,* 25, 265, 1985.

9. **Anderson, R.,** Effect of Dipel 4L on zooplankton of a small pond, Abbott Laboratories, unpublished report, 1981.

10. **Laszlo, S.,** The effect of *Bacillus thuringiensis* on the arthropod fauna of a pond (chiefly insects), *Novenyved. Plant Prot.,* 15, 251, 1979.

11. **Eidt, D. C.,** Toxicity of *Bacillus thuringiensis* var. *kurstaki* to aquatic insects, *Can. Entomol.,* 117, 829, 1985.

12. **Lacey, L. A. and Mulla, M. S.,** Evaluation of *Bacillus thuringiensis* as a biocide of blackfly larvae (Diptera:Simuliidae), *J. Invertebr. Pathol.,* 30, 46, 1977.

13. **Lacey, L. A., Mulla, M. S., and Dulmage, H. T.,** Some factors affecting the pathogenicity of *Bacillus thuringiensis* Berliner against blackflies, *Environ. Entomol.,* 7, 583, 1978.

14. **Eco-Analysts, Inc.,** The effects of *Bacillus thuringiensis* on Trichoptera, in *Environmental Monitoring Reports from the 1981 Maine Spruce Budworm Supression Project,* Maine Forest Service, Augusta, ME, 1981, 134.

15. **Feng, S. Y.,** Experimental bacterial infections in the oyster *Crassostrea virginica, J. Invertebr. Pathol.,* 8, 505, 1966.

16. **Moulinier, C.,** Study on the action of *Bacillus thuringiensis* var. *kurstaki* Serotype 3a,3b on the formation and growth of *Crassostrea gigas* larvae, Laboratory of Medical Parasitology and Mycology, University of Bordeaux, Bordeaux, France, 1986.

17. **Alzieu, C., deBarjac, H., and Maggi, P.,** Tolerance of marine fauna to *Bacillus thuringiensis,* Science et Pêche., *Bull. Inst. Pêches Marit. Maroc,* 250, 11, 1975.

18. **DeLucca, A., Simonson, J., and Larson, A.,** *Bacillus thuringiensis* distribution in soils of the United States, *Can. J. Microbiol.,* 27, 865, 1981.

19. **Saleh, S. M., Harris, R. F., and Allen, O. N.,** Fate of *Bacillus thuringiensis* in soil: effect of soil pH and organic amendment, *Can. J. Microbiol.,* 16, 677, 1970.

20. **Pruett, C., Burges, H., and Wyborn, C.,** Effect of exposure to soil on potency and spore viability of *Bacillus thuringiensis, J. Invertebr. Pathol.,* 35, 168, 1980.

21. **West, A. W.,** Fate of the insecticidal, proteinaceous parasporal crystal of *Bacillus thuringiensis* in soil, *Soil Biol. Biochem.,* 16, 357, 1984.

22. **Smirnoff, W. A. and Heimpel, A. M.,** Notes on the pathogenicity of *Bacillus thuringiensis* var. *thuringiensis* Berliner for the earthworm, *Lumbricus terrestris* Linnaeus, *J. Insect Pathol.,* 3, 403, 1961.

23. **Benz, G. and Altwegg, A.,** Safety of *Bacillus thuringiensis* for earthworms, *J. Invertebr. Pathol.,* 26, 125, 1975.

24. **Atlavinyte, O., Galvelis, A., Daciulyte, J., and Lagauskas, A.,** Effects of entorobacterin on earthworm activity, *Pedobiologia,* 23, 372, 1982.

25. **Brand, R., Pinnock, D., Jackson, K., and Milstead, J.,** Methods for assessing field persistence of *Bacillus thuringiensis* spores, *J. Invertebr. Pathol.,* 25, 199, 1975.

26. **Salama, H., Foda, M., Zaki, F., and Khalafallah, A.,** Persistence of *Bacillus thuringiensis* Berliner spores in cotton cultivations, *Z. Angew. Entomol.,* 95, 321, 1983.

27. **Pozsgay, M., Fast, P., Kaplan, H., and Carey, P.,** The effect of sunlight on the protein crystals from *Bacillus thuringiensis* var. *kurstaki* HD1 and NRD12: a Raman spectroscopic study, *J. Invertebr. Pathol.,* 50, 246, 1987.

28. **Yousten, A. A.,** Effect of the *Bacillus thuringiensis* δ-endotoxin on an insect predator which has consumed intoxicated cabbage looper larvae, *J. Invertebr. Pathol.,* 21, 312, 1973.

29. **Workman, R. B.,** Pesticides toxic to striped earwig, an important insect predator, *Proc. Fl. State Hortic. Soc.,* 90, 401, 1977.

30. **Hamed, A. R.,** Effects of *Bacillus thuringiensis* on parasites and predators of *Yponomeuta evonymellus* (Lep., Yponomeutidae), *Z. Angew. Entomol.,* 87, 294, 1978-79.

31. **Elsey, K. D.,** *Jalysus spinosus* effect of insecticide treatments on this predator of tobacco pests, *Environ. Entomol.,* 2, 240, 1973.

32. **Jensen, R.,** Comparison of various insecticides including DIPEL WP for control of the velvetbean caterpillar *(Anticarsia gemmatalis)* and the green cloverworm *(Plathypena scabra)* on soybeans and the effect on non-target species, Abbott Laboratories Exp. No. D911-1582, 1974.

33. **Harding, J., Wolfenbarger, D., Dupnik, T., and Fuchs, T.,** Large scale tests comparing *Bacillus thuringiensis* with methyl-parathion for cotton insect control, field test report, Abbott Laboratories, 1972.

34. **Wallner, W. and Surgeoner, G.,** Control of oakleaf caterpillar, *Heterocampa manteo,* and the impact of controls on non-target organisms, unpublished study, Abbott Laboratories, 1974.

35. **Salama, H. and Zaki, F.,** Interaction between *Bacillus thuringiensis* Berliner and the parasites and predators of *Spodoptera littoralis* in Egypt, *Z. Angew. Entomol.,* 94, 425, 1983.

36. **Salama, H., Zaki, F., and Sharaby, A.,** Effect of *Bacillus thuringiensis* Berl. on parasites and predators of the cotton leafworm *Spodoptera littoralis* (Boisd.), *Z. Angew. Entomol.,* 94, 498, 1982.

37. **Obadofin, A. A. and Finlayson, D. G.,** Interactions of several insecticides and a carabid predator (*Bembidion lampros* (Hrbst.)) and their effects on *Hylemya brassicae* (Bouché), *Can. J. Plant Sci.,* 57, 1121, 1977.

38. **Wilkinson, J. D., Biever, K. D., and Ignoffo, C. M.,** Contact toxicity of some chemical and biological pesticides to several insect parasitoids and predators, *Entomophaga,* 20, 113, 1975.

39. **Johnson, A.,** *Bacillus thuringiensis* and tobacco budworm control on flue-cured tobacco, *J. Econ. Entomol.,* 67, 755, 1974.

40. **Asquith, D.,** Response of the predaceous black lady beetle *Stethorus punctum* (LeConte), to apple orchard insecticide treatments, Abbott Laboratories Exp. No. D986-4007, 1975.

42. **Buckner, C. H., Kingsbury, P. D., McLeod, B. B., Mortenstern, K. L., and Roy, D. G. H.,** Impact of aerial treatment on non-target organisms, Algonquin Park, Ontario, and Spruce Woods, Manitoba. Section F, in Evaluation of Commercial Preparations of *Bacillus thuringiensis* with and without Chitinase Against Spruce Budworm, Information Report CC-X-59, Chemical Control Institute, 1974.

43. **Kazakova, S. and Dzhunusov, K.,** The effect of Bitoxibacillin-202 on certain orchard insects in the Issyk-kul' depression, *Rev. Appl. Entomol. Ser. A,* 65, 5987, 1977.

44. **Hassan, S. A.,** Results of the laboratory testing of a series of pesticides on egg parasites of the genus *Trichogramma* (Hymenoptera, Trichogrammatidae), *Nachrichtenbl. Detsch. Pflanzenschutzdienstes* (Braunschweig), 35, 21, 1983.

45. **Horn, D. J.,** Selective mortality of parasitoids and predators of *Myzus persicae* on collards treated with malathion, carbaryl, or *Bacillus thuringiensis, Entomol. Exp. Appl.,* 34, 208, 1983.

46. **Dunbar, D. M., Kaya, H. K., Doane, C. C., Anderson, J. F., and Weseloh, R. M.,** Aerial application of *Bacillus thuringiensis* against larvae of the elm spanworm and gypsy moth and effects on parasitoids of the gypsy moth, Bulletin 735, Connecticut Experiment, 1972.

47. **Fusco, R. A.,** Field evaluation of a commercial preparation of *Bacillus thuringiensis,* DIPEL 4L, progress report, Gypsy Moth Pest Management Methods Development Project, Pennsylvania Bureau of Forestry, 1980.

48. **Krieg, A. and Franz, J.,** Experiments on the control of wax moths with bacteria, *Naturwissenschaften,* 46, 22, 1959.

49. **Fisher, R. and Rosner, L.,** Toxicology of the microbial insecticide Thuricide, *J. Agric. Food Chem.,* 7, 686, 1959.

50. **Wilson, W.,** Observations on the effects of feeding large quantities of *Bacillus thuringiensis* Berliner to honey bees, *J. Insect Pathol.,* 4, 269, 1962.

51. **Martouret, D. and Euverte, G.,** The effect of *Bacillus thuringiensis* Berliner preparations on the honey bee under conditions of forced feeding, *J. Insect Pathol.,* 6, 198, 1964.

52. **Krieg, A. and Herfs, W.,** The effects of *Bacillus thuringiensis* on honey bees, *Entom. Exp. Appl.,* 6, 1, 1963.

53. **Cantwell, G. E., Knox, D. A., Lehnert, T., and Michael, A. S.,** Mortality of the honey bee, *Apis mellifera,* in colonies treated with certain biological insecticides, *J. Invertebr. Pathol.,* 8, 228, 1966.

54. **Krieg, A.,** About toxic effects of cultures of *Bacillus cereus* and *Bacillus thuringiensis* on honey bees *(Apis mellifera), Z. Pflanzenkr. Pflanzenschutz,* 80, 483, 1973.

55. **Krieg, A., Hassan, S., and Pinsdorf, W.,** Comparison of the effect of the variety *israelensis* with other varieties of *B. thuringiensis* on non-target organisms of the order Hymenoptera:*Trichogramma cacoeciae* and *Apis mellifera, Anz. Schaedlingsk. Pflanz. Umweltschutz,* 53, 81, 1980.

56. **Burges, H. D.,** Control of the wax moth *Galleria mellonella* on beecomb by H-serotype of *Bacillus thuringiensis* and the effect of chemical additives, *Apidologie,* 8, 155, 1977.

57. **Burges, H. D. and Bailey, L.,** Control of the greater and lesser wax moths (*Galleria mellonella* and *Achroia grisella*) with *Bacillus thuringiensis, J. Invertebr. Pathol.,* 11, 184, 1968.

58. **Cantwell, G. E. and Shieh, T. R.,** Certan—A new bacterial insecticide against the greater wax moth, *Am. Bee J.,* 121, 424, 1981.

59. **Burges, H. D.,** Leaching of *Bacillus thuringiensis* spores from foundation beeswax into honey and their subsequent survival, *J. Invertebr. Pathol.,* 28, 393, 1976.

60. **Hassan, S. and Krieg, A.,** *Bacillus thuringiensis* preparations harmless to the parasite *Trichogramma cacoeciae* (Hym.:Trichogrammatidae), *Z. Pflanzenkr. Pflanzenschutz,* 82, 515, 1975.

61. **Thoms, E. and Watson, T.,** Effect of Dipel (*Bacillus thuringiensis*) on the survival of immature and adult *Hyposoter exiguae* (Hymenoptera:Ichneumonidae), *J. Invertebr. Pathol.,* 47, 178, 1986.

62. **Muck, O., Hassan, S., Huger, A., and Krieg, A.,** Effects of *Bacillus thuringiensis* Berliner on the parasitic hymenopteran *Apanteles glomeratus* L. (Braconidae) and *Pimpla turionellae* (L.) (Ichneumonidae), *Z. Angew. Entomol.,* 92, 303, 1981.

63. **Dunbar, J. P. and Johnson, A. W.**, *Bacillus thuringiensis:* effects on the survival of a tobacco budworm parasitoid and predator in the laboratory, *Environ. Entomol.*, 4, 352, 1975.

64. **Temerak, S. A.**, Detrimental effects of rearing a braconid parasitoid on the pink borer larvae inoculated by different concentrations of the bacterium, *Bacillus thuringiensis* Berliner, *Z. Angew. Entomol.*, 89, 315, 1980.

65. **Weseloh, R. M. and Andreadis, T. G.**, Possible mechanism for synergism between *Bacillus thuringiensis* and the gypsy moth (Lepidoptera:Lymantriidae) parasitoid, *Apanteles melanoscelus* (Hymenoptera:Braconidae), *Ann. Entomol. Soc. Am.*, 75, 435, 1982.

66. **Wallner, W. E., DuBois, N. R., and Grinberg, P. S.**, Alteration of parasitism by *Rogas lymantriae* (Hymenoptera:Braconidae) in *Bacillus thuringiensis*-stressed gypsy moth (Lepidoptera:Lymantriidae) hosts, *J. Econ. Entomol.*, 76, 275, 1983.

67. **Weseloh, R. M., Andreadis, T. G., Moore, R. E. B., Anderson, J. P., DuBois, N. R., and Lewis, F. B.**, Field confirmation of a mechanism causing synergism between *Bacillus thuringiensis* and the gypsy moth parasitoid, *Apanteles melanoscelus*, *J. Invertebr. Pathol.*, 41, 99, 1983.

68. **Hamel, D. R.**, The effects of *Bacillus thuringiensis* on parasitoids of the western spruce budworm, *Choristoneura occidentalis* (Lepidoptera:Tortricidae), and the spruce coneworm, *Dioryctria reniculelloides* (Lepidoptera:Pyralidae), in Montana, *Can. Entomol.*, 109, 1409, 1977.

69. **Morris, O. N., Armstrong, J. A., and Hildebrand, M. J.**, Aerial field trials with a new formulation of *Bacillus thuringiensis* against the spruce budworm, *Choristoneura fumiferana* (Clem.), Report CC-X-144, Chemical Control Research Institute, Ottawa, Ontario, Can., 1977.

70. **Morris, O. N., Hildebrand, M. J., and Armstrong, J. A.**, Preliminary field studies on the use of additives to improve deposition rate and efficacy of commercial formulations of *Bacillus thuringiensis* applied against the spruce budworm, *Choristoneura fumiferana* (Lepidoptera:Tortricidae), Report FPM-X-32, Forest Pest Management Institute, Sault Ste. Marie, Ontario, 1980.

71. **Ahmad, S., O'Neill, J. R., Mague, D. L., and Nowalk, R. K.**, Toxicity of *Bacillus thuringiensis* to gypsy moth larvae parasitized by *Apanteles melanoscelus*, *Environ. Entomol.*, 7, 73, 1978.

72. **Kaya, H. K. and Dunbar, D. M.**, Effect of *Bacillus thuringiensis* and carbaryl on an elm spanworm egg parasite *Telonomus alsophilae*, *J. Econ. Entomol.*, 65, 1132, 1972.

73. **Biache, G.**, Effects of *Bacillus thuringiensis* on *Pimpla instigator* (Ichneumonidae-Pimplinae), *Ann. Soc. Entomol. Fr.*, 11, 609, 1975.

74. **Krieg, A.**, The effects of *Bacillus thuringiensis* preparations on spider mites (Tetranychidae), *Anz. Schaedlingskd. Pflanz. Umweltschutz*, 45, 169, 1972.

75. **Horsburgh, R. and Cobb, L.**, Effect of Dipel WP and Dipel + Guthion tank mix combination for control of variegated leafroller, tufted apple bud moth, and red-banded leaf roller larvae on apples in a full season and late season program in Virginia, Experiment Number D-986-4240, Abbott Laboratories, 1981.

76. **Weires, R. W. and Smith, G. L.**, Apple mite control, N.Y. St. Agric. Exp. Sta. Rep., Hudson Valley, NY, 1977.

77. **Bond, R. P. M., Boyce, C. B. C., and French, S. J.**, A purification and some properties of an insecticidal exotoxin from *Bacillus thuringiensis* Berliner, *Biochem. J.*, 114, 447, 1969.

78. **Farkaš, J., Šebesta, K., Horská, K., Samek, Z., Dolejš, L., and Sorm, F.**, Structure of thuringiensin, the thermostable exotoxin from *Bacillus thuringiensis, Coll. Czech. Chem. Commun.*, 42, 909, 1977.

79. **Cantwell, G. E., Heimpel, A. M., and Thompson, M. J.**, The production of an exotoxin by various crystal-forming bacteria related to *Bacillus thuringiensis* var. *thuringiensis* Berliner, *J. Insect Pathol.*, 6, 466, 1964.

80. **DeBarjac, H., Burgerjon, A., and Bonnefoi, A.**, The production of heat-stable toxin by nine serotypes of *Bacillus thuringiensis, J. Invertebr. Pathol.*, 8, 537, 1966.

81. **Krieg, A.**, A taxonomic study of *Bacillus thuringiensis* Berliner, *J. Invertebr. Pathol.*, 12, 366, 1968.

82. **Mohd-Salleh, M. B. and Beegle, C. C.**, Fermentation media and production of exotoxin by three varieties of *Bacillus thuringiensis, J. Invertebr. Pathol.*, 35, 75, 1980.

83. **Rogoff, M. H., Ignoffo, C. M., Singer, S., Gard, I., and Prieto, A. P.**, Insecticidal activity of thirty-one strains of *Bacillus* against five insect species, *J. Invertebr. Pathol.*, 14, 122, 1969.

84. **Šebesta, K., Horská, K., and Vaňkova, J.**, Inhibition of *de novo* RNA synthesis by the insecticidal exotoxin of *Bacillus thuringiensis* var. *gelechiae, Coll. Czech. Chem. Commun.*, 34, 1786, 1969.

85. **Šebesta, K. and Horská, K.**, Inhibition of DNA-dependent RNA polymerase by the exotoxin of *Bacillus thuringiensis* var. *gelechiae, Biochim. Biophys. Acta*, 169, 281, 1968.

86. **Šebesta, K. and Horská, K.**, Mechanism of inhibition of DNA-dependent RNA polymerase by exotoxin of *Bacillus thuringiensis, Biochim. Biophys. Acta*, 209, 357, 1970.

87. **Beebee, T., Korner, A., and Bond, R. P. M.**, Differential inhibition of mammalian ribonucleic acid polymerases by an exotoxin from *Bacillus thuringiensis, Biochem. J.*, 227, 619, 1972.

88. **Mackedonski, V. V., Nikolaev, N., Šebesta, K., and Hadjiolov, A. A.**, Inhibition of ribonucleic acid biosynthesis in mice liver by the exotoxin of *Bacillus thuringiensis, Biochim. Biophys. Acta*, 272, 56, 1972.

89. **Kim, Y. T., Gregory, B. G., and Ignoffo, C. M.,** The beta-exotoxins of *Bacillus thuringiensis.* III. Effects on *in vivo* synthesis of macromolecules in an insect system, *J. Invertebr. Pathol.,* 20, 46, 1972.

90. **Bond, R. P. M., Boyce, C. B. C., Rogoff, M. H., and Shieh, T. R.,** The thermostable exotoxin of *Bacillus thuringiensis, in Microbial Control of Insects and Mites,* Burges, H. D. and Hussey, N. W., Eds., Academic Press, London, 1971, 275.

91. **Faust, R. M.,** The *Bacillus thuringiensis* beta-exotoxin: current status, *Bull. Entomol. Soc. Am.,* 19, 153, 1973.

92. **Burrato, B. and Majors, K.,** Acute toxicity evaluation of ABG-6162A (beta-exotoxin formulation) in rabbits, Study number TE85-288, unpublished study for Abbott Laboratories, 1986.

93. **Majors, K. R. and Tekeli, S.,** Acute intravenous toxicity evaluation of thuringiensin (beta-exotoxin) in rats and rabbits, Study number T87-039, unpublished study for Abbott Laboratories, 1987.

94. **Carlberg, G.,** Biological effects of the thermostable beta-exotoxin produced by different serotypes of *Bacillus thuringiensis,* Thesis, University of Helsinki, 1973.

95. **Briggs, J. D.,** Reduction of adult house-fly emergence by the effects of *Bacillus* spp. on the development of immature forms, *J. Insect. Pathol.,* 2, 418, 1960.

96. **Dunn, P. H.,** Control of houseflies in bovine feces by a feed additive containing *Bacillus thuringiensis* var. *thuringiensis* Berliner, *J. Insect Pathol.,* 2, 13, 1960.

97. **Burgerjon, A., Grison, P., and Kachkouli, A.,** Activity of the heat-stable toxin of *Bacillus thuringiensis* Berliner in *Locusta migratoria* (Linnaeus) (Locustidae, Orthoptera), *J. Insect Pathol.,* 6, 381, 1966.

98. **Hitchings, D. L.,** *Bacillus thuringiensis,* a reproduction inhibitor for southern armyworm, *J. Econ. Entomol.,* 60, 596, 1967.

99. **Ignoffo, C. M. and Gregory, B.,** Effects of *Bacillus thuringiensis* beta-exotoxin on larval maturation, adult longevity, fecundity and egg viability in several species of Lepidoptera, *Environ. Entomol.,* 1, 269, 1972.

100. **Burges, H. D.,** Teratogenicity of the thermostable beta exotoxin of *Bacillus thuringiensis* in *Galleria mellonella, J. Invertebr. Pathol.,* 26, 419, 1975.

101. **Wolfenbarger, D. A., Guerra, A. A., Dulmage, H. T., and Garcia, R. D.,** Properties of the beta-exotoxin of *Bacillus thuringiensis* IMC 10,001 against the tobacco budworm, *J. Econ. Entomol.,* 65, 1245, 1972.

102. **Burgerjon, A., Biache, G., and Cals, P.,** Teratology of the Colorado potato beetle, *Leptinotarsa decemlineata,* as provoked by larval administration of the thermostable toxin of *Bacillus thuringiensis, J. Invertebr. Pathol.,* 14, 274, 1969.

103. **Vaňková, J.,** House-fly susceptability to *Bacillus thuringiensis* var. *israelensis* and a comparison with the activity of other insecticidal bacterial preparations, *Acta Entomol. Bohemoslov.,* 78, 358, 1981.

104. **Tremblay, F. L. J., Huot, L., and Perron, J. M.,** Penetration of the thermostable exotoxin of *Bacillus thuringiensis* into the egg chorion of *Acheta domesticus, Entomol. Exp. Appl.,* 15, 397, 1972.

105. **Ignoffo, C. M. and Gregory, B.,** Effects of *Bacillus thuringiensis* beta-exotoxin on larval maturation, adult longevity, fecundity and egg viability in several species of Lepidoptera, *Environ. Entomol.,* 1, 269, 1972.

106. **Majors, K. R. and Burrato, B.,** Acute oral toxicity evaluation of ABG-6146 (beta-exotoxin formulation) in mice and rats, Study number T83-437, unpublished study for Abbott Laboratories, 1984.

107. **Glaza, S. M.,** Oral LD50 in rats with thuringiensin tech methanol precipitate: 51571BD, Sample number 61002424. unpublished study by Hazelton Laboratories Americas, Inc. for Abbott Laboratories, 1986.

108. **Schmid, E. and Benz, G.,** Oral and parenteral toxicity of *Bacillus thuringiensis* 'exotoxin' and its inactivation in larvae of *Galleria mellonella, Experientia,* 25, 96, 1969.

109. **Sebesta, K., Horska, K., and Vankova, J.,** Isolation and properties of the insecticidal exotoxin of *Bacillus thuringiensis* var. *gelechiae* Auct., *Coll. Czech. Chem. Commun.,* 34, 891, 1969.

110. **Sebesta, K. and Horska, K.,** The fate of exotoxin from *Bacillus thuringiensis* in mice, *Coll. Czech. Chem. Commun.,* 38, 2533, 1973.

111. **Vaňková, J., Horská, K., and Šebesta, K.,** The fate of exotoxin of *Bacillus thuringiensis* in *Galleria mellonella* caterpillars, *J. Invertebr. Pathol.,* 23, 209, 1974.

112. **Beebee, T. J. C. and Bond, R. P. M.,** Effect of an exotoxin from *Bacillus thuringiensis* on deoxyribonucleic acid-dependent ribonucleic acid polymerase in nuclei from adult *Sarcophaga bullata, Biochem. J.,* 136, 9, 1973.

113. **Beebee, T. J. C. and Bond, R. P. M.,** Effect of the exotoxin of *Bacillus thuringiensis* on normal and ecdysone-stimulated ribonucleic acid polymerase activity in intact nuclei from the fat-body of *Sarcophaga bullata* larvae, *Biochem. J.,* 136, 1, 1973.

114. **Campbell, D. P., Dieball, D. E., and Brackett, J. M.,** Rapid HPLC assay for the beta-exotoxin of *Bacillus thuringiensis, J. Agric. Food Chem.,* 35, 156, 1987.

115. **Argauer, R. J., Cantwell, G. E., and Faust, R. M.,** A new heat-stable exotoxin produced by *Bacillus thuringiensis* subsp. *morrisoni,* in press, 1987.

116. **Forbis, A. D.**, Acute toxicity of A-56570 to *Daphnia magna*, Static Acute Bioassay Report 31098, unpublished study by Analytical Biochemistry Laboratories, Inc. for Abbott Laboratories, 1983.

117. **Dieball, D. E. and Jovanovich, A. P.**, A study of the hydrolysis of thuringiensin at pH 5, 7 and 9, Project Number 95-268-62, unpublished Study for Abbott Laboratories, 1987.

118. **Dieball, D. E.**, A study of adsorptive/desorptive properties of thuringiensin (beta-exotoxin, ABG-6162A) with four reference soils, Project Number 79-0030-62, unpublished study for Abbott Laboratories, 1985.

119. **Benz, G.**, On the chemical nature of the heat-stable exotoxin of *Bacillus thuringiensis*, *Experientia*, 22, 81, 1966.

120. **Pan-Agricultural Labs, Inc.**, Effect of two thuringiensin formulations on two-spotted spider mite and a predator on cotton, unpublished study for Abbott Laboratories, 1985.

121. **Schuster, M.**, Evaluation of thuringiensin (ABG-6162A) for control of tarnished plant bugs and cotton fleahoppers plus effect on beneficial arthropods on cotton in the Texas Blacklands Region, unpublished study for Abbott Laboratories, 1986.

122. **Collins, R.**, Control of a variety of insect groups and species with exotoxin, unpublished study for Abbott Laboratories, 1984.

123. **Grau, P.**, The effect of thuringiensin on the pest and beneficial arthropods in alfalfa, unpublished study for Abbott Laboratories, 1984.

124. **Sanford, K. H.**, Evaluation of ABG-6162A against the wintermoth, *Operophtera brumata* on apple in Canada, unpublished study for Abbott Laboratories, 1985.

125. **Tanigoshi, L.**, Laboratory and field tests on the effects of thuringiensin on adult and nymph *Lygus* bugs with additional data on nabid bugs and pea leaf weevil, unpublished study for Abbott Laboratories, 1985.

126. **Burgerjon, A. and Biache, G.**, Alimentation au laboratorie de *Perillus bioculatus* Fabr. avec des larves de *Leptinotarsa decemlineata* Say intoxiquées par la toxine thermostable de *Bacillus thuringiensis* Berliner, *Entomophaga*, 11, 279, 1966.

127. **Cantwell, G.**, Evaluation of thuringiensin (ABG-6162A) effects on parasites and predators of the Colorado potato beetle and the Mexican bean beetle in a greenhouse feeding study, unpublished study for Abbott Laboratories, 1984a.

128. **Cantwell, G., Knox, D., and Michael, A.**, Mortality of honey bees, *Apis mellifera* Linnaeus, fed exotoxin of *Bacillus thuringiensis* var. *thuringiensis* Berliner, *J. Insect Pathol.*, 6, 532, 1964.

129. **Cantwell, G. E., Knox, D. A., Lehnert, T., and Michael, A. S.**, Mortality of honey bee, *Apis mellifera*, in colonies treated with certain biological insecticides, *J. Invertebr. Pathol.*, 8, 228, 1966.

130. **Krieg, A.**, Toxic effect of *Bacillus cereus* and *Bacillus thuringiensis* cultures on the honey bee *(Apis mellifera)*, *Z. Planzenkr. Pflanzenschutz*, 80, 483, 1973.

131. **Vandenberg, J. and Shimanuki, H.**, Two commercial preparations of the beta-exotoxin of *Bacillus thuringiensis* influence the mortality of caged adult honey bees, *Apis mellifera* (Hymenoptera:Apidae), *Environ. Entomol.*, 15, 166, 1986.

132. **Atkins, L.**, A toxicity test with thuringiensin against honey bee worker adults, unpublished study for Abbott Laboratories, 1985.

133. **Mayer, D.**, The effect ot thuringiensin on adult honey bees *(Apis mellifera)* and leafcutter bees *(Megachile rotundata)* in a laboratory mortality bioassay, unpublished study for Abbott Laboratories, 1985a.

134. **Mayer, D.**, The effect of thuringiensin on honey bees *(Apis mellifera)*, alfalfa leafcutter bees *(Megachile rotundata)*, lygus bugs *(Lygus* spp.), pea aphids *(Acrythosiphon pisum)* and predators on blooming alfalfa, unpublished study for Abbott Laboratories, 1985b.

135. **Mayer, D.**, The effect of thuringiensin on adult and larval honey bees *(Apis mellifera)*, in pollen shedding sweet corn, unpublished study for Abbott Laboratories, 1985c.

136. **Cantwell, G.**, Evaluation of thuringiensin (ABG-6162A) and the CPB egg parasite, *Edovum puttleri*, for control of Colorado potato beetle on tomatoes, unpublished study for Abbott Laboratories, 1984b.

137. **Cantwell, G.**, Evaluation of thuringiensin (ABG-6162A) and the MBB larval parasite, *Pediobius foveolatus*, for control of Mexican bean beetle on snap beans, unpublished study for Abbott Laboratories, 1984c.

138. **Grannett, J.**, Effect of thuringiensin on the *Trioxys pallidus*, a hymenopterous parasite of the walnut aphid, *Chromaphis juglandicola*, unpublished Study for Abbott Laboratories, 1985.

139. **Purcell, M.**, Effect of thuringiensin on the walnut aphid, *Cromaphis juglandicola* and its parasite, *Trioxys pallidus*, unpublished study for Abbott Laboratories, 1985.

140. **Patterson, C., Potts, M., and Rodriguez, J. G.**, Laboratory evaluation of Abbott thuringiensin, unpublished study for Abbott Laboratories, 1986.

141. **Haverty, M. I.**, Sensitivity of selected non-target insects to the carrier of Dipel 4L in the laboratory, *Environ. Entomol.*, 11, 337, 1982.

Chapter 12

SAFETY OF *BACILLUS THURINGIENSIS* SSP. *ISRAELENSIS* AND *BACILLUS SPHAERICUS* TO NONTARGET ORGANISMS IN THE AQUATIC ENVIRONMENT

L. A. Lacey and M. S. Mulla

TABLE OF CONTENTS

I. INTRODUCTION

Mosquitoes and blackflies transmit the causal agents of some of the most debilitating diseases of mankind, including malaria, onchocerciasis, several arboviruses, lymphatic filariasis, and a diversity of other parasitic diseases of humans and other vertebrates. Selective control of the insect vectors would enable reduction of morbidity and mortality caused by these diseases with minimal impact on the environment. Prior to the discovery of highly efficacious and selective bacterial pathogens of vector insects, the impact on the aquatic environment and safety of available and potential microbial control candidates were reviewed by Singer[80] and other researchers at a workshop/symposium sponsored by the U.S. Environmental Protection Agency (EPA).[9] In addition to recommendations for the study of persistence and safety of candidate microbial control agents, the participants recommended continued search for and development of microorganisms for management of aquatic pests. Within the past 15 years, isolations of efficacious bacteria with larvicidal activities comparable to those of some of the conventional chemical insecticides have been made. Extensive field trials indicate that certain isolates of *Bacillus sphaericus* (mostly serotype 5a,5b) and *Bacillus thuringiensis* spp. *israelensis* (= serotype H-14) offer selective control of mosquito and blackfly larvae with little impact on nontarget organisms (NTOs).

The commercial development of *B. thuringiensis* spp. *israelensis* rapidly followed its discovery by Goldberg and Margalit.[30] Its efficacy against several species of pest and vector mosquitoes and blackflies is documented in various habitats worldwide.[26,40,45,46,59,86] The speed with which commercial products of this agent were registered in the U.S. was due to the high degree of selectivity of the toxin of this microorganism and the existing registration and safety record of its agricultural counterparts, strains of *B. thuringiensis* with selective activity toward Lepidoptera.

Although *B. sphaericus* has a considerably narrower host range, it also has good potential as a mosquito larvicide.[17,45,59,89] Several isolates have demonstrated outstanding larvicidal activity against a number of mosquito species, and, under certain circumstances, prolonged residual activity and recycling of the bacterium have been observed.[12,19,35,48,60,69,70] Recent isolation and favorable field testing of highly efficacious strains have renewed commercial interest in *B. sphaericus*.

Widespread public and regulatory agency acceptance of these two microbial control agents will be preceded by rigorous safety testing under a variety of environmental conditions. Although *B. thuringiensis* ssp. *israelensis* is registered for use in several countries and a variety of studies indicate a lack of deleterious effects on the vast majority of NTOs, its use, particularly in lotic (running water) habitats, continues to generate a certain amount of controversy and some opposition. The major concerns are its safety to humans, beneficial NTOs, and its overall effect on the food web, particularly to fisheries. *Bacillus sphaericus* has not yet been registered for use.

In this chapter we will review the available data on short- and long-term effects of both of these microbial control agents on NTOs in aquatic habitats.

II. IMPACT OF *B. THURINGIENSIS* SSP. *ISRAELENSIS* ON NONTARGET INVERTEBRATE ORGANISMS IN THE LOTIC HABITATS

Of the two microbial control agents reviewed in this chapter, only *B. thuringiensis* ssp. *israelensis* exhibits larvicidal activity against blackflies, hence it is the only one used extensively in lotic habitats. Although *B. sphaericus* is active against mosquito species that are found at the margins of streams (e.g., *Anopheles minimus*, and others), no exhaustive studies on the effects of this bacterium on NTOs in the lotic environment have yet been conducted.

A fortuitous association of biotic factors govern the selectivity of *B. thuringiensis* ssp. *israelensis* for blackfly larvae. The combination of filter feeding behavior of the target simuliid larvae, their relatively high gut pH, and the proper complement of proteolytic enzymes enables capture, dissolution and activation of the toxin-containing parasporal inclusion bodies. The absence of any of these factors greatly reduces or nullifies the susceptibility of a given species to the deleterious effects of *B. thuringiensis* ssp. *israelensis*. For example, filter feeding Ephemeroptera, Trichoptera, and others may actually ingest the toxin, but without subsequent activation it remains harmless to these organisms. On the other hand, certain lotic Nematocera may be intrinsically susceptible to the bacterial toxin, but insufficient amounts of the toxin may be ingested due to the mode of feeding (e.g., predator, scraper, gatherer, etc.).

A. METHODS FOR MEASURING THE IMPACT ON LOTIC NTOS
1. Laboratory Studies
A variety of bioassay techniques have been developed for the laboratory evaluation of larvicidal activity of insecticides used for blackfly control, including *B. thuringiensis* ssp. *israelensis*. These have been reviewed by Lacey et al.[47] and Walsh.[84] Unfortunately, their utility for determining the acute impact of the same larvicides on NTOs is somewhat limited. Laboratory evaluations of the effect of this microorganism on nontarget fauna from lotic ecosystems are only useful if the bioassay procedure enables feeding rates and behavior comparable to those observed in the natural habitat. For example, data generated in assays of *B. thuringiensis* ssp. *israelensis* against lotic organisms conducted in still water at abnormally low temperatures are of limited value.[85] Even a relatively simple bioassay system that permits attachment and normal feeding can generate useful information on susceptibility for certain functional groups. For example, bioassay systems in which current is created with air bubbles,[44] magnetic stirrers[15,32] or spinning bottles[33,47] may be adequate for species that attach to relatively smooth surfaces and feed within the boundary layer. However, these systems may not provide the diversity of conditions found in natural streams that would enable bioassay of *B. thuringiensis* ssp. *israelensis* against a wide range of NTOs. The more sophisticated recirculating trough systems used by Gaugler et al.[27] and Kurtak et al.[38] could, with the inclusion of natural substrates, better simulate diversified field conditions.

2. Field Studies
Unidirectional flow-through systems, or mini-gutters, operated at streamside using pumped or gravity-fed stream water provide a realistic simulation of the habitat with the convenience of small size, rapid set-up, and repeatability.[31,42,87] Small trough systems containing natural substrates anchored in the stream or river that is being treated offer several of the advantages of the mini-gutter system, excepting repeatability, under more natural conditions. Netting placed at the downstream end of the troughs allows precise measurement of drift and mortality.[21,29,81]

The most accurate assessment of the impact of insecticides on NTOs, especially the effects of repeated applications, are conducted *in situ*. Several sampling techniques have been employed for the censusing of invertebrate populations before and after treatment with insecticides.[83] These include sampling of drift, use of colonizable artificial substrates, use of the Surber sampler and other devices, counts from natural substrates, and lab rearing/holding field-exposed organisms. Many early field trials of *B. thuringiensis* ssp. *israelensis* against blackfly larvae also included preliminary assessment of possible negative impact on nontarget fauna. Studies conducted in various stream types in both tropical and temperate climates demonstrated little or no noticeable effect on several taxa of lotic insects (literature or specific products reviewed by Cibulsky and Fusco,[14] Gaugler and Finney,[26] and Knutti and Beck[36]).

B. SHORT-TERM IMPACT ON LOTIC NTOS

Table 1 lists the results of several evaluations of the short-term impact of *B. thuringiensis* ssp. *israelensis* on lotic NTOs. In most cases, few adverse effects are apparent for the majority of the nontarget fauna being monitored. In some instances an increase in NTOs was observed after application of *B. thuringiensis* ssp. *israelensis*[28,56] (also cited by D. Molloy, personal communication). The apparent increase may simply be due to reduced competition for space through removal of the target simuliid species enabling expanded utilization of substrate not previously available. One of the more detailed assessments of short-term impact to date revealed neither any increase in drift of 27 nontarget families, nor significant reduction in the population densities of 75 nontarget genera, after treatment of a third-order, cold water trout stream with operational dosages of the Vectobac®-AS formulation. (D. Molloy, personal communication).

Some lotic larval Nematocera, however, are adversely affected by exposure to *B. thuringiensis* ssp. *israelensis*. Those most affected are in the superfamily Culicoidea which comprises the Culicidae, Simuliidae, Chironomidae, Blepharoceridae, and several other families found in both lotic and lenitic habitats. Increased drift and/or decrease in population density of certain nontarget members of this superfamily have resulted from treatment with operational dosages of *B. thuringiensis* ssp. *israelensis*. The Chironomidae appear to be the most commonly affected family of NTOs.[8,11,20,21,29,57,72,76,82,88] In addition to effects on some chironomids, Back et al.[8] reported increased mortality and drift in blepharocerid larvae that were exposed to high concentrations of Teknar®. No adverse effects of the Vectobac-AS formulation on chironomid nor blepharocerid larvae were observed when it was applied at a lower rate (Table 1).[28]

Increased drift has also been observed for other taxa besides the Nematocera. Significant increase in the drift of Ephemeroptera and Trichoptera shortly after application of *B. thuringiensis* ssp. *israelensis* has been reported by Gibon et al.,[29] Troubat et al.,[82] Pistrang and Burger,[72] Dejoux et al.,[21] and de Moor and Car.[57] Considering the lack of demonstrated susceptibility of these taxa to *B. thuringiensis* ssp. *israelensis* toxin, it is possible that an increase in drift rates may be due to the temporary increase in particulates during treatment periods, especially immediately downstream of application points. Adjuvants may also contribute to the irritability and subsequent drift of benthic NTOs. Dejoux (personal communication) attributed the adverse effects of Teknar on certain caddisflies in treated streams of Ivory Coast to the xylene component of that formulation. Although increased drift may not always be indicative of direct lethal effect of the bacterial toxin, any unnatural augmentation in drifting behavior will place the drifting organism at greater risk of predation. Dejoux et al.[21] reported that the increase in drift following application of chemical insecticides was far greater than that induced by *B. thuringiensis* ssp. *israelensis* formulations.

Mortality in non-nematoceran NTOs following treatment with formulations of *B. thuringiensis* ssp. *israelensis* is less frequently reported than increased drift. The high level of mortality in Ephemeroptera reported by Car and de Moor[11] was due, at least in part, to stressful handling. The same authors reported an unexplained decrease in the population density of the snail, *Burnupia* sp., following treatment with Teknar. Other studies of the effect of *B. thuringiensis* ssp. *israelensis* on other lentic and nonriverine Mollusca have not demonstrated any adverse effects.[25,58]

One of the greatest concerns of regulatory agencies and the public in general is the effect that *B. thuringiensis* ssp. *israelensis* formulations might have on fish. To date, concentrations of formulations based on this microorganism and used for blackfly control have not been reported to harm fish or other vertebrates. Extraordinarily high concentrations of the Teknar formulation, however, have produced deleterious effects in brook trout (*Salvelinus fontinalis;*, but Fortin et al.[23] conclude that these are not caused by the endotoxin but by xylene, a constituent of the Teknar WDC formulation. Gibbs et al.[28] studied the effects of

TABLE 1
Short-Term Effects of *Bacillus thuringiensis* ssp. *israelensis* on Lotic Nontarget Organisms under Natural Conditions

Major taxa of predominant NTOs	Formulation/ concentration	Sampling method	Impact on NTOs	Location	Ref.
Ephemeroptera, Trichoptera, Lepidoptera, Odonata, Mollusca, Hirudinea	Primary powder (0.2 mg/1/10 min)	Drift samples in stream trough	No adverse effect	Ivory Coast	20
Trichoptera, Plecoptera, Coleoptera, Ephemeroptera, Odonata	Aqueous suspension (10^5 spores per ml/1 min)	Counts from natural substrates	No adverse effect	Newfoundland, Canada	16
Chironomidae, Ephemeroptera, Trichoptera	Teknar® WDC (1.6 mg/1/10 min)	Drift samples, in stream trough	Increased drift, some reduction of chironomids	Ivory Coast	29
Chironomidae, Trichoptera, Ephemeroptera, Plecoptera, Elmidae	Primary powder (0.5 mg/1/15 min)	Surber sampler, counts from natural substrates	No adverse effect	New York	56
Chironomidae, Ephemeroptera, Trichoptera	Teknar® WDC (1.5 mg/1/10 min)	Artificial substrates	No adverse effect	Ivory Coast	42
Chironomidae, Ephemeroptera, Trichoptera	Teknar® WDC (1.6 mg/1/10 min)	Drift samples in stream trough	Increased drift, some reduction of chironomids	Ivory Coast	82
Ephemeroptera, Trichoptera, Plecoptera, Coleoptera, Chironomidae, Dixidae	Teknar® WDC (2 mg/1/15 min) or primary powder (0.2 mg/1/15 min)	Surber samples	No adverse effect	New Zealand	13
Chironomidae	Teknar® WDC (1.6 mg/1/10 min) or primary powder (3.0 mg/1/10 min)	Counts from natural substrates	No adverse effect	South Africa	10
Chironomidae, Ephemeroptera, Plecoptera, Trichoptera, Gastropoda, Platyhelminthes	Teknar® WDC (1.6 mg/1/10 min or 2.3 mg/ 1/7 min)	Drift samples, counts from natural substrates	Mortality in Ephemeroptera, some reduction of chironomids (Tanytarsini), reduction in *Burnupia* (Gastropoda)	South Africa	11
Ephemeroptera, Trichoptera, Plecoptera, Chironomidae	Teknar® WDC (10 mg/1/10 min)	Drift samples, "kick" samples	Some reduction of chironomid larvae. Increase in drifting by Ephemeroptera (2 spp.), Trichoptera (2 spp.)	New Hampshire	72

TABLE 1 (continued)
Short-Term Effects of *Bacillus thuringiensis* ssp. *israelensis* on Lotic Nontarget
Organisms under Natural Conditions

Major taxa of predominant NTOs	Formulation/ concentration	Sampling method	Impact on NTOs	Location	Ref.
Chironomidae	0.7—2.0 mg/l operational dosage (application time not given)	Counts on natural and artificial substrates	Some reduction of chironomids at 17 × operational dosage	Federal Republic of Germany	76
Chironomidae, Blepharoceridae, Trichoptera, Plecoptera, Ephemeroptera	Teknar® WDC (5.86 mg/1/15 min)	Artificial substrates and drift nets	Increased drift in Blepharoceridae, reduction in two genera of chironomids, other taxa not adversely affected	Quebec	8
Chironomidae and other Nematocera, Ephemeroptera, Lepidoptera (Pyralidae), Trichoptera	Teknar® WDC (1.6 mg/1/10 min) or primary powder (0.2 mg/1/10 min)	Drift samples, in stream troughs and Surber samples	Increased drift, some reduction of Hydropsychidae (Trichoptera)	Ivory Coast	21
Ephemeroptera, Plecoptera, Trichoptera, Chironomidae, and other Nematocera, Coleoptera	Vectobac®-AS (10 mg/1/1-5 min)	Artificial substrates and drift nets	No significant adverse effects	Maine	28
Chironomidae, Ephemeroptera, Trichoptera	Teknar® WDC (1.6 mg/1/10 min)	Drift samples, counts on natural substrates	Slight increase in drift, some reduction of chironomids (Tanytarsini)	South Africa	57
Coleoptera, Diptera, Ephemeroptera, Pleocoptera, Trichoptera	Vectobac®-AS (1 mg/1/10—30 min)	Drift nets, counts from natural substrates, "portable invertebrate sampling box"	No adverse effect	New York	personal communication

operational dosages (10 mg/l for 1 to 5 min) of Vectobac-AS on the feeding behavior of *S. fontinalis* and slimy sculpins, *Cottus cognatus*. Although there was a slight increase in the ingestion of blackfly larvae in posttreatment samples (including check streams), it was not possible to attribute any change to the Vectobac treatments.

C. LONG-TERM IMPACT ON LOTIC ECOSYSTEMS

Monitoring of long-term impact of *B. thuringiensis* ssp. *israelensis* on lotic ecosystems will enable determination of chronic effects of the microbial toxin on susceptible NTOs as well as its effect on the structure of the food web. Any intervention that disrupts or restructures the aquatic community could possibly have deleterious effects on higher trophic levels. The more diversified the food web (in terms of number of organisms and species at each trophic level), the less likely is a major catastrophic effect due to complete or partial removal of a given species or species group (e.g., simuliids or certain species of the Chironomidae).

Molloy (personal communication) reports no long-term deleterious effect on the nontarget insect community in a third-order stream in the Adirondack Mountains (New York) after 2 years of continuous blackfly control using *B. thuringiensis* ssp. *israelensis* exclusively. The largest ongoing blackfly control effort, the Onchocerciasis Control Programme (OCP) in West Africa's Volta River Basin, actively monitors the impact of larvicides on a wide range of NTOs.[22,37] As the long-term impact of *B. thuringiensis* ssp. *israelensis* on the aquatic ecosystems of the OCP is reported in another chapter of this book, all reference to this particular topic has been avoided in the present chapter.

III. IMPACT OF *B. THURINGIENSIS* SSP. *ISRAELENSIS* AND *B. SPHAERICUS* ON NTOS IN LENITIC HABITATS

The number of studies conducted on the impact of these two bacteria in lenitic habitats far outnumbers those in lotic habitats. In the early evaluations of *B. thuringiensis* ssp. *israelensis* against NTOs, it became evident that aside from some effects on chironomids and certain other species of the Culicoidea, the preponderance of NTOs remains unaffected.[24,25,49,51,54,55,64,65,79,86] Here we will focus on the effects of *B. thuringiensis* ssp. *israelensis* and *B. sphaericus* on selected groups of lenitic NTOs representing various trophic levels in the food chain.

A. ODONATA

Immature odonates, or dragonfly (Anisoptera) and damselfly (Zygoptera) naiads, are aquatic in nature and are found abundantly in many permanent and semipermanent larval mosquito habitats. The naiads are predacious on mosquito larvae and a variety of other macroinvertebrates. For example, naiads of the dragonfly *Tarnetrum corruptum* consumed about 36 fourth instar mosquito larvae per individual in 4 d in laboratory tests.[6] Similar rates of consumption were exhibited by naiads of the damselfly *Enallagma civile*, consuming about 28 mosquito larvae per individual per 4 d.

Mosquito larvae that were intoxicated with extremely high dosages of *B. thuringiensis* and *B. sphaericus*, were offered to *T. corruptum* and *E. civile*. The prey larvae had been exposed to 1 gm of bacterial preparations per l (1000-fold maximum larvicidal rate) and offered to the naiads soon after signs of intoxication were apparent. The preparations of pathogens used in these tests were an aqueous suspension (AS) of *B. thuringiensis* ssp. *israelensis* (ABG-6145, 600 ITU/mg; Abbott Laboratories, North Chicago, IL) and the AS of *B. sphaericus* strain 2362 (BSP-1, 2×10^7 spores per ml; Solvay and Co., Brussels). In these experiments the naiads ingested measurable amounts of each toxin, partly dissolved and activated in the prey's intestine. They also ingested large numbers of resting and germinating spores present in the intestinal tract of mosquito larvae.[5,7,18,35,70] There was no significant difference in the rate of consumption of untreated mosquito larvae and larvae treated with *B. thuringiensis* ssp. *israelensis* and *B. sphaericus*. The duration of development of the dragonfly and damselfly naiads, from the time of exposure to emergence, was essentially the same in the checks and treated individuals.[6] Sebastien and Brust[78] also observed no adverse effects of *B. thuringiensis* ssp. *israelensis* on odonates.

Field trials of both agents against larval mosquito populations showed similar trends for dragonfly and damselfly naiads in treated and check mesocosms.[59,60,62,64,65] In another study in a wetland marsh, application of *B. thuringiensis* ssp. *israelensis* at a rate of 0.8 kg/ha caused no noticeable reduction in the nymphs of several species of Anisoptera and Zygoptera.[68] From the data obtained in laboratory and field tests, it is evident that both microbial agents applied at larvicidal or even much higher rates have no noticeable adverse effects on odonate naiads.

B. EPHEMEROPTERA

Ephemeroptera are among the most abundant and early invaders of freshly flooded

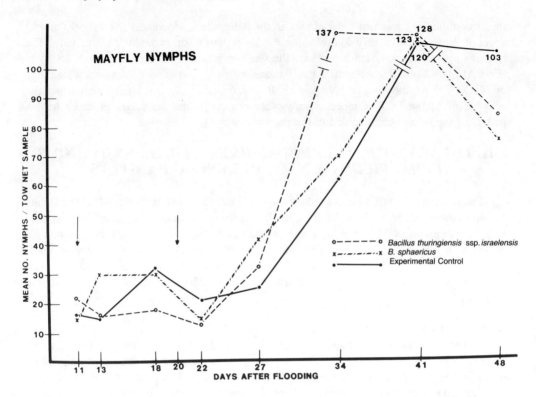

FIGURE 1. Population trends of mayfly nymphs in untreated experimental larval mosquito pond habitats and those treated at larvicidal rates with *Bacillus thuringiensis* ssp. *israelensis* and *B. sphaericus* as dictated by the density of mosquito larvae. Arrows indicate application of treatments.

biotopes of mosquito larvae. They are primarily herbivores feeding on algae and other plant material and play an important role in the ecology of permanently and temporarily flooded areas. As they frequently play a key role in primary production, it is important that these organisms are not adversely affected by interventions used in vector control programs.

In the course of field trials of *B. thuringiensis* ssp. *israelensis* and *B. sphaericus* in California against mosquito larvae, extensive quantitative observations were made on their effects on mayfly naiads, mostly *Callibaetis pacificus*.[62-65] In another study on the effects of *B. thuringiensis* ssp. *israelensis* (Vectobac, Abbott Laboratories) and *B. sphaericus* 2362 (primary powder, Abbott Laboratories), using larvicidal rates of 0.56 and 0.22 kg/ha, respectively, no noticeable effects were recorded for the mayfly naiad, *C. pacificus* (Mulla, unpublished data). The population trends subsequent to flooding of dry ponds were similar in treated and untreated ponds (Figure 1). Two applications of both entomopathogens were required for mosquito control at a time when naiad mayfly populations were low in all treated and check ponds. By the 27th day postflooding, the naiad population began to increase; it reached peak population 34 to 42 d postflooding.

In a study in Florida where *B. thuringiensis* ssp. *israelensis* (ABG-6108) was tested against larval chironomids in the field, there were no adverse effects on *Baetis* sp. at the highest dosage of 10 kg/ha.[2] This dosage is approximately 20-fold the larvicidal dosage for mosquitoes. From the field data obtained in subtropical regions in the U.S., (from California and Florida) it is evident that *B. thuringiensis* ssp. *israelensis* and *B. sphaericus* had no noticeable effects on mayfly species encountered at even much higher rates of application than necessary for mosquito control. Similar results are reported from more temperate climes. In a wetland marsh in California, a *B. thuringiensis* ssp. *israelensis* formulation applied at

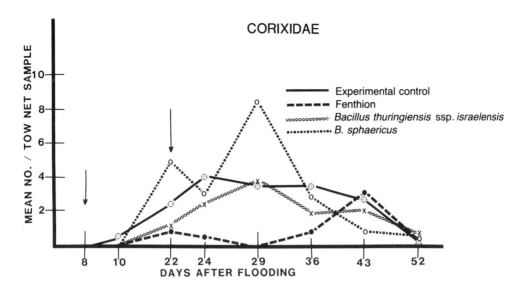

FIGURE 2. Population trends of corixids (mostly *Corisella decolor*) in larval mosquito habitats treated with 0.56 kg/ha of *Bacillus thuringiensis* spp. *israelensis* and 0.25 kg/ha of *B. sphaericus*. Arrows indicate application of treatments.

the larvicidal rate of 0.8 kg/ha did not affect *Callibaetis* adversely.[68] The ephemeropteran *Cloeon* sp. found in larval habitats of *Aedes vexans* in the Federal Republic of Germany was not impacted by *B. thuringiensis* ssp. *israelensis* treatments and it suffered no mortality in laboratory tests at rates as high as 180 ppm.[77]

C. HETEROPTERA

Members of the families Corixidae and Notonectidae are quite common in mosquito larval habitats. Corixids are omnivores feeding primarily on detritus, plant material, and small aquatic organisms including small mosquito larvae. Notonectids, on the other hand, are exclusively carnivorous, feeding on a variety of macroinvertebrates including mosquito larvae.

1. Corixidae

In a field study in California, corixids (*Corisella decolor* and other species) were found to be unaffected by mosquito larvicidal rates of *B. thuringiensis* ssp. *israelensis* (0.56 kg/ ha) and *B. sphaericus* (0.25 kg/ha) when two treatments were applied to mosquito larval habitats (Figure 2). Corixid populations were low and essentially displayed similar trends in treated and untreated mesocosms (Mulla, unpublished data). Similarly low population levels of *Corisella* spp. prevailed before and after treatment with 0.8 kg/ha of a *B. thuringiensis* ssp. *israelensis* formulation.[68] Corixids collected from mosquito larval habitats on the upper Rhine were not affected either after exposure to 180 ppm of *B. thuringiensis* ssp. *israelensis* for 48 h.[77]

2. Notonectidae (Laboratory)

To assess the impact of *B. thuringiensis* ssp. *israelensis* and *B. sphaericus* on notonectids, Aly and Mulla[6] fed *Culex quinquefasciatus* larvae intoxicated with extremely high dosages of the pathogens to field-collected fourth to fifth instar *Notonecta undulata*. The larvae intoxicated with *B. thuringiensis* ssp. *israelensis* (ABG-6145 600 ITU/mg Abbott Laboratories) and *B. sphaericus* (BSP-2 strain 2362, 2 × 10^7 spores per ml; Solvay and Co.) were fed at the rate of ten larvae per predator per day for 4 d. The predators consumed the majority of the offered prey. Thereafter they were fed unintoxicated larvae and observed for 15 to 17 d.

Adult notonectids that were offered intoxicated mosquito larvae and observed for 11 d, showed no mortality due to consumption of prey positive for *B. thuringiensis* ssp. *israelensis* or *B. sphaericus*. Similarly, nymphs offered intoxicated prey for 4 d and observed for 17 d suffered no mortality. Developmental rates, oviposition, and egg viability were the same in treated and untreated populations.[6]

Studies conducted by Olejníček and Maryskova[71] showed that mosquito larvae intoxicated by *B. thuringiensis* ssp. *israelensis* caused no marked mortality in *Notonecta glauca* to which they were fed. In laboratory experiments using field-collected notonectids, 10% mortality occurred in checks, while 11% mortality occurred in the nymphs consuming intoxicated larvae. Schnetter et al.[77] found no mortality in *N. glauca* exposed for 48 h to 180 ppm of *B. thuringiensis* ssp. *israelensis* without offering intoxicated prey.

Effects of two strains of *B. sphaericus* (1593M and 2362) were assessed on field populations of *Anisops bouvieri* brought into the laboratory, and the LC_{50} was found to be in excess of 500-fold the larvicidal rates of the two strains.[52]

3. Notonectidae (Field)

Adverse effects of *B. thuringiensis* ssp. *israelensis* and *B. sphaericus* have not been noted on notonectids in several studies by Mulla[60,62] and Mulla et al.[63-65] In field trials of *B. thuringiensis* ssp. *israelensis* and *B. sphaericus* conducted during three seasons, natural populations of *Notonecta unifasciata, N. undulata,* and *Buenoa* sp. were low during the treatments. Thereafter, though, their populations grew rapidly following similar trends in check and untreated ponds.[62]

In a field study conducted in Florida, Purcell[73] noted reduction in population of *N. indica* and *Buenoa elegans* after application of *B. thuringiensis* ssp. *israelensis* but attributed this to flying activity of these bugs. In these studies 1-d posttreatment counts were compared with pretreatment numbers without concomitant censusing of untreated check populations. *Notonecta* spp. prevailed in low numbers in a wetland marsh in California. They were not adversely affected by the application of 0.8 kg/ha of a formulation of *B. thuringiensis* ssp. *israelensis*,[68] and in fact their populations increased posttreatment. In simulated field habitats, Sebastien and Brust[78] observed about 50% reduction in *Notonecta* species exposed to 12 ppm of an Abbott formulation of *B. thuringiensis* ssp. *israelensis,* but a reduction was not observed after treatment with a Biochem formulation, notwithstanding that the potency of the latter was five- to tenfold the former. In the absence of an experimental control population and because of the discrepancy between formulations, the claim that one formulation killed *Notonecta* is not entirely supportable. The authors concluded that the notonectid adults in their study are quite mobile; they tend to migrate when prey density decreases or the habitat becomes unsuitable.

D. COLEOPTERA

Members of the families Dytiscidae (diving beetles) and Hydrophylidae (water scavenger beetles) are important predators and key regulators of mosquito larvae. They invade and occupy an aquatic habitat within 1 to 2 weeks after flooding. The young and mature larvae, which reach peak populations in 2 to 3 weeks postflooding, feed on mosquito larvae and other aquatic organisms.

During studies on the evaluation of *B. thuringiensis* ssp. *israelensis* and *B. sphaericus,* no adverse effects were noted on larvae of these beetles.[60-65] Further detailed studies over a three-season period showed that *B. thuringiensis* ssp. *israelensis* and *B. sphaericus* treatment at mosquito larvicidal rates had no noticeable effects on larvae of these two families.[62]

Substantial numbers of beetle larvae prevailed during the time of the second application for mosquito control. Population trends in check and treated populations were essentially similar. Another field study in California clearly showed the lack of adverse effects on beetle larvae by *B. thuringiensis* ssp. *israelensis* after application of 0.8 kg/ha in a wildlife marsh.[68] Laboratory exposure of several species of field-collected aquatic beetles to 180 ppm of *B. thuringiensis* ssp. *israelensis* caused little or no mortality in adult beetles.[77] Based on the available data we conclude that both entomopathogens have no noticeable adverse impact on dytiscid or hydrophilid beetle larvae at mosquito larvicidal rates.

E. DIPTERA/CHIRONOMIDAE

Chironomid midges are closely related to mosquitoes and are some of the most abundant and ubiquitous aquatic Diptera breeding in lotic and lenitic situations. Their larvae are found in the benthos as well as on algae, rooted, and floating plants and other substrates. The larvae of some species are filter feeders, while others are browsers and predators. Midges utilize temporary as well as permanent bodies of water; most mosquito larval habitats support ample populations of them. When emerging in large numbers from habitats near human populations they may constitute a nuisance problem,[1] but by and large they are beneficial, playing an important role in the food web. Therefore, in vector control programs where midges are not a pest problem attempts are made to administer larvicides that are innocuous to midge larvae.

Studies on the effects of *B. thuringiensis* ssp. *israelensis* and *B. sphaericus* on chironomid midges have been somewhat limited. Since the feeding behavior (e.g., filter feeders, browsers, predators, etc.) varies from one group or species to another, it is expected that the various groups will have entirely different levels of susceptibility, if any, to the two microbial entomopathogens.

1. Laboratory Studies (*B. thuringiensis* ssp. *israelensis*)

Research to date indicates that some chironomid midge larvae demonstrate low susceptibility to *B. thuringiensis* ssp. *israelensis*, while none have been found to be affected by *B. sphaericus*. Ali[2] and Ali et al.[3] conducted studies on larvae of four chironomine species (*Glyptotendipes paripes*, *Chironomus crassicaudatus*, *C. decorus*, and *Tanytarsus* spp.) in the laboratory in Florida. Test methods employed were those reported by Mulla and Khasawinah.[66] Four preparations of *B. thuringiensis* ssp. *israelensis* were tested against these midges with and without provision of larval food. The susceptibility of the various species to several preparations of *B. thuringiensis* ssp. *israelensis* is shown in Table 2, along with LC$_{90}$ values in mg/l (ppm) given for *Aedes aegypti*. From this table it is quite clear that midge larvae were less susceptible to *B. thuringiensis* ssp. *israelensis*, being 13- to 75-fold more tolerant than mosquito larvae to the various preparations. Provision of food to the larvae increased their tolerance, a situation closely simulating field conditions. It should be noted that 1st instar larvae of *G. paripes* were 10- to 14-fold more susceptible than the third instar larvae. These research data clearly documented the low level of susceptibility of midge larvae (especially third instars) to *B. thuringiensis* ssp. *israelensis* formulations. Using field-collected larvae of *Chironomus thummi*, Schnetter et al.[77] observed complete mortality after exposure to 1.8 ppm of *B. thuringiensis* ssp. *israelensis* for 48 h without food. The lethal dose was approximately five- to tenfold the mosquito larvicidal concentrations in that area.

In a preliminary study conducted in California in tubs placed in screen houses, Garcia et al.[24] induced low to high levels of mortality in some nematocerous Diptera, including a *Dixa* sp. (Dixidae), *Palpomyia* sp. (Ceratopogonidae), several taxa of Chironomidae (genera and species not given) at dosages of *B. thuringiensis* ssp. *israelensis* preparations which were fifty- to several hundredfold greater than concentrations used for mosquito control. No dosage-response lines were produced, nor was mortality among experimental controls pro-

TABLE 2
Susceptibility of Chironomid Midges and Mosquito
Larvae to Preparations of *B. thuringiensis* ssp.
israelensis

Species and microbial preparation	Instar	No food	With food
Glyptotendipes paripes (midge)			
IPS-78 (LY)	3rd	13.1	29.7
R-153-78 (WP)	1st	—	1.3
	3rd	9.8	13.3
ABG-6108 (WP)	1st	—	2.8
	3rd	32.4	38.6
SAN-402 (WDC)	1st	—	3.7
	3rd	23.6	37.7
Chironomus crassicaudatus (midge)			
IPS-78 (LY)	3rd	10.5	15.9
R-153-78 (WP)	3rd	4.9	6.8
ABG-6108 (WP)	3rd	47.0	—
SAN-402 (WDC)	3rd	28.3	36.5
C. decorus (midge)			
IPS-78 (LY)	3rd	8.6	10.8
R-153-78 (WP)	3rd	4.6	8.2
ABG-6108 (WP)	3rd	30.8	34.7
SAN-402 (WDC)	3rd	27.0	—
Aedes aegypti (mosquito)[a]			
IPS-78 (LY)	3rd	0.19	0.51
R-153-78 (WP)	3rd	0.13	0.32
ABG-6108 (WP)	3rd	0.88	2.01
SAN-402 (WDC)	3rd	0.99	2.39

[a]   24-h lethal concentration.

From Ali, A., Baggs, R. A., and Stewart, J. P., *J. Econ. Entomol.*, 74, 672, 1981. With permission.

vided. This study indicates that certain *B. thuringiensis* ssp. *israelensis* preparations can induce mortality in some nematoceran species relatively closely related to mosquitoes, but not in other taxa of NTOs.

2. Field Studies (*B. thuringiensis* ssp. *israelensis*)

Ali[2] evaluated *B. thuringiensis* ssp. *israelensis* (ABG-6108, 1000 ITU/mg; Abbott Laboratories) against chironomid midges in experimental ponds and a golf course pond in Florida at the rates of 1, 2, 3, 4, and 10 kg/ha. In these studies, he obtained a high level of control of chironomine species (Chironomini and Tanytarsini) at the higher rates. At the highest rate (10 kg/ha), 88% control of chironomine larvae was achieved 7 d after treatment, however, the level of control was only 59% 3 d after treatment. At 2 and 4 kg/ha, maximum control did not exceed 61 and 72%, respectively. It should be noted that *B. thuringiensis* ssp. *israelensis* is quite slow acting on chironomines which could be due to the low susceptibility of the older instars.[3] Although the younger larvae may be killed, the high level of reduction is not noted until the older larvae pupate and emerge as adults. This period of development may require 3 to 7 d. It was also noted that larval midge populations began to recover 3 to 4 weeks after treatment even at the highest dosage applied.

In California, Miura et al.,[54] using mosquito larvicidal rates of *B. thuringiensis* ssp. *israelensis* (Teknar, SAN 402 WDCI) at 0.25 kg/ha, showed no reduction in field populations of chironomid midges following treatment. Holding field-collected midge larvae in the laboratory did cause some mortality but this may have been due to unfavorable conditions to which these insects were exposed. Reduction (40 to 75%) in some chironomid species, however, was noted by Mulligan and Schaefer[68] after application of a *B. thuringiensis* ssp. *israelensis* formulation to a wildlife marsh.

Recent field studies in California (Mulla and Chaney, unpublished data 1987), using two commercial preparations of *B. thuringiensis* ssp. *israelensis* showed similar results and trends as those observed in Florida.[2,3] The studies were initiated to ascertain whether *B. thuringiensis* ssp. *israelensis* (provided by Abbott Laboratories) were employed by treating ponds (30 m² × 30 cm deep) with two rates of Vectobac AS (600 ITU/mg) and Vectobac WP. Each treatment, including experimental controls, was replicated three times. Benthic larval samples were taken with a scoop prior to and 3 d following treatment and weekly thereafter. Larvae were segregated and counted according to the procedures of Mulla et al.[67] and percent reduction was calculated. The larval population consisted mostly of chironomines (*C. decorus* and *C. fulvipilus*) and some tanypodines, mainly *Paralauterborniella elachistus*. The results of these field studies showed that Vectobac AS with a low potency yielded no more than 32 to 57% reduction of larvae 7 and 14 d posttreatment at 22 kg/ha. On the other hand, the Vectobac WP formulation yielded almost complete control of chironomines 7 and 14 d posttreatment at both rates of 5.6 and 11 kg/ha (Table 3). Due to cool weather, no recovery of larvae was noted in the treated ponds for the duration of the experiment. The dosages of Vectobac WP employed were equal to 20 to 40 times the mosquito larvicidal rates in the same habitats.[59,65]

To test *B. thuringiensis* ssp. *israelensis* further in the field against diverse groups of chironomid midge larvae, ABG-6164 (Abbott Laboratories, primary powder) was evaluated in mesocosms in Riverside, CA. The material was applied at 1.1, 2.75, and 5.6 kg/ha to mesocosms (27 m² × 30 cm deep) and devoid of vegetation. The midge fauna was more diverse, with species of *Chironomus, Dicrotendipes, Tanytarsus, Paralauterborniella,* and *Procladius* predominating. These and other genera showed marked differences in their susceptibility to *B. thuringiensis* ssp. *israelensis*. The *Chironomus* spp. were markedly reduced at all dosages, with 100% reduction at the 5.6-kg/ha rate. Populations cf *Paralauterborniella* and *Tanytarsus* spp. were also reduced but not at satisfactorily high levels. *Dicrotendipes* and *Procladius* spp. were not affected, in fact, their populations increased markedly in the treated mesocosms. Similarly, Rogatin and Baizhanov[75] noted sufficient reduction of chironomids in their studies to recommend the use of *B. thuringiensis* ssp. *israelensis* for the control of certain species.

Field data from studies in California and Florida, justified the conclusion that chironomine midges will not be affected by mosquito larvicidal dosages of *B. thuringiensis* ssp. *israelensis* preparations. These midges are reduced or controlled at rates as high as 20- to 40-fold the mosquito larvicidal rates. It is evident that different species and groups of midges have different levels of susceptibility; some are almost completely refractory to this microorganism while others show low to moderate levels of susceptibility. Whether *B. thuringiensis* ssp. *israelensis* can be used in chironomid midge control programs will depend on the type of larval habitats and the economic base of affected communities.

3. Laboratory Studies (*B. sphaericus*)

Using field-collected larvae of *C. crassicaudatus* and *G. paripes*, Ali and Nayar[4] found these species to be completely insensitive to experimental primary powders of *B. sphaericus* 1593 and 2362. They calculated an LC_{50} of >50 ppm, an extremely high concentration, being more than 10,000-fold the toxic dose for the mosquito *Culex quinquefasciatus*. Similarly, Mathavan and Velpandi[52] found field-collected *Chironomus* larvae to be totally insensitive to this microbial agent. The LC_{50} values were 250 ppm for the two strains (1593M and 2362), being 50,000- and 250,000-fold the mosquito larvicidal concentrations. Therefore, it is quite evident that chironomid midges are totally refractory to this entomopathogen.

TABLE 3
Efficacy of *Bacillus thuringiensis* ssp. *israelensis* and *Bacillus sphaericus* against Chironomid Larvae in Outdoor Ponds[90]

Material and formulation	Rate (kg/ha)	Pre number	Mean number larvae per sample[a] and % reduction posttreatment					
			3 d post		7 d post		14 d post[b]	
			Number	% reduction	Number	% reduction	Number	% reduction
Vectobac® AS (600 IU/mg)	11	85	63	7	59	37	90	13
	22	67	49	9	50	32	35	57
Vectobac® AS primary powder (5500 IU/mg)	5.6	95	39	49	13	88	2	98
	11.0	76	29	52	5	94	0	100
Bacillus sphaericus 2362 BSP-2 liquid (2×10^7 spores/g)	22	75	72	0	59	28	57	38
B. sphaericus 2362 (ABG-6184) primary powder (7.6×10^{10} spores per g)	11	112	84	6	136	0	101	26
Experimental control	—	110	88	—	121	—	134	—

[a] Benthic mud samples were taken with a long-handled scoop (scoop dimensions: 15 cm^2 × 3 cm deep). Samples were processed according to procedures in References 66 and 67.

[b] Level of populations and % control in the various treatments remained unchanged 21 and 28 d after treatment. Once the larvae were killed, there was no further larval recruitment due to the cool weather in December when water temperature was mean min 49°, mean max 63° F.

4. Field Studies (*B. sphaericus*)

At mosquito larvicidal rates, this pathogen did not adversely impact lenitic populations of chironomid midges studied over a period of 48 d, during which time two applications were made.[62] Mulla and Chaney (1987, unpublished data) conducted a field study on the effectiveness of *B. sphaericus* against chironomid midges in replicated mesocosms in the lower desert region of California. Application of *B. sphaericus* 2362 strain, both AS (BSP-2, Solvay and Co.) and powder (ABG-6184, Abbott Laboratories) at the rates of 22 and 11 kg/ha, respectively, produced little or no reduction of chironomids (Table 3). These rates are 50- to 100-fold those larvicidal for mosquitoes.[62] Therefore, *B. sphaericus* can be considered as having no activity against midges.

F. DIPTERA/CULICIDAE/*TOXORHYNCHITES SSP.*

Mosquito larvae of the genus *Toxorhynchites* are predacious, feeding almost exclusively on larvae of other container-breeding mosquitoes. The adults of *Toxorhynchites* mosquitoes offer the added benefit of not taking blood meals. Consequently, numerous attempts have been made to employ them for biological control of container-breeding mosquitoes such as *Aedes aegypti*. In integrated control efforts, combinations of larvicides, adulticides, predators, and environmental manipulation can assert effective, compatible, and complementary control on pest and vector mosquitoes. Before combining such agents as *B. thuringiensis* ssp. *israelensis*, *B. sphaericus*, and others with *Toxorhynchites*, it will be necessary to assess their deleterious effects on *Toxorhynchites*. Laboratory tests indicate no effect of *B. thuringiensis* ssp. *israelensis* on larvae of *Toxorhynchites* spp. in the absence of prey and relative tolerance to operational dosages of the bacterium in the presence of prey.[39,41,50]

Laboratory tests on the effect of *B. sphaericus* on *Toxorhynchites* revealed that most species of the latter were not susceptible to extremely high concentrations of the bacterium in the presence of prey.[39] Although *Toxorhynchites rutilus* was susceptible to serotype 5a,5b isolates of *B. sphaericus*,[39] it was not affected by high concentrations of the 2297 isolate (serotype 25) in the presence of prey.[43]

The lack of effect or reduced effect of *B. thuringiensis* ssp. *israelensis* and *B. sphaericus* on the Toxorhynchitinae may enable combined use of these agents. The ability of *Toxorhynchites* to withstand prolonged periods of starvation and to tolerate operational dosages of the two microbial control agents may enhance their complementarity. As concentrations of bacterial toxins decline to a level permitting reinvasion and development of prey larvae, surviving *Toxorhynchites* larvae could again assume a controlling role.

G. CRUSTACEA

This group of organisms contains a large number of freshwater species, many of which cohabit with mosquitoes. Few investigators have reported observations on Crustacea subjected to treatments with either reported observations on Crustacea subjected to treatments with either *B. thuringiensis* ssp. *israelensis* or *B. sphaericus* strains. In a laboratory study, Reish et al.[74] found the marine amphipod *Elasmopus bampo* to be quite tolerant to *B. thuringiensis* ssp. *israelensis* (Bactimos®) with a 96-h LC_{50} in the range of 12.8 mg/l. This concentration was over 100 to 200 times the mosquito larvicidal dosage of this formulation.

In laboratory studies in India, two strains of *B. sphaericus* (1593 M and 2362) were tested against field-collected populations of two crustacean species. The cladoceran, *Daphnia similis*, and the anostracan (fairy shrimp), *Streptocephalus dichotomus*, were not affected by extremely high concentrations of the two strains of the agent.[52] The 1593 M strain affected *S. dichotomus* at about 2500-fold the larvicidal concentration for *Culex quinquefasciatus*, while strain 2362 had some effects at 15,000 times the larvicidal concentration. Likewise, *D. similis* was not affected except by very high concentrations of 4000- and 27,000-fold the larvicidal concentration of the two strains, respectively.

Crustaceans found in the larval habitats of *Aedes vexans* in the upper Rhine Valley were collected and exposed in the laboratory for 48 h to 180 ppm of *B. thuringiensis* ssp. *israelensis* by Schnetter et al.[77] *Daphnia magna, Cyclops* sp. and *Rivulogammarus pulex* suffered no mortality at this extremely high concentration. However, 57 and 100% mortality was observed at 18 and 180 ppm, respectively, for the anostracan, *Chirocephalus grubei*. The lower concentration of 18 ppm is about 100-fold the larvicidal dosage used for the control of local mosquitoes.

In a Californian study on freshwater Crustacea, Mulligan and Schaefer[68] observed populations of some of these organisms in a habitat treated with a formulation of *B. thuringiensis* ssp. *israelensis* at the rate of 0.8 kg/ha. The cladoceran (*Simocephalus* sp.), ostracod (*Cypris* sp.), and copepod (*Cyclops* sp.) were not adversely affected. In another investigation in California, Mulla[62] studied the impact of two treatments at mosquito larvicidal rates of *B. thuringiensis* ssp. *israelensis* and *B. sphaericus* against some native crustaceans. Populations of the conchostracan *Eulimnadia texana* followed similar trends as measured by tow-net samples both in experimental control and treated mesocosms. Populations of the dominant cladoceran, *Moina rectirostris,* also grew, peaked, and declined in a similar manner both in the treated and untreated mesocosms.

Laboratory studies on the impact of *B. thuringiensis* ssp. *israelensis* and *B. sphaericus* were conducted by Holck and Meek[34] using the swamp crawfish, *Procambarus clarkii*. The LC_{50} values were 75 and 103 mg/l (ppm) for *B. sphaericus* (BSP-1 formulation) and *B. thuringiensis* ssp. *israelensis* (Bactimos formulation), respectively, which are approximately 1000-fold the mosquito larvicidal rates of these two entomopathogens.

H. MISCELLANEOUS NTOs

Preliminary studies on the effects of *B. thuringiensis* ssp. *israelensis* on a wide variety of other NTOs revealed very few adverse reactions to the bacterium.[24,25] However, the impact of larvicidal microbial agents has been quantitatively assessed on relatively few other aquatic organisms. Studies by Mathavan and Velpandi[52] on the effects of two strains of *B. sphaericus* (2362 and 1593 M) revealed that *Tubifex tubifex* were insensitive, having an LC_{50} in the range of 5- to 20-thousand fold that of larvicidal rates. Similarly, *Laccotrephes maculatus* (Nepidae) were completely insensitive with an LC_{50} in the range of 10- to 50,000-fold the mosquito larvicidal rate.[52] Mathavan et al.,[53] however, demonstrated reduced feeding, delayed maturity, and other deleterious effects in *Laccotrephes griseus* after long-term exposure to sublethal doses of *B. sphaericus*. Mollusca, Trichoptera, Turbellaria (*Dugesia tigrina*), and Amphibia (*Triturus vulgaris* and *Rana temporaria*) were not affected by 48-h exposures to 180 ppm of *B. thuringiensis* ssp. *israelensis* in tests conducted by Schnetter et al.[77]

IV. CONCLUSIONS

From the available data derived from a myriad of studies conducted in a wide diversity of habitats it is evident that primary powders and formulations of both *B. thuringiensis* ssp. *israelensis* and *B. sphaericus* offer highly selective control of vector and pest Nematocera with minimal adverse impact on the environment.

ACKNOWLEDGMENTS

We thank several of our colleagues for sending their published and unpublished data on the effects of *Bacillus* on NTOs. We are grateful for the bibliographic assistance of Ms. Ione Auston and for preparation of the manuscript by Ms. Cynthia Lacey.

REFERENCES

1. **Ali, A.,** Nuisance chironomids and their control: a review, *Bull. Entomol. Soc. Am.,* 26, 3, 1980.
2. **Ali, A.,** *Bacillus thuringiensis* serovar. *israelensis* (ABG-6108) against chironomids and some nontarget aquatic invertebrates, *J. Invertebr. Pathol.,* 38, 264, 1981.
3. **Ali, A., Baggs, R. A., and Stewart, J. P.,** Susceptibility of some Florida chironomid midges and mosquitoes to various formulations of *Bacillus thuringiensis* serovar. *israelensis, J. Econ. Entomol.,* 74, 672, 1981.
4. **Ali, A. and Nayar, J. K.,** Efficacy of *Bacillus sphaericus* Neide against larval mosquitoes (Diptera: Culicidae) and midges (Diptera: Chironomidae) in the laboratory, *Fla. Entomol.,* 69, 685, 1986.
5. **Aly, C.,** Germination of *Bacillus thuringiensis* var. *israelensis* spores in the gut of *Aedes* larvae (Diptera: Culicidae), *J. Invertebr. Pathol.,* 45, 1, 1985.
6. **Aly, C. and Mulla, M. S.,** Effect of two microbial insecticides on aquatic predators of mosquitoes, *Z. Angew. Entomol.,* 103, 113, 1987.
7. **Aly, C., Mulla, M. S., and Federici, B. A.,** Sporulation and toxin production by *Bacillus thuringiensis* var. *israelensis* in cadavers of mosquito larvae (Diptera: Culicidae), *J. Invertebr. Pathol.,* 46, 251, 1985.
8. **Back, C., Boisvert, J., Lacoursière, J. O., and Charpentier, G.,** High-dosage treatment of a Quebec stream with *Bacillus thuringiensis* serovar. *israelensis* efficacy against black fly larvae (Diptera: Simuliidae) and impact on non-target insects, *Can. Entomol.,* 117, 1523, 1985.
9. **Bourquin, A. W., Ahearn, D. G., and Meyers, S. P., Eds.,** Impact of the Use of Microorganisms in the Aquatic Environment, Ecological Research Series, Environmental Protection Agency Publication EPA-660/3-75-001, 1975.
10. **Car, M.,** Laboratory and field trials with two *Bacillus thuringiensis* var. *israelensis* products for *Simulium* (Diptera: Nematocera) control in a small polluted river in South Africa, *Onderstepoort J. Vet. Res.,* 51, 141, 1984.
11. **Car, M. and De Moor, F. C.,** The response of Vaal River drift and benthos to *Simulium* (Diptera: Nematocera) control using *Bacillus thuringiensis* var. *israelensis* (H-14), *Onderstepoort J. Vet. Res.,* 51, 155, 1984.
12. **Charles, J. F. and Nicolas, L.,** Recycling of *Bacillus sphaericus* 2362 in mosquito larvae: a laboratory study, *Ann. Inst. Pasteur Paris,* 137B, 101, 1986.
13. **Chilcott, C. N., Pillai, J. S., and Kalmakoff, J.,** Efficacy of *Bacillus thuringiensis* var. *israelensis* as a biocontrol agent against larvae of Simuliidae (Diptera) in New Zealand, *N. Z. J. Zool.,* 10, 319, 1983.
14. **Cibulsky, R. J. and Fusco, R. A.,** Recent experiences with Vectobac for black fly control: an industrial perspective on future developments, in *Black Flies: Ecology, Population Management, and Annotated World List,* Kim, K. C. and Merritt, R. W., Eds., Pennsylvania State University Press, University Park, PA, 1987, 419.
15. **Colbo, M. H. and Thompson, B. H.,** An efficient technique for laboratory rearing of *Simulium verecundum* S. and J. (Diptera: Simuliidae), *Can. J. Zool.,* 56, 507, 1978.
16. **Colbo, M. H. and Undeen, A. H.,** Effect of *Bacillus thuringiensis* var. *israelensis* on non-target insects in stream trials for control of Simuliidae, *Mosq. News,* 40, 368, 1980.
17. **Davidson, E. W.,** *Bacillus sphaericus* as a microbial control agent for mosquito larvae, in *Integrated Mosquito Control Methodologies,* Laird, M. and Miles, J., Eds., Vol. 2, Academic Press, New York, 1985, 213.
18. **Davidson, E. W., Singer, S., and Briggs, J. D.,** Pathogenesis of *Bacillus sphaericus* strain SS II-1 infections in *Culex pipiens quinquefasciatus* (= *Culex pipiens fatigans*) larvae, *J. Invertebr. Pathol.,* 25, 179, 1975.
19. **Davidson, E. W., Urbina, M., Payne, J., Mulla, M. S., Darwazeh, H., Dulmage, H. T., and Correa, J. A.,** Fate of *Bacillus sphaericus* 1593 and 2362 spores used as larvicides in the aquatic environment, *Appl. Environ. Microbiol.,* 47, 125, 1984.
20. **Dejoux, C.,** Recherches préliminaires concernant l'action de *Bacillus thuringiensis israelensis* de Barjac sur la faune d'invertébrés d'un cours d'eau tropical, mimeographed document WHO/VBC/79.721, World Health Organization, 1979.
21. **Dejoux, C., Gibon, F. M., and Yameogo, L.,** Toxicité pour la faune non-cible de quelques insecticides nouveaux utilisés en milieu aquatique tropical. IV. Les *Bacillus thuringiensis* var. *israelensis, Rev. Hydrobiol. Trop.,* 18, 31, 1985.
22. **Elouard, J. M. and Fairhurst, C. P.,** Ten years of surveillance of the rivers treated with antiblackfly insecticides by the Onchocerciasis Control Programme: medium and long-term impact on the invertebrates, *Chemosphere,* in press, 1988.
23. **Fortin, C., Lapointe, D., and Charpentier, G.,** Susceptibility of brook trout (*Salvelinus fontinalis*) fry to a liquid formulation of *Bacillus thuringiensis* serovar. *israelensis* (Teknar®) used for blackfly control, *Can. J. Fish. Aquat. Sci.,* 43, 1667, 1986.

24. **Garcia, R., Des Rochers, B., and Tozer, W.,** Studies on the toxicity of *Bacillus thuringiensis* var. *israelensis* against organisms found in association with mosquito larvae, *Proc. Calif. Mosq. Vect. Cont. Assoc.,* 48, 33, 1980.

25. **Garcia, R., Des Rochers, B., and Tozer, W.,** Studies on *Bacillus thuringiensis* var. *israelensis* against mosquito larvae and other organisms, *Proc. Calif. Mosq. Vect. Control Assoc.,* 49, 25, 1981.

26. **Gaugler, R. and Finney, J.,** A review of *Bacillus thuringiensis* var. *israelensis* (serotype 14) as a biological control agent of black flies (Simuliidae), in *Biological Control of Black Flies (Diptera: Simuliidae) with Bacillus thuringiensis var. israelensis (serotype 14), A Review with Recommendations for Laboratory and Field Protocol, Vol. 12(4),* Molloy, D., Ed., Misc. Publ. Entomological Society of America, 1982.

27. **Gaugler, R., Molloy, D., Haskins, T., and Rider, G.,** A bioassay system for the evaluation of black fly (Diptera: Simuliidae) control agents under simulated stream conditions, *Can. Entomol.,* 112, 1271, 1980.

28. **Gibbs, K. E., Brautigam, F. C., Stubbs, C. S., and Zibilske, L. M.,** Experimental applications of *B. t. i.* for larval black fly control: persistence and downstream carry, efficacy, impact on non-target invertebrates and fish feeding, *Tech. Bull. Maine Agric. Exp. Sta.,* 123, 1986.

29. **Gibon, F.-M., Elouard, J.-M., and Troubat, J.-J.,** Action du *Bacillus thuringiensis* var. *israelensis* sur les invertébrés aquatiques. I. Effets d'un traitement expérimental sur la Maraoué, *Rapp. Lab. Hydrobiol. O.R.S.T.O.M., Bouaké,* 38, 1980.

30. **Goldberg, L. J. and Margalit, J.,** A bacterial spore demonstrating rapid larvicidal activity against *Anopheles sergentii, Uranotaenia unguiculata, Culex univitattus, Aedes aegypti* and *Culex pipiens, Mosq. News,* 37, 355, 1977.

31. **Guillet, P., Escaffre, H., and Prud'hom, J.-M.,** L'utilisation d'une formulation à base de *Bacillus thuringiensis* H14 dans la lutte contre l'onchocercose en Afrique de l'Ouest. I. Efficacité et modalité d'application, *Cah. O.R.S.T.O.M. Sér. Entomol. Méd. Parasitol.,* 20, 175, 1982.

32. **Guillet, P., Hougard, J.-M., Doannio, J., Escaffre, H., and Duval, J.,** Evaluation de la sensibilité des larves du complexe *Simulium damnosum* à la toxine de *Bacillus thuringiensis* H 14. I. Méthodologie, *Cah. O.R.S.T.O.M. Sér. Entomol. Med. Parasitol.,* 23, 241, 1985.

33. **Hembree, S. C., Frommer, R. L., and Remington, M. P.,** A bioassay apparatus for evaluating larvicides against black flies, *Mosq. News,* 40, 647, 1980.

34. **Holck, A. R. and Meek, C. L.,** Dose-mortality responses of crawfish and mosquitoes to selected pesticides, *J. Am. Mosq. Control Assoc.,* 3, 407, 1987.

35. **Karch, S. and Coz, J.,** Recycling of *Bacillus sphaericus* in dead larvae of *Culex pipiens* (Diptera: Culicidae), *Cah. O.R.S.T.O.M. Ser. Entomol. Med. Parasitol.,* 24, 41, 1986.

36. **Knutti, H. J. and Beck, W. R.,** The control of black fly larvae with Teknar®, in *Black Flies: Ecology, Population Management, and Annotated World List,* Kim, K. C. and Merritt, R. W., Eds., Pennsylvania State University Press, University Park, PA, 1987, 409.

37. **Kurtak, D. C., Grunewald, J., and Baldry, D. A. T.,** Control of black fly vectors of Onchocerciasis in Africa, in *Black Flies: Ecology, Population Management, and Annotated World List,* Kim, K. C. and Merritt, R. W., Eds., Pennsylvania State University Press, University Park, PA, 1987, 341.

38. **Kurtak, D., Jamnback, H., Meyer, R., Ocran, M., and Renaud, P.,** Evaluation of larvicides for the control of *Simulium damnosum* s.l. (Diptera: Simuliidae) in West Africa, *J. Am. Mosq. Control Assoc.,* 3, 201, 1987.

39. **Lacey, L. A.,** Larvicidal activity of *Bacillus* pathogens against *Toxorhynchites* mosquitoes (Diptera: Culicidae), *J. Med. Entomol.,* 20, 620, 1983.

40. **Lacey, L. A.,** *Bacillus thuringiensis* serotype H-14 (Bacteria), in *Biological Control of Mosquitoes,* Chapman, H. C., Ed., *Am. Mosq. Control. Assoc. Bull.,* 6, 132, 1985.

41. **Lacey, L. A. and Dame, D. A.,** The effect of *Bacillus thuringiensis* var. *israelensis* on *Toxorhynchites rutilus rutilus* (Diptera: Culicidae) in the presence and absence of prey, *J. Med. Entomol.,* 19, 593, 1982.

42. **Lacey, L. A., Escaffre, H., Philippon, B., Sékétéli, A., and Guillet, P.,** Large river treatment with *Bacillus thuringiensis* (H-14) for the control of *Simulium damnosum* s.l. in the Onchoceriasis Control Programme, *Z. Tropenmed. Parasitol.,* 33, 97, 1982.

43. **Lacey, L. A., Lacey, C. M., Peacock, B., and Thiery, I.,** Mosquito host range and field activity of *Bacillus sphaericus* isolate 2297 (serotype 25), *J. Am. Mosq. Control Assoc.,* 4, 51, 1988.

44. **Lacey, L. A. and Mulla, M. S.,** A new bioassay unit for evaluating larvicides against blackflies, *J. Econ. Entomol.,* 70, 453, 1977.

45. **Lacey, L. A. and Undeen, A. H.,** Microbial control of black flies and mosquitoes, *Annu. Rev. Entomol.,* 31, 265, 1986.

46. **Lacey, L. A. and Undeen, A. H.,** The biological control potential of pathogens and parasites of black flies, in *Black Flies: Ecology, Population Management, and Annotated World List,* Kim, K. C. and Merritt, R. W., Eds., Pennsylvania State University Press, University Park, PA, 1987, 327.

47. **Lacey, L. A., Undeen, A. H., and Chance, M. M.,** Laboratory procedures for the bioassay and comparative efficacy evaluation of *Bacillus thuringiensis* var. *israelensis* (Serotype 14), in *Biological Control of Black Flies (Diptera: Simuliidae) with Bacillus thuringiensis var. israelensis (Serotype 14), A Review with Recommendations for Laboratory and Field Protocol,* Vol. 12(4), Molloy, D., Ed., Misc. Publ. Entomological Society of America, 1982, 19.

48. **Lacey, L. A., Urbina, M. J., and Heitzman, C. M.,** Sustained release formulations of *Bacillus sphaericus* and *Bacillus thuringiensis* (H-14) for control of container breeding *Culex quinquefasciatus, Mosq. News,* 44, 26, 1984.

49. **Larget, I. and deBarjac, H.,** Spécificité et principe actif de *Bacillus thuringiensis* var. *israelensis, Bull. Soc. Pathol. Exot.,* 74, 216, 1981.

50. **Larget, I. and Charles, J. F.,** Étude de l'activité larvicide de *Bacillus thuringiensis* variété *israelensis* sur les larves de Toxorhynchitinae, *Bull. Soc. Pathol. Exot.,* 75, 121, 1982.

51. **LeBrun, P. and Vlayen, P.,** Étude de la bioactivité comparée et des effets secondaires de *Bacillus thuringiensis* H 14, *Z. Angew. Entomol.,* 91, 15, 1981.

52. **Mathavan, S. and Velpandi, A.,** Toxicity of *Bacillus sphaericus* strains to selected target and non-target aquatic organisms, *Indian J. Med. Res.,* 80, 653, 1984.

53. **Mathavan, S., Velpandi, A., and Johnson, J. C.,** Sub-toxic effects of *Bacillus sphaericus* 1593 M on feeding, growth and reproduction of *Laccotrephes griseus* (Hemiptera: Nepidae), *Exp. Biol.,* 46, 149, 1987.

54. **Miura, T., Takahashi, R. M., and Mulligan, F. S., III,** Effects of the bacterial mosquito larvicide *Bacillus thuringiensis* serotype H-14 on selected aquatic organisms, *Mosq. News,* 40, 619, 1980.

55. **Miura, T., Takahashi, R. M., and Mulligan, F. S., III,** Impact of the use of candidate bacterial mosquito larvicides on some selected aquatic organisms, *Proc. Calif. Mosq. Vector Control Assoc.,* 49, 45, 1982.

56. **Molloy, D. and Jamnback, H.,** Field evaluation of *Bacillus thuringiensis* var. *israelensis* as a black fly biocontrol agent and its effect on nontarget stream insects, *J. Econ. Entomol.,* 74, 314, 1981.

57. **de Moor, F. C. and Car, M.,** A field evaluation of *Bacillus thuringiensis* var. *israelensis* as a biological control agent for *Simulium chutteri* (Diptera: Nematocera) in the middle Orange River, *Onderstepoort J. Vet. Res.,* 53, 43, 1986.

58. **Moulinier, Cl., Mas, J. P., Moulinier, Y., deBarjac, H., Giap, G., and Couprie, B.,** Étude de l'innocuité de *Bacillus thuringiensis* var. *israelensis* pour les larves d'huître, *Bull. Soc. Pathol. Exot.,* 74, 381, 1981.

59. **Mulla, M. S.,** Field evaluation and efficacy of bacterial agents and their formulations against mosquito larvae, in *Integrated Mosquito Control Methodologies,* Vol. 2, Laird, M. and Miles, J. W., Eds., Academic Press, San Diego, CA, 1985, 227.

60. **Mulla, M. S.,** Efficacy of the microbial agent *Bacillus sphaericus* Neide against mosquitoes (Diptera: Culicidae) in southern California, *Bull. Soc. Vector Ecol.,* 11, 247, 1986.

61. **Mulla, M. S.,** Role of BTI and *Bacillus sphaericus* in mosquito control programs, in *Fundamental and Applied Aspects of Invertebrate Pathology,* Sampson, A., Vlak, J. M., and Peters, D., Eds., Foundation 4th Int. Colloq. Invertebrate Pathology, Wageningen, 1987, 494.

62. **Mulla, M. S.,** Activity, field efficacy and use of *Bacillus thuringiensis* (H-14) against mosquitoes, in *Bacterial Control of Mosquitoes and Blackflies,* de Barjac, H. and Sutherland, D. J., Eds., Rutgers University Press, New Brunswick, NJ, in press, 1988.

63. **Mulla, M. S., Darwazeh, H. A., Davidson, E. W., and Dulmage, H. T.,** Efficacy and persistence of the microbial agent *Bacillus sphaericus* for the control of mosquito larvae in organically enriched habitats, *Mosq. News,* 44, 166, 1984.

64. **Mulla, M. S., Darwazeh, H. A., Davidson, E. W., Dulmage, H. T., and Singer, S.,** Larvicidal activity and field efficacy of *Bacillus sphaericus* strains against mosquito larvae and their safety to nontarget organisms, *Mosq. Mews,* 44, 336, 1984.

65. **Mulla, M. S., Federici, B. A., and Darwazeh, H. A.,** Larvicidal efficacy of *Bacillus thuringiensis* serotype H-14 against stagnant-water mosquitoes and its effects on nontarget organisms, *Environ. Entomol.,* 11, 788, 1982.

66. **Mulla, M. S. and Khasawinah, A. M.,** Laboratory and field evaluation of larvicides against chironomid midges, *J. Econ. Entomol.,* 62, 37, 1969.

67. **Mulla, M. S., Norland, R. L., Fanara, D. M., Darwazeh, H. A., and McKean, D. W.,** Control of chironomid midges in recreational lakes, *J. Econ. Entomol.,* 64, 300, 1971.

68. **Mulligan, F. S., III and Schaefer, C. H.,** Integration of a selective mosquito control agent *Bacillus thuringiensis* serotype H.14, with natural predator populations in pesticide-sensitive habitats, *Proc. Calif. Mosq. Vector Control Assoc.,* 49, 19, 1982.

69. **Mulligan, F. S., III, Schaefer, C. H., and Miura, T.,** Laboratory and field evaluation of *Bacillus sphaericus* as a mosquito control agent, *J. Econ. Entomol.,* 71, 774, 1978.

70. **Nicolas, L., Dossou-Yovo, and Hougard, J.-M.,** Persistence and recycling of *Bacillus sphaericus* 2362 spores in *Culex quinquefasciatus* breeding sites in West Africa, *Appl. Microbiol. Biotechnol.,* 25, 341, 1987.

71. **Olejníček, J. and Maryskova, B.,** The influence of *Bacillus thuringiensis* var. *israelensis* on the mosquito predator *Notonecta glauca, Folia Parasitol. (Prague),* 33, 279, 1986.

72. **Pistrang, L. A. and Burger, J. F.,** Effect of *Bacillus thuringiensis* var. *israelensis* on a genetically-defined population of blackflies (Diptera: Simuliidae) and associated insects in a montane New Hampshire stream, *Can. Entomol.,* 116, 975, 1984.

73. **Purcell, B. H.,** Effects of *Bacillus thuringiensis* var. *israelensis* on *Aedes taeniorhynchus* and some non-target organisms in the salt marsh, *Mosq. News,* 41, 476, 1981.

74. **Reish, D. J., LeMay, J. A., and Asato, S. L.,** The effect of BTI (H-14) and methoprene on two species of marine invertebrates from southern California estuaries, *Bull. Soc. Vector Ecol.,* 10, 20, 1985.

75. **Rogatin, A. B. and Baizhanov, M.,** Laboratory study of the effect of an experimental series of a bacterial preparation of *Bacillus thuringiensis* serotype 14 on various groups of hydrobionts, *Izv. Akad. Nauk. Kaz. SSR Ser. Biol.,* D6, 22, 1984 (in Russian).

76. **Rutschke, J. and Grunewald, J.,** The control of blackflies (Diptera: Simuliidae) as cattle pests with *Bacillus thuringiensis* H-14, *Zentralbl. Bakteriol. Parasitenkd. Infektionskr. Hyg. Abt. 1 Ref.,* 258, 413, 1984.

77. **Schnetter, W., Engler, S., Morawcsik, J., and Becker, N.,** Wirksamkeit von *Bacillus thuringiensis* var. *israelensis* gegen Stechmückenlarven und Nontarget-Organismen, *Mitt. Dtsch. Ges. Allg. Angew. Entomol.,* 2, 195, 1981.

78. **Sebastien, R. J. and Brust, R. A.,** An evaluation of two formulations of *Bacillus thuringiensis* var. *israelensis* for larval mosquito control in sod-lined simulated pools, *Mosq. News,* 41, 508, 1981.

79. **Sinègre, G., Gaven, B., and Jullien, J. L.,** Sécurité d'emploi du sérotype H-14 de *Bacillus thuringiensis* pour la faune non-cible des gîtes à moustiques du littoral Méditerranéen Français, *Parassitologia (Rome),* 22, 205, 1980.

80. **Singer, S.,** Use of bacteria for control of aquatic insect pests, in *Impact of the Use of Microorganisms in the Aquatic Environment. Ecological Research Series,* Bourquin, A. W., Ahearn, D. G., and Meyers, S. P., Eds., Environmental Protection Agency Publication EPA-660/3-75-001, 1975, 5.

81. **Troubat, J.-J.,** Dispositif à gouttières multiples destiné à tester *in situ* la toxicité des insecticides vis a vis des invertébrés benthiques, *Rev. Hydrobiol. Trop.,* 14, 149, 1981.

82. **Troubat, J.-J., Gibon, F.-M., Wongbe, A. I., and Bihoum, M.,** Action du *Bacillus thuringiensis* Berliner H14 sur les invertébrés aquatiques. II. Effets du'un épandage sur le cycle de dérive et les densités d'insectes benthiques, *Rapp. Lab. Hydrobiol., O.R.S.T.O.M. Bouaké,* 48, 1982.

83. **Undeen, A. H. and Lacey, L. A.,** Field procedures for the evaluation of *Bacillus thuringiensis* var. *israelensis* (serotype 14) against black flies (Simuliidae) and nontarget organisms in streams, in *Biological Control of Black Flies (Diptera: Simuliidae) with Bacillus thuringiensis var. israelensis (Serotype 14), A Review with Recommendations for Laboratory and Field Protocol,* Vol. 12(4), Molloy, D., Ed., Misc. Publ. Entomological Society of America, 1982, 25.

84. **Walsh, J. F.,** The feeding behaviour of *Simulium* larvae, and the development, testing and monitoring of the use of larvicides, with special reference to the control of *Simulium damnosum* Theobald s.l. (Diptera: Simuliidae): a review, *Bull. Entomol. Res.,* 75, 549, 1985.

85. **Weiser, J. and Vankova, J.,** Toxicity of *Bacillus thuringiensis israelensis* for black-flies and other freshwater invertebrates, *Proc. Int. Colloq. Invertebr. Pathol.,* Prague, 8, 243, 1978.

86. **WHO,** Data sheet on the biological control agent *Bacillus thuringiensis* serotype H-14 (de Barjac, 1978), mimeographed document, WHO/VBC/79.750, Rev. 1, VBC/BCDS/79.01, World Health Organization, 1979.

87. **Wilton, D. P. and Travis, B. V.,** An improved method for simulated stream tests of blackfly larvicides, *Mosq. News,* 25, 118, 1965.

88. **Yameogo, L.,** Modification des entomocénoses d'un cours d'eau tropical soumis à traitement antisimulidien avec *Bacillus thuringiensis* var. *israelensis,* mimeograph, Mém. d'ingenieur de l'Université de Ouagadougou, Upper Volta, 1980.

89. **Yousten, A. A.,** *Bacillus sphaericus:* Microbiological factors related to its potential as a mosquito larvicide, *Adv. Biotechnol. Proc.,* 3, 315, 1984.

Chapter 13

SAFETY CONSIDERATIONS IN THE USE OF *BACILLUS POPILLIAE*, THE MILKY DISEASE PATHOGEN OF SCARABAEIDAE

F. D. Obenchain and B-J. Ellis

TABLE OF CONTENTS

I. INTRODUCTION

The literature on *Bacillus popilliae*-group pathogens and the milky disease of scarabaeid beetle larvae (generally called white grubs) contains more than 300 citations,[1,2] including major reviews covering aspects of bacterial systematics, host specificity, pathogenesis, physiological control of growth and sporulation, problems associated with *in vitro* culture, and epizootiology,[3-9] along with reviews on host biology and the applied use of *B. popilliae* as an insecticide.[10-16] Many of these same references discuss briefly the safety and environmental implications associated with this group of presumedly obligate pathogens. These issues are more specifically addressed in several general reviews on the safety of microbial insecticides as a class.[17-22] Most of the safety tests supporting the registration of *B. popilliae*-based microbial insecticides in the U.S. have not been published. Nearly all have been performed on *in vivo*-produced spores or spore formulations. This review will report on the details of those safety studies and examine their relevance to a new formulation of *in vitro*-produced *B. popilliae* spores which has recently been registered by the U.S. Environmental Protection Agency (EPA).[23]

II. HISTORICAL REVIEW

A. THE MILKY DISEASE STATE

The name "milky disease" describes the milk-white color of hemolymph taken from scarabaeid larvae when they are heavily infected with sporulating stages of a *B. popilliae*-group pathogen. Although several species were previously identified on the basis of differences in morphological structure or host preferences, the most recent taxonomic review recognized a single species with a number of subspecies.[8]

Vegetative stages of these forms are characterized as aerobic, spore-forming rods which grow *in vitro* only on complex media. Vegetative growth is catalase-negative, facultatively anaerobic, and restricted to temperatures between 17 and 34°C. Oval spores bear a number of longitudinal ridges and develop centrally within a markedly swollen sporangium which does not elongate during spore morphogenesis. Refractory parasporal bodies, numbering 0 to 2 as characteristic of the subspecies, may also develop within the sporangium during morphogenesis. Mature sporangia, containing spores and parasporal bodies (when the latter are present), accumulate in the hemolymph of infected grubs to concentrations[7] as high as 2 to 3 × 10^{10} sporangia per ml. These sporangia do not autolyse to release free spores or parasporal crystals into the hemolymph. The cause of death in infected grubs may be physiological starvation,[20] since no toxic compounds appear to be involved. All of these latter characteristics are recognized as central evolutionary relationships between the closely adapted entomopathogens in the *B. popilliae* group and their susceptible scarabaeid hosts. These characteristics are equally important to the efficacy and environmental safety of *B. popilliae* microbial insecticides.

B. DISCOVERY AND APPLICATION

Although the Japanese beetle, *Popillia japonica,* may have been accidentally introduced into the U.S. (near Riverton, NJ) as recently as 1912, its obvious potential for economic damage and its rapid spread had made it a target for a cooperative federal-state research program by 1917.[12] Between 1922 and 1933, beetle larvae from several widely separated field sites were found to be infected by a number of unidentified bacteria and fungi to the extent that 50% of the grubs collected in the fall for experimental work during the winter were lost to disease. Three syndromes were recognized from the general appearance of the dead grubs. In the first group, the grubs were covered with fungal hyphae. Those in the second group were dark in color (red, green, brown, or black) and appeared to be infected with several micro-organisms, while grubs in the third group were white and filled with a spore-forming organism in nearly pure culture.

In a five-year period, beginning in 1934, the milky disease bacteria responsible for the white color were studied intensively by Dutky[3,24] and co-workers. The putative disease organisms were described as two closely related species; *Bacillus popilliae*, which formed a single parasporal crystal during sporulation, and *B. lentimorbus* (now considered to be a subspecies of *B. popilliae*) which did not form parasporal crystals. Their role as causative agents of milky disease was confirmed by serial passage of spore suspensions heated to 80°C for 10 min. Not all of Koch's Postulates were satisfied at that time since the two subspecies were said to grow poorly or not at all on most culture media and those cultures that grew failed to sporulate. *In vivo*-produced spores, however, were shown to remain viable for many years in soil. Early observations on the infectivity and yield (1000-fold increase over spores used in the inoculum) from injected, field-collected grubs formed the basis for a method of mass production.[25] Spore powders, formulated from a dried slurry of calcium carbonate and homogenated infected grubs, were extended with talc to a final concentration of 1×10^8 spores per g and subjected to initial field tests for efficacy and to preliminary safety tests with birds.

When the powder formulation was applied to turf infested with Japanese beetle grubs (typically in spots at the rate of 1.67 g per spot at 1.22 m intervals) localized epizootics of milky disease were initiated.[25-27] Based on these data, together with no indication of germination, growth, or toxic effects of *B. popilliae* fed to starlings and chickens and no allergic reactions among workers who had handled infected grubs or produced spore powders, Dutky applied to several government agencies (Bureau of Entomology and Plant Quarantine, Bureau of Animal Industry, Bureau of Plant Industry, and the Apiculture Division, in the U.S. Department of Agriculture (USDA), and to the U.S. Public Health Service) for approval to use government-produced spore formulations in a federal-state cooperative colonization program. Government approval was obtained and in a 14-year period, between 1939 and 1952, approximately 83,600 kg of powder (produced from 4 to 5 million injected, field-collected grubs) was applied to 194,000 different sites in 14 eastern states (U.S.) and the District of Columbia and to a total of more than 42,000 ha. Commercial production of the spore powder, under nonexclusive government license of Dutky's 1941 control method[26] and 1942 *in vivo* propagation[25] patents, began in the mid-1940s and still continues under those basic protocols.

C. HOST-RANGE AND PATHOGEN SPECIFICITY/VIRULENCE

Attempts to establish *B. popilliae* infections in various non-scarabaeid insects have been unsuccessful. Dutky[3] reported that spores injected into larval *Musca domestica* were gradually sequestered by fatbody tissues prior to pupation but reappeared when larval tissues lysed during metamorphosis. He also reported that spores and vegetative rods had no effects on the development of larval *Galleria mellonella*. Aizawa[28] failed to infect *Bombyx mori* larvae with either *B. popilliae* ssp. *popilliae* or *B. popilliae* ssp. *lentimorbus* in the laboratory. Natural milky disease infections, as diagnosed by the accumulation of sporulated bacilli with unlysed, swollen sporangia, have been reported only from beetles in the family Scarabaeidae.

Numerous laboratory studies have been performed on the infectivity of *B. popilliae* subspecies to various scarabaeid hosts, both by injection and feeding of sporulated and/or vegetative forms.[3] Only those which used the *per os* route are relevant to the situation in the field, however, and actual field data are scant. From his work on *B. popilliae* ssp. *rhopaea*, Milner[8,29] suggested that the effective host range of a subspecies is probably confined to a few species within a single genus. Field and laboratory studies in the U.S., on *B. popilliae* strains associated with native and introduced scarabaeid pests, only partially support that generalization. Tashiro[16,30] found that the commercial strain of *B. popilliae*, used for Japanese beetle control, was equally infective for grubs of the European chafer, *Rhizotrogus (Amphimallon) majalis*, at least in the laboratory. By contrast, standard doses of the ''DeByrne'' strain of *B. popilliae* ssp. *popilliae* and the ''Amphimallon'' strain of *B. popilliae* ssp. *lentimorbus*, both isolated from field sites with established epizootics among European chafer grubs, were only 1 to 10% as infective for Japanese beetle grubs as they were for chafers in laboratory tests. Studies on strains associated with *Melolontha*,[31,32] *Ataenus*,[33] and *Cyclocephala* grubs,[34,35] however, showed highly restricted host ranges supporting Milner's hypothesis.

The closely adapted *B. popilliae* strains of the Japanese beetle and the European chafer were isolated only 24 and 12 years, respectively, after the introduction of these pests into the U.S.,[12,15] and in the case of the chafer, only 6 years after those field sites were inoculated with the commercial Japanese beetle strain. Summarizing studies conducted between 1974 and 1984, Chang and Wan[36] note that following the introduction of commercial spore powder from the U.S., more than 39 *B. popilliae* strains have been isolated from 14 Chinese species of scarabaeid grubs in 7 genera. While host specificity is also a function of the defense mechanisms and nutrition of the host, characteristics of the pathogen strains must be better understood before the taxonomic situation can be clarified. Serological studies,[37-39] generation of plasmid profiles[40] or study of plasmid exchange systems,[41] and determination of complex biochemical parameters have been hampered by difficulties associated with the *in vitro* growth and sporulation of *B. popilliae*. Recent developments, discussed below, should stimulate such studies in the future, both in support of new or improved product development and of our increased appreciation of differences in host specificity and virulence between various isolates and potential environmental consequences of their use as pesticides.

D. *IN VITRO* GROWTH AND SPORULATION

There are nearly 90 published studies on various aspects of *in vitro* growth and sporulation in this fastidious group of entomopathogens and these works have been extensively reviewed.[4-9,42-44] One center of such investigations was the USDA Northern Regional Research Laboratories (NRRL) in Peoria, IL. Beginning in the early 1960s, an investment of 50 to 60 man-years of work led to the development of techniques for the extensive vegetative growth of a small number of *B. popilliae* isolates on complex solid or liquid media. Limited production of mature sporangia, containing spore and parasporal bodies, was also achieved but only in a few derivative strains. *In vitro*-grown vegetative stages were fully infective to Japanese beetle grubs by injection, but spores were less infective by injection and showed little activity *per os*.[45,46] At the end of this 14-year period, government funding was diverted to other priorities and *in vitro* studies wound down. In retrospect, it is ironic that the isolates most studied (NRRL B-2309 and derivatives) were descendents of the ''DeByrne'' strain which was highly adapted to the European chafer, while efficacy bioassays were performed on Japanese beetle grubs.

Formulations of *B. popilliae* vegetative stages in encapsulated form have been subjected to limited efficacy and safety testing.[47-49] The lack of infectivity of these preparations to Japanese beetle grubs has been attributed to the absence of parasporal bodies[9] but *Melolontha* grubs showed no toxic responses when they were fed or injected with parasporal bodies of their adapted strain.[43] Although solubilized parasporal bodies produced mortality when injected into Japanese beetle grubs,[50] this is not a normal route of exposure and the role of parasporal bodies in the infective process of *B. popilliae* has not been established.

The first step in the *in vitro* culture of *B. popilliae* or any other entomopathogen is the isolation of the organism from the diseased host. Recently, Krieg[51] reiterated a widely recognized concern when he warned that the identity of a presumptive *Bacillus* pathogen should be confirmed on the basis of Koch's Postulates since saprophytic sporeformers may secondarily invade the diseased host and feign the true disease organism. In a recent study, forming the basis of patented composition claims,[52] it was demonstrated that the *in vitro* sporulation process of *B. popilliae* differs from the *in vivo* process. In preferred liquid media, sporulation of established laboratory strains or new field isolates involved the autolysis of mature sporangia with the release of free spores. The latter were infective in *per os* bioassays, giving rise to the typical sporangium-bearing spores in their Japanese beetle host. These same free spores, though infective *per os*, were not infective in parallel injection bioassays. It now appears likely that earlier workers may not have recognized true *B. popilliae* isolates when those isolates sporulated with autolysis of the sporangia and when injection bioassays (more easily performed than feeding assays) of sporangium-free spores failed to produce milky disease. Future isolation of potentially useful *B. popilliae* strains from diseased specimens would be facilitated by the development of serological and other identification techniques which could be substituted for feeding bioassays in the process of screening out saprophytic sporeformers.

III. SAFETY DATA

A. REGISTRATION REQUIREMENTS FOR USE ON PASTURE GRASS

When the U.S. Federal Insecticide, Fungicide, and Rodenticide Act was passed in 1947, the registration of commercial *B. popilliae* spore powder was continued under a "grandfather" clause and additional safety data were not requested. In 1968, however, the newly established EPA withdrew the registered use of these products on pastures, pending submission of data establishing the safety of *B. popilliae* residues on forage grasses.[53] Data to satisfy those requirements were generated in a series of studies performed by a USDA team headed by Dr. A. M. Heimpel and submitted through the Inter-regional Research Project No. 4 (Rutgers University) to the EPA as Pesticide Petition 6E1692 in 1977. EPA requested further follow-up data on various aspects of the submitted tests and an additional mouse subcutaneous pathogenicity test on representative commercial products. Following submission of test data in those additional areas, Title 40 of the Code of Federal Regulations was amended by addition of Sect. 180.1076, establishing an exemption from the requirements of a residue tolerance for "viable spores of the microorganism *B. popilliae*" when used as an insecticide on pasture and rangeland forage.

B. PRODUCT AND RESIDUE ANALYSIS DATA

Table 1 summarizes the existing data base supporting the registration of commercial *B. popilliae*-based microbial insecticides in the U.S. End-use products which have been tested include those manufactured by Fairfax Laboratories, Clinton Corners, New York (Japidemic® and Doom®),[54,55,58,59,62] and by the Ringer Corporation, formerly Reuter Laboratories, Manassas, Virginia (Milky Spore and Grub Attack®).[55,56,60] The remaining end-use product tests were performed on products of government manufacture.

Thompson and Heimpel[54] performed a series of microbiological screens on production samples for detection of extraneous biotypes, including coliform bacteria, anaerobic or aerobic bacterium (other than *B. popilliae*), or species of the *Shigella-Salmonella* complex. No coliform or anaerobic bacteria were found, but a single lot contained a nonpathogenic bacterium classified as *Aerobacter* sp. Aerobic sporeformers (mostly *Bacillus subtilis*, *B. cereus*, and *B. megaterium*) averaged 4×10^4/g, and aerobic nonsporeformers averaged 2.1×10^4/g. No members of the *Shigella-Salmonella* complex were detected. It was concluded that commercial products manufactured according to Dutky's *in vivo* protocols[25] contained no extraneous bacteria harmful to man and that bacterial contaminant counts were comparable to those allowed in U.S. "Extra Grade" powdered milk products. Subcutaneous injection of commercial products[55] or sporangium-free *in vitro*-produced spores at 3 to 4

TABLE 1
Summary of Safety Data on *Bacillus popilliae* and *B. popilliae* Formulations

U.S. EPA Testing Requirements (EPA Guideline no.)	Test animal (number)	Formulation	Dose	Test results	Date	Ref.
Product analysis testing						
Formation of unintentional ingredients (151-22)	Microbial screen	End product	—	No human or animal pathogens	1974	54
Product batch analysis — subcutaneous injection (40 CFR§180.1076(3))	Mouse (3 × 5)	End product	1×10^6	No effects	1981	55
	Mouse (5)	ATCC-53256 pri. powder	3.2×10^6	No effects	1988	23, 56
	Mouse (5)	ATCC-53256 pri. powder	4.5×10^6	No effects	1988	23, 56
Residue analysis (153-4)	Soil bioassay	End product	$2 \times 10^6/\text{ft}^2$	Present after 25 years	1967	57
Tier I toxicological data						
Acute oral (152-30)	Rat (20)	End product	$5 \times 10^7/\text{d}$ (×21)	No effects	1973	58
	Rat (20)	Spore	$5 \times 10^7/\text{d}$ (×21)	No effects	1973	58
	Monkey (4)	End product	$2.5 \times 10^8/\text{d}$ (×21)	No effects	1973	58
	Monkey (4)	Spore	$2.5 \times 10^8/\text{d}$ (×21)	No effects	1973	58
Acute dermal (152-31)	Guinea pig (10)	Spore	1×10^6	No mortality; no serum antibodies.	1977	59
IV, IC, IP (152-33)	Mouse (5 × 7)	NRRL-B2309 rods	2.74×10^8 — 4.26×10^9	$IPLD_{50} = 3.1 \times 10^9$ cells	1971	49
Primary dermal irritation (152-34)	Guinea pig (6)	End product	5×10^7	Very slight irritation; complete reversal in 7 d	1983	60
Primary eye irritation (152-35)	Rabbit (10)	Spore	1×10^6	No irritation	1977	61
Hypersensitivity incidents (152-37)	Human (8)	Spore and end product	Production exposure	No occupational sensitization	1973	62

	Spore and end product	Production exposure			
Immune response (152-38)					
Human (8)			No serum antibodies detected	1973	62
Tier I nontarget data					
Avian oral (154-16)					
Starling (−)	Spore	$3 \times 10^{10}/d \ (\times 1)$	No effects; passed viable spores	1938	63
Chicken (−)	Spore	$3 \times 10^{10}/d \ (\times 33)$	No effects; passed viable spores	1938	63
Avian injection (154-17)					
Chicken (−)	Spore	1.1×10^9	No effects	1938	63
Non-target plant (154-22)					
Pastures	End product	$2.06 \times 10^9/\text{ft}^2$	No negative effects on turf	1975	64
Non-target insect (154-23)					
Tiphia	Milky host	—	Reduced parasite survival	1943	65
Neodiprion	Spore	2.1×10^9 per shoot	No infectivity	1942	66
Musca	Spore	—	No effect after injection	1963	3
Galleria	Spore and rods	—	No effect on host after injection	1963	3
Bombyx	Spore	—	No infection on injection	1967	28

times required doses[56] produced no mortality in test mice and no signs of gross pathology at necropsy. These studies demonstrate that established quality control protocols generate end-use formulations which present no appreciable hazards to man or animals. Residue analyses, showing levels of *B. popilliae* spores 25 years after colonization,[57] lack statistical precision, but provide no indication of potential threat to man, animals, or the environment. In fact, the rise and fall in spore levels appears to be directly related to the density of grub infestations and this is one of *B. popilliae's* most environmentally attractive features.

C. TIER I TOXICOLOGICAL DATA

Table 1 is in the form of a modified data matrix, relating existing test data to the current EPA data registration requirements. Tests performed prior to 1981 may not follow the protocols indicated by the EPA Guideline Numbers, but in many cases these tests followed protocols far more strict than those currently adopted. As an example, data on oral toxicity came from 21 d stomach intubation studies on rats and monkeys, using commercial end-use product and preparations of *in vivo* spores from hemolymph slides.[58] The test animals, composed of equal numbers of both sexes, were observed for mortality, any behavioral changes associated with treatment, and extensive physical, clinical, and serological parameters during the test and recovery periods. Half of the test animals were killed 21 d following the test period and the rest were killed 7 d later; following death, tissues of major circulatory, respiratory, digestive, and glandular organ systems were examined for histopathological changes. In both rats and monkeys there were no differences in the behavior or appearance of test and control animals during or following the 21 d treatment period. There were no deaths or signs of infection, and physical, clinical, and serological parameters were comparable. Serum samples, tested in another laboratory, showed no sign of antibody production to the test articles. There were no gross changes in any of the organ systems at necropsy and no significance was attributed to a relative increase in the weight of male spleens in monkeys treated with end-use product or an absolute increase in the weight of seminal vesicles and relative weight increase of female gonads in monkeys treated with spores. Similarly, no significance was attributed to a 29% reduction in absolute weight of thyroids at the 3-week killing of male rats treated with product and a 24% reduction in those killed at 4 weeks post-treatment. There were no histopathological changes noted in nine tissues taken from all animals and the investigator concluded that "no unambiguous differences were noted between treated and control groups in the course of the study."

Protocols for studies performed to determine acute dermal toxicity,[59] primary dermal irritation,[60] and primary eye irritation (with no wash-out)[61] more closely follow current EPA Guidelines. A suspension of *in vivo* spores in distilled water produced no mortality, no dermal irritation, and induced no serum antibodies when applied at the rate of 1×10^6 spores to two abraded and two unabraded sites on the backs on ten guinea pigs. Application of 0.5 g of end-use product (containing 5.0×10^7 *B. popilliae* spores), moistened with 0.5 ml of 0.9% saline, was also made to each of two abraded and two unabraded sites on the backs of six guinea pigs and covered for 24 h, then wiped free of test material. The animals exhibited well-defined erythema, with very slight edema at 24 h and very slight erythema at 72 h. The primary irritation index was 1.6 on the Draize Scale and all animals were free of irritation within 4 to 6 d after test article application. No signs of eye irritation were observed among ten rabbits at 24, 48, and 72 h, or at 7 and 14 d post-instillation following application of 1×10^6 spores from hemolymph slides in 0.1 ml of distilled water; *B. popilliae* was not considered an eye irritant.

Heimpel and Hrubant[62] supervised the clinical, medical, and serological investigation of eight employees of Fairfax Biological Laboratories with combined full or part-time employment totaling 85 years in the *in vivo* manufacturing process of *B. popilliae* product. All individuals (five women and three men) were given routine physical examinations and chest

X-rays, and clinical blood tests and urinalyses were performed on three occasions during the spring production season (May through July) in 1971. Three sets of serum samples were also taken from the volunteers and tested for antibody production by agglutination, precipitin ring, and agar diffusion tests against (1) antigens prepared from vegetative stages of *B. popilliae* laboratory strains (NRRL B-2309, NRRL B-2309S, NRRL B-2309L, and NRRL B-3329F), (2) antigens prepared from hemolymph of field-collected milky Japanese beetle grubs, and (3) antigens from hemolymph of Japanese beetle grubs that had been injected and bled at the laboratory while the serum samples were being drawn. The human sera examined contained no antibodies to any of the three antigen preparations and none of the medical findings, physical or clinical, were unusual for the age and sex of each volunteer. It was concluded that there was no evidence of illness associated with occupational exposure to *B. popilliae* and that continued exposure does not induce antibody production in the human.

A laboratory investigation on the potential pathogenicity of *B. popilliae* vegetative stages to mammals was performed with 14 h log-phase rods of NRRL B-2309 and a putatively nonpathogenic laboratory strain of *B. subtilis* by Welsh and McMahon.[49] Rods were harvested by centrifugation, washed in 0.1% tryptone to remove extracellular debris, resuspended in tryptone, serially diluted for the injection series, and analyzed for viable cell count by a plating technique. Doses of the rod dilutions were injected i.p. (0.5 ml) and subdermally (0.25 ml) into groups of seven mice. Mortality occurred in mice receiving viable rods in doses of 2.52×10^9 and higher ($LD_{50} = 3.1 \times 10^9$). Mice died within 24 to 48 h of injection in both experimental series. Intraperitoneal injection with 2×10^{10} heat-killed rods did not produce mortality and mice injected subdermally with 1×10^{10} *B. popilliae* rods did not die but showed localized tissue necrosis. The claim of these workers that *B. popilliae* organisms were isolated from these necrotic lesions 5 to 7 d after injection can not be explained and is the only observation of concern. The high doses and route of infection are unlikely to occur in nature and do not suggest any danger to man or animals.

D. TIER I NONTARGET DATA

Grubs of the Japanese beetle, with or without infections of *B. popilliae,* are commonly eaten by birds such as starlings and chickens[12,13] and it was appropriate that the first safety tests were performed on these species.[63] Starlings were tested for acute oral toxicity effects at a single spore dose (3×10^{10}) which is equivalent to consumption of 10 to 15 milky-diseased third-instar grubs. There was no indication that the *B. popilliae* spores germinated in the digestive tract of the starlings or persisted in their feces. Chickens were fed the same number of spores for each of 33 consecutive d or injected with 0.5×10^9 spores per kg of body weight (estimated by Ignoffo[19]), also with no sign of spore germination and no negative effects to the chickens.

Limited nontarget plant safety data comes from the observations of Ladd[64] on pasture treated with commercial *B. popilliae* product at the rate of 2000 lb/acre (200 times rate recommended on the label); he reported that there were no noticeable negative effects on the growth of pasture grasses in treated areas which had received about 2×10^9 spores per ft². That observation is not surprising since a single, optimally infected milky grub might contain that number of mature sporangia at its death. Other nontarget data from the literature show the lack of infectivity of *B. popilliae* among a variety of beneficial or non-scarabaeid pest insect species from a range of insect orders (Table 1).

E. DATA REQUIREMENTS OF THE EPA REREGISTRATION PROGRAM

The EPA reregistration program was initiated in 1980 following a U.S. congressional mandate to systematically review groups of registered pesticides, to analyze the adequacy of the data base supporting those registrations, to identify "data-gaps", and to require that any such missing or inadequate data are provided by the pesticide registrants. A review of the *B. popilliae* data base has been initiated by the EPA and current registrants have been requested to provide additional nontarget data involving freshwater fish (Guideline No. 154-19), freshwater invertebrates (154-20), nontarget plant studies (154-22), nontarget insect studies (154-23), and honeybee testing (154-24). Completion of these studies is anticipated in the next several years.

IV. DISCUSSION

Although product supply has been limited by the scale and efficiency of *in vivo* production techniques, *B. popilliae* microbial pesticide formulations have been used for close to 50 years without raising any significant environmental safety issues. Questions about ways to improve product cost and effectiveness have been addressed in a number of ways. Laboratory studies by Sharpe and Detroy[67,68] demonstrated that the crystalline endotoxin of *B. thuringiensis* ssp. *galleriae* caused a fatal septicemia when spore paste was fed to Japanese beetle grubs. Sharpe also demonstrated that the *B. thuringiensis* ssp. *galleriae* endotoxin facilitated the infection of grubs with *in vitro*-produced *B. popilliae* spores (NRRL B-2309). He further suggested that encapsulated *B. thuringiensis* crystalline endotoxin might be combined with spores or vegetative cells of *B. popilliae* to enhance their infectivity in the field. Laboratory bioassays and field tests of a formulation containing both *B. popilliae* and *B. thuringiensis* ssp. *kurstaki* showed no such enhancement, however.[23]

The recent achievement of *B. popilliae* sporulation in liquid media at yields of 2×10^9 sporangium-free spore per ml by Ellis et al.[52] should sharply increase the supply of product. A formulation containing 6.4×10^8 sporangium-free spores per g has laboratory and field infectivity equivalent to the *in vivo* formulation developed 50 years ago by Dutky and this formulation has been registered by the EPA under the Grub Attack® trademark. It is anticipated that additional *B. popilliae* products based on subspecies or strains with a different spectrum of host infectivity may be developed and registered in the near future.

In his 1980 review on risk analysis in the registration of bacterial pesticides, Burges[21] noted that considerations must be made beyond those addressed in safety testing, even when all tests for apparent safety hazards are negative. *B. popilliae* spores and spore formulations neither kill nor infect test animals and they do not induce occupational disease in production workers. There is no evidence of toxin production in the disease mechanisms of natural or artificial (injected) scarabaeid hosts and the pathogen does not grow at temperatures above 35.5°C. For that reason, *B. popilliae* is not able to infect man or higher animals. Moreover, spores are applied as an inoculum; an increase in spore numbers in the soil occurs only in the presence of susceptible, targeted pests. Still, the persistence of spores in the soil provides for long-term pest control and this reduces the need to resort to residual chemicals. All existing commercial formulations, whether based on *in vivo* or on *in vitro* production methods, contain strains of *B. popilliae* which are highly adapted to the Japanese beetle. These strains are characterized by their low virulence and high spore yield in infected hosts. There appear to be no tangible safety or environmental risks associated with the use of any of these *B. popilliae*-based insecticides.

It seems feasible that future products, based on lines of *B. popilliae* grown *in vitro*, could contain more virulent strains which would kill targeted grubs faster but produce lower numbers of spores in infected hosts. Presuming no differences in the relevant safety data, the predictable drawback to such rapidly acting products would be the need to reapply in order to maintain grub control from year to year. The choice between that type of product and current commercial products, containing the less virulent, spore-productive strains, would then be based on their cost effectiveness.

REFERENCES

1. **Klein, M. G., Johnson, C. H., and Ladd, T. L., Jr.,** A bibliography of the milky bacteria (*Bacillus* spp.) associated with the Japanese beetle, *Popillia japonica* and closely related Scarabaeidae, *Bull. Entomol. Soc. Am.,* 22(3), 305, 1976.

2. **Klein, M.,** *Bacillus popilliae* — Prospects and problems, in *Fundamental and Applied Aspects of Invertebrate Pathology,* Samson, R. A., Vlak, J. M., and Peters, D., Eds., Proc. 4th Int. Colloq. Invertebr. Pathol., Wageningen, 1986, 534.

3. **Dutky, S. R.,** The milky diseases, in *Insect Pathology: An Advanced Treatise,* Vol. 2, Steinhaus, E. A., Ed., Academic Press, New York, 1963, 75.

4. **Rhodes, R. A.,** Symposium on microbial insecticides. II. Milky disease of the Japanese beetle, *Bacteriol. Rev.,* 29, 373, 1965.

5. **Rhodes, R. A.,** Milky disease of the Japanese beetle, in *Proc. of Joint U.S.—Japan Seminar on Microbial Control of Insect Pests,* Fukuoka, Japan, April 1967, publ. 1968, 85.

6. **Steinkraus, K. H. and Tashiro, H.,** Milky-disease bacteria, *Appl. Microbiol.,* 15, 325, 1967.

7. **St. Julian, G. and Bulla, L. A., Jr.,** Milky disease, in *Current Topics in Comparative Pathobiology,* Vol. 2, Cheng, T. C., Ed., Academic Press, New York, 1973, 57.

8. **Milner, R. J.,** Identification of the *Bacillus popilliae* group of insect pathogens, in *Microbial Control of Pests and Plant Diseases 1970—1980,* Burges, H. D., Ed., Academic Press, London, 1981, 45.

9. **Klein, M.,** Advances in the use of *Bacillus popilliae* for pest control, in *Microbial Control of Pests and Plant Diseases 1970—1980,* Burges, H. D., Ed., Academic Press, London, 1981, 183.

10. **Hurpin, B.,** Recherches épizootiologiques sur la maladie laiteuse à *Bacillus popilliae* "*Melolontha,*" *Ann. Epiphyt.,* 18, 127, 1967.

11. **Hurpin, B.,** Ecologie des scarabeides et lutte biologique contre les vers blacs, *Bull. Soc. Ecol.,* 2, 122, 1971.

12. **Fleming, W. E.,** *Biological Control of the Japanese Beetle, Tech. Bull. No. 1383,* U.S. Department of Agriculture, 1968.

13. **Fleming, W. E.,** *Biology of the Japanese Beetle, Tech. Bull. No. 1449,* U.S. Department of Agriculture, 1972.

14. **Fowler, M.,** Milky disease (*Bacillus* spp.) occurrence and experimental infection in larvae of *Costelytra zealandica* and other Scarabaeidae, *N. Z. J. Zool.,* 1, 97, 1974.

15. **Tashiro, H., Gyrisco, G. G., Gambrell, F. L., Fiori, B. J., and Breitfeld, H.,** Biology of the European Chafer *Amphimallon majalis* (Coleoptera Scarabaeidae) in Northeastern United States, *N.Y. Agric. Exp. Stn. Tech. Bull.,* 828, 1969.

16. **Tashiro, H.,** Coleopteran/scarabaeid Pests, in *Turfgrass Insects of the United States and Canada,* Comstock Publishing, Cornell University Press, Ithaca, NY, 1987, chapters 10 through 15.

17. **Bailey, L.,** The safety of pest-insect pathogens for beneficial insects, in *Microbial Control of Insects and Mites,* Burges, H. D. and Hussey, N. W., Eds., Academic Press, London, 1971, chap. 23.

18. **Heimpel, A. M.,** Safety of insect pathogens for man and vertebrates, in *Microbial Control of Insects and Mites,* Burges, H. D. and Hussey, N. W., Eds., Academic Press, New York, 1971, 469.

19. **Ignoffo, C. M.,** Effects of entomopathogens on vertebrates, *Ann. N.Y. Acad. Sci.,* 217, 141, 1973.

20. **Bulla, L. A., Jr., Rhodes, R. A., and St. Julian, G.,** Bacteria as insect pathogens, *Annu. Rev. Microbiol.,* 29, 163, 1975.

21. **Burges, H. D.,** Risk analysis in the registration of pesticidal bacteria: pathogenicity and toxicological aspects, *Environ. Prot. Biol. Forms Control Pest Organ. Ecol. Bull. (Stockholm),* 31, 81, 1980.

22. **Burges, H. D.,** Safety, safety testing and quality control of microbial pesticides, in *Microbial Control of Pests and Plant Diseases 1970—1980,* Burges, H. D., Ed., Academic Press, New York, 1981, 738.

23. **Obenchain, F. D.,** unpublished information, 1988.

24. **Dutky, S. R.,** Two new spore-forming bacteria causing milky diseases of the Japanese beetle larvae, *J. Agric. Res.,* 61, 57, 1940.

25. **Dutky, S. R.,** Process for propagating bacteria, U.S. Patent 2,293,890, 1942.

26. **Dutky, S. R.,** Method for the control of Japanese beetle, U.S. Patent 2,258,319, 1941.

27. **White, R. T. and Dutky, S. R.,** The effect of the introduction of milky diseases on populations of Japanese beetle larvae, *J. Econ. Entomol.,* 33, 306, 1940.

28. **Aizawa, K.,** *Mushi,* 39, 97, 1967. (as cited in Reference 17)

29. **Milner, R. J.,** A new variety of milky disease, *Bacillus popilliae* var. *rhopaea,* from *Rhopaea verreauxi, Aust. J. Biol. Sci.,* 27, 235, 1974.

30. **Tashiro, H.,** Susceptibility of European chafer and Japanese beetle larvae to different strains of milky disease organisms, *J. Econ. Entomol.,* 50, 350, 1957.

31. **Hurpin, B.,** Sur la virulence de *Bacillus popilliae* Dutky pour les larves du hanneton commun *Melolontha melolontha, C. R. Soc. Biol.,* 151, 1833, 1957.

32. **Hurpin, B.,** Étude de diverses souches de maladie laiteuse sur les larves de *Melolontha melolontha* L. et sur celles de quelques espèces voisines, *Entomophaga,* 4, 233, 1959.

33. **Kawanishi, C. Y., Splittstoesser, C. M., Tashiro, H., and Steinkraus, K. H.,** *Ataenius spretulus,* a potentially important turf pest, and its associated milky disease bacterium, *Environ. Entomol.,* 3, 177, 1974.

34. **Klein, M. G. and Scoles, L. E.,** *Proc. North Cent. Branch Entomol. Soc. Am.,* 33, 42, 1978.

35. **Warren, G. W. and Potter, D. A.,** Pathogenicity of *Bacillus popilliae (Cyclocephala* strain) and other milky disease bacteria in grubs of the southern masked chafer (Coleoptera: Scarabaeidae), *J. Econ. Entomol.,* 76, 69, 1983.

36. **Chang, S. and Wan, Y.,** Studies on milky disease bacteria in China, *Sinozoologia,* 10, 219, 1986.

37. **Hrubant, G. R. and Rhodes, R. A.,** Agglutinability of sporeforming insect pathogens with antiglobulins to milky disease bacteria, *J. Invertebr. Pathol.,* 11, 371, 1968.

38. **Luthy, P. and Krywienczyk, J.,** Serological comparison of three milky disease isolates, *J. Invertebr. Pathol.,* 19, 163, 1972.

39. **Krywienczyk, J. and Luthy, P.,** Serological relationship between three varieties of *Bacillus popilliae, J. Invertebr. Pathol.,* 23, 275, 1974.

40. **Faust, R. M., Spizizen, J., Gage, V., and Travers, R. S.,** Extrachromosomal DNA in *Bacillus thuringiensis* var. *kurstaki,* var. *finitimus,* var. *sotto,* and in *Bacillus popilliae, J. Invertebr. Pathol.,* 33, 233, 1979.

41. **Bakhiet, N. and Stahly, D. P.,** Studies on transfection and transformation of protoplasts of *Bacillus larvae, Bacillus subtilis* and *Bacillus popilliae, Appl. Environ. Microbiol.,* 49, 577, 1985.

42. **Wyss, C.,** Sporulationveruche mit drei Varietaten von *Bacillus popilliae* Dutky, *Zentralbl. Bakteriol. Parasitenkd. Infektionskr. Hyg.,* Abt. 2, 126, 461, 1971.

43. **Luthy, P.,** Zur Bakteriologischen schädlingsbekampfung: die entomopathogenen *Bacillus* Arten, *Bacillus thuringiensis* und *Bacillus popilliae, Vierteljahresschr. Naturforsch. Ges. Zuerich,* 120, 81, 1975.

44. **Bulla, L. A., Jr., Costilow, R. N., and Sharpe, E. S.,** Biology of *Bacillus popilliae* pathogen of scarabaeid beetles, *Adv. Appl. Microbiol.,* 23, 1, 1978.

45. **Schwartz, P. H., Jr. and Sharpe, E.,** Infectivity of spores of *Bacillus popilliae* produced on a laboratory medium, *J. Invertebr. Pathol.,* 15, 126, 1970.

46. **Sharpe, E. S., St. Julian, G., and Crowell, C.,** Characteristics of a new strain of *Bacillus popilliae* sporogenic *in vitro, Appl. Microbiol.,* 19, 681, 1970.

47. **Cloran, J. and McMahon, K. J.,** Use of coatings to protect lyophilized *Bacillus popilliae* from moisture, *Appl. Microbiol.,* 26, 502, 1973.

48. **Hepper, G. R. and McMahon, K. J.,** Viability of lyophilized *Bacillus popilliae* in tung oil pellets coated with paraffin and rubber, *N. D. Res. Rep.,* 54, 1, 1974.

49. **Welsh, R. D. and McMahon, K. J.,** Pathogenicity of vegetative cells of *Bacillus popilliae* for mice, *Trans. Kans. Acad. Sci.,* 72, 509, 1971.

50. **Weiner, B. A.,** Isolation and partial characterization of the parasporal body of *Bacillus popilliae, Can. J. Microbiol.,* 24, 1557, 1978.

51. **Krieg, A.,** The genus *Bacillus:* insect pathogens, in *The Prokaryotes: A Handbook on Habitats, Isolation and Identification of Bacteria,* Vol. 2, Starr, M. P., et al., Eds., Springer-Verlag, New York, 1981, chap. 136.

52. **Ellis, B.-J., Mehta, R., and Obenchain, F. D.,** *In vitro* method for producing infective bacterial spores and spore-containing insecticidal compositions, U.S. Patent No. 4,824,671, 1989.

53. **Heimpel, A. M.,** Safety of insect pathogens, in *Proceedings of the Summer Institute on Biological Control of Plant Insects and Diseases,* Maxwell, F. G. and Harris, F. A., Eds., University Press of Mississippi, Jackson, 1974, 628.

54. **Thompson, J. V. and Heimpel, A. M.,** Microbiological examination of the *Bacillus popilliae* product called Doom, *Environ. Entomol.,* 3, 182, 1974.

55. **Wolfe, G. W. and Mense, M. A.,** Pathogenicity test in mice; *Bacillus popilliae,* final report, Hazelton Laboratory of America Inc., Vienna, VA, 1981.

56. **Reagan, E. L.,** Acute subcutaneous injection study of *Bacillus popilliae* spore preparation, RLI-88F1-FTP in CD-1 mice, Food and Drug Research Laboratories, Waverly, NY, 1988.

57. **Ladd, T. L., Jr. and McCabe, P. J.,** Persistence of spores of *Bacillus popilliae,* the causal organism of type A milky disease of Japanese beetle larvae, *J. Econ. Entomol.,* 60, 493, 1967.

58. **Vos, B. J.,** Safety evaluation by repeated oral administration of *Bacillus popilliae* to rats and monkeys, U.S. Environmental Protection Agency, Pesticide Petition No. 6E1692, Woodward Research Corp., Herndon, VA, 1973.

59. **Reno, F. E.,** Acute dermal toxicity in guinea pigs; *Bacillus popilliae,* in spore form, final report, supplement to U.S. Environmental Protection Agency Pesticide Petition No. 6E1692, Hazelton Laboratory of America Inc., Vienna, VA, 1977.

60. **Auletta, C. S., Daly, I. W., and Lamb, C.,** Primary dermal irritation study in guinea pigs, Bio/dynamics Inc., East Millstone, NJ, 1983.

61. **Reno, F. E.,** Acute eye irritation potential study in rabbits, supplement to U.S. Environmental Protection Agency Pesticide Petition No. 6E1692, Hazelton Laboratory of America Inc., Vienna, VA, 1977.
62. **Heimpel, A. M. and Hrubant, G. G.,** Medical examination of humans exposed to *Bacillus popilliae* and *Popillia japonica* during production of commercial milky disease spore dust, *Environ. Entomol.,* 2, 793, 1973.
63. **Dutky, S. R.,** personal communications, 1972; cited in Reference 19.
64. **Ladd, T. L., Jr.,** U.S. Environmental Protection Agency Pesticide Petition No. 6E1692, personal communications, 1975.
65. **White, R. T.,** Effect of milky disease on *Tiphia* parasites of Japanese beetle larvae, *J. N.Y. Entomol. Soc.,* 51, 213, 1943.
66. **Hawley, I. M. and Dobbins, T. N.,** Leconte's sawfly not susceptible to the type A milky disease organism, Report of Moorestown, NJ Station, August, 1942, Hadley, C. H., Ed., U.S. Environmental Protection Agency Pesticide Petition No. 6E1692, 1977.
67. **Sharpe, E. S.,** Toxicity of the parasporal crystal of *Bacillus thuringiensis* to Japanese beetle larvae, *J. Invertebr. Pathol.,* 27, 421, 1976.
68. **Sharpe, E. S. and Detroy, R. W.,** Susceptibility of Japanese beetle larvae to *Bacillus thuringiensis:* associated effects of diapause, midgut pH, and milky disease, *J. Invertebr. Pathol.,* 34, 90, 1979.

Chapter 14

ENVIRONMENTAL RISKS OF GENETICALLY ENGINEERED ENTOMOPATHOGENS

J. R. Fuxa

It has been recognized that ecology-epizootiology is a central issue in the development and release of genetically engineered microorganisms (GEMs). It is just as likely to be a bottleneck for this new industry as are the laboratory-related and industrial problems. One major reason for so sudden an interest in ecology is risk assessment of these agents. Risk is an estimate of the probability and severity of harm,[19] or a prediction of the likelihood of something going wrong in a certain set of circumstances.[10] Risk assessment is the process of obtaining quantitative or qualitative measures of risk levels.[5]

There are three major environmental concerns about releasing GEMs. The first is that they might have unexpected and deleterious properties outside the laboratory. Some scientists are concerned that recombinant DNA is so distinct from other manipulations that there will be unexpected problems,[17] just as there were with DDT.[2] They point out that genes from almost any organism can be isolated and introduced into any other.[21] Exchange between organisms from different genera will be unpredictable. Genetic exchange in nature is most likely within genera. Double, triple, and quadruple mutants probably never occur in nature. However, they can now be produced in the laboratory.[1] GEMs may have unusual evolutionary potential due to their added genetic variability and possibly their unusual capacity for reproductive isolation.[18] The recipient organism could become pathogenic to nontarget organisms (NTOs); this is particularly a concern with entomopathogens because they are intended to increase virulence, host range, or survivability.[9] Or, they could transfer the gene(s) to another nonpathogen (or other nonpathogens) and make it (them) pathogenic.[17]

Other scientists believe that released GEMs will be more predictable. They argue that there has been no hazard peculiar to recombinant-DNA in work with innumerable genes in many kinds of organisms or in hundreds of laboratories.[17] There is no evidence that hazards of recombinant-DNA organisms are inherently different from those of unmodified organisms or those modified by other techniques.[17] It has been pointed out that, at least with certain types of parental organisms, only one or a few genes are added to up to tens of thousands; thus it is essentially the same organism[3] and may even be safer because an added gene is well characterized[15] and implanted with precision.[18] A National Academy of Science (NAS) committee concluded that the organism to be released and its target environment, not the method by which it was modified, should be the basis for risk assessment.[17] Another response to some of these concerns is that the addition of functional genetic material is not the same as random disruption of DNA or random mutation, even though random mutations can be beneficial and have been used in crop plant breeding.[18] In answer to the concern about pathogenicity to NTOs, it has been mentioned that one gene cannot make that much difference. Pathogenicity results from an array of characteristics and therefore from many more than one gene.[3] Exchanges between distantly related organisms are likely to be *less* harmful because one gene cannot make much difference in changing a nonpathogen to a pathogen or in increasing fitness in the environment.[4]

A second major category of concern is that a released GEM could cause ecological disruptions. The arguments against release center around the concern that the released organism would be unpredictable in the environment and would itself become a pest species. Previous introductions to the U.S., including the gypsy moth (*Porthetria dispar*), the starling (*Sturnus* v. *vulgaris*), and the Kudzu vine (the Southeast Asian *Pueraria lobata*), have caused disruptions. Some of these have been very destructive;[20,23] according to one author, 71 of 854 introductions have caused extinction of a resident species, which is only one of many potential problems.[18] According to another, 75% of domesticated animals have caused pest problems, usually somewhere other than the place where they were domesticated or where wild ancestors lived.[23] Of the world's 18 worst weeds, only one is not used somewhere in agriculture.[23] GEMs could become "super species" with new survival modes and advantages,[18] even with a change in only one or a few genes. Examples include influenza viral strains and resistance of insects and bacteria to pesticides and antibiotics, respectively.[14,20] In an argument especially pertinent to entomopathogens, widespread and persistent mortality agents are very likely to drive the evolution of resistance in the target or other pest species;[22] this is a valid criticism of some of the proposed strategies for release, which seem almost to be designed to induce resistance.[7]

Those in favor of release point out that our entire agricultural system is based on novel "introductions".[14] Major population shifts have been observed, but they reflect changes in the environment rather than or solely due to characteristics of the introduced organism.[4,17] It usually is difficult even to establish an introduced species though this is hard to estimate since failures are seldom reported.[18] By one estimate only 1 of 1000 to 1 of 100 introductions[23] has led to destructive invasions.[17] Domesticated organisms in particular have been harmless or unable to survive, and have a long history of safety.[13] Many of the critical generalizations are not entirely appropriate: some introduced pests were not deliberately introduced;[14] if the recombinant is reintroduced into the area native to the parentals, this is not analogous to the introduction of a non-native species;[17] and successful invasions have been limited mostly to organisms with high reproductive potential and dispersal rates and broad ecological niches. More complex ecosystems have been less susceptible to invasion.[22] With regard to the "super species" argument, the probability that random genetic change will produce a super species is very low; organisms are limited by environment, not by the lack of a gene or small group of genes.[14] Like hybrids, the GEMs will be less fit.[18] The energetic burden of synthesizing additional macromolecules and disruption of normal processes will reduce fitness.[11] The influenza and resistance examples represent very specific genetic changes which will differ from the ones being added by man.[3]

The third major area of concern is the unintended transfer of the genetic material to other organisms. If this happened, then the first two areas of concern resurface: an unintended recipient could have unforeseen properties (e.g., pathogenicity in a nonpathogen) or cause ecological disruptions. There is evidence that genetic transfer can occur. Conjugation, transduction, and transformation occur among bacteria, though perhaps at a lower frequency in the field than in the laboratory. There would be barriers to such transfers in the field, such as concentrations of organisms not being high enough, appropriate species or strains not being present, and barriers to expression; but these barriers would be expected to limit, not halt, such transfers.[16]

On the other hand, known gene transfers between very different organisms have not resulted in new super species, and they usually put the recipient at a disadvantage.[3] Past research indicates that transfers of large segments of genetic material rarely persist in a population unless there is strong selection pressure.[17] A new environment, or a changed one, is what causes populations to explode.[3] Even when genes for pathogenicity are on mobile genetic elements, there is a low probability of dissemination to related bacteria with an array of genes for pathogenicity, and an even lower probability of transfer to unrelated species with such an array.[17]

The potential hazards of using genetically engineered entomopathogens must be balanced against the potential benefits. Basically, genetic engineering is expected to bolster microbial control of insects and thus reduce the array of hazards or harmful side effects associated with chemical insecticides.

No single technology will suffice for risk assessment of GEMs. Only three types of studies can be used prior to release: laboratory experiments with GEMs, release of GEMs into microcosms, and biology and ecology of the parental microorganisms.[7,8] Results of these three types of study cannot be extrapolated to draw conclusions or make predictions about the actual releases of a GEM; therefore it will be important to have all three types of study to minimize the danger of extrapolation. For example, there has been much argument in favor of microcosms to study the fate of a released GEM. But the fate of a GEM released into the environment, which is a critical factor in risk assessment, has the potential to be affected by too many complex factors and their interactions to be duplicated in microcosms.

The probability of environmental harm is a product of six probabilities: release, survival, multiplication, dissemination, transfer of genetic information, and harm.[12] Several of these (survival, multiplication, dissemination) are ecologically important characteristics of an organism.

Thus, the ecology of the entomopathogens that will be parental to GEMs, particularly their survival, dispersal, and population growth, is important to risk assessment. These ecological parameters have been reviewed recently for artificial releases of viruses, bacteria, and fungi.[7,8] Persistence of entomopathogens generally is measured in units of days when they remain exposed to sunlight or to certain other environmental factors. Baculoviruses are likely to persist for years if they reach the soil reservoir. Their dispersal and population growth after release is variable and dependent on factors such as transmission, habitat, and virulence. The introduced population can remain localized and decline, or it can spread over wide areas. The bacteria also persist in soil. However, the rather sketchy data available indicate that only *Bacillus popilliae* and occasionally *B. sphaericus* populations grow after release, and none of the bacteria are known for their long-range dispersal. Fungi can persist for months or 1 to 2 years in soil or host cadavers. Their dispersal and population growth are variable and not well understood, though transport of conidia by wind gives them high potential capability for dispersal. A few models of general epizootiology and specific host-pathogen systems have been published for entomopathogens, but none of them is suitable for prediction of fate of a released pathogen. The information in the literature is heavily biased toward results of successful introductions. Thus this information represents the capabilities of these organisms, not the probabilities of what might happen after release.

Data concerning effects on the environment after release of entomopathogens are scanty but somewhat similar for the three pathogen groups.[7,8] The major demonstrated effects reflect the purpose of release: increase in numbers of pathogen units in the environment and decrease in host-insect numbers and damage to crop plants or other resources. Widespread use of a pathogen or toxin is likely to induce resistance in host populations in many, but not all cases, though there is only one example from microbial control of insects. Isolates of viruses and bacteria tested to date have caused no direct harm to nontarget organisms. Certain fungal isolates can cause allergic reactions, occasional infections in certain reptiles, and infections in certain nontarget arthropods; but they too are generally considered safe. Viruses and bacteria have been demonstrated to indirectly harm invertebrate parasitoids and predators by removing their food source, the host insect; but they rarely have as severe an effect as chemical insecticides on populations of beneficial arthropods.

Thus, it is clear that if a GEM is derived from these parental entomopathogens and if it is not intentionally disabled in some manner, it will have the potential to survive, multiply, and spread. Based only on these previous studies, however, it would be difficult to estimate the probabilities of survival, multiplication, and dissemination of even the parental organisms, much less extrapolate to predict what would happen with GEMs. If, in addition, one adds genetic transfer and harm to risk assessment, it is clear why releases must be evaluated on a case-by-case basis until experience is gained with them.

At present only a qualitative approach to risk assessment of GEMs is feasible; a quantitative one is not. Any attempts at prediction must be viewed as crude approximations.[14] Risk at least for the near future must be determined from opinions of experts in the field.[10] We will not be able to predict risk from statistics for previous deaths or other problems, as is done for automobile deaths or cancer, or even the risk of a city being destroyed by a meteorite.[10] However, qualitative statements should be sufficient for risk assessment.[21]

Overall, GEMs are perceived as having low probability of causing environmental harm but potentially severe consequences if harm occurs.[4,6] Zero risk will be unachievable for the foreseeable future.[5] On the other hand, society cannot insist on zero risk in this endeavor just as it does not do so in others.[14] Care will be necessary in risk assessment of GEMs, particularly as experience is gained in the initial cases. But if care is exercised, most introductions will have low risk of environmental damage.[17,18] Even the severest critics of GEMs conclude that there will be few or no problems in the vast majority of cases.[14]

Excessive concern over the risks of introducing GEMs into the environment could seriously hamper innovation and a whole new U.S. industry, including its competitiveness with similar industries in other nations. On the other hand, just one ecological disaster would likely result in severe, adverse public reaction to the entire concept of using GEMs for insect control. On a positive note, there are two or three points of general agreement among groups on both sides of the issue of whether GEMs should be released, the most important being that everybody wants to avoid any ecological disasters associated with release.

REFERENCES

1. **Alexander, M.,** Spread of organisms with novel genotypes, in *Biotechnology and the Environment,* Teich, A. H., Levin, M. A., and Pace, J. H., Eds., American Association for the Advancement of Science, Washington, D.C., 1985, 115.
2. **Brill, W. J.,** Safety concerns and genetic engineering in agriculture, in *Biotechnology. The Renewable Frontier,* Koshland, D. E., Jr., Ed., Westview Press, Boulder, CO, 1986, 321.
3. **Brill, W. J.,** Why engineered organisms are safe, *Issues Sci. Technol.,* 4, 44, 1988.
4. **Davis, B. D.,** Bacterial domestication: underlying assumptions, *Science,* 235, 1329, 1332, 1987.
5. **Fiksel, J. R. and Covello, V. T.,** The suitability and applicability of risk assessment methods for environmental applications of biotechnology, in *Biotechnology Risk Assessment,* Fiksel, J. F. and Covello, V. T., Eds., Pergamon Press, New York, 1986, 1.
6. **Florio, J. J.,** Regulation in biotechnology, in *Biotechnology. Implications for Public Policy,* Panem, S., Ed., The Brookings Institution, Washington, D.C., 1985, 41.
7. **Fuxa, J. R.,** Risk assessment of genetically-engineered entomopathogens: effects of microbial control agents on the environment including their persistence and dispersal, Environmental Science and Engineering Fellows Report, American Association for the Advancement of Science/Environmental Protection Agency, Washington, D.C., 1987.
8. **Fuxa, J. R.,** Fate of released entomopathogens with reference to risk assessment of genetically engineered microorganisms, *Bull. Entomol. Soc. Am.,* in press.
9. **Karny, G. M., Perpick, J. G., and Levin, M. A.,** Environmental aspects of biotechnology; a discussion, in *Biotechnology and the Environment,* Teich, A. H., Levin, M. A., and Pace, J. H., Eds., American Association for the Advancement of Science, Washington, D.C., 1985, 179.
10. **King, S. R.,** Economic impacts of biotechnology, in *Biotechnology and the Environment,* Teich, A. H., Levin, M. A., and Pace, J. H., Eds., American Association for the Advancement of Science, Washington, D.C., 1985, 29.
11. **Lenski, R. E. and Nguyen, T. T.,** Stability of recombinant DNA and its effects on fitness, in *Planned Release of Genetically Engineered Organisms (Trends in Biotechnology/Trends in Ecology and Evolution Special Publication),* Hodgson, J. and Sugden, A. M., Eds., Elsevier, Cambridge, 1988, S18.
12. **Levin, M. A.,** Review of environmental risk assessment studies sponsored by EPA, *Recomb. DNA Tech. Bull.,* 5, 177, 1982.
13. **Levin, S. A.,** Safety standards for the environmental release of genetically engineered organisms, in *Planned Release of Genetically Engineered Organisms (Trends in Biotechnology/Trends in Ecology and Evolution Special Publication),* Hodgson, J. and Sugden, A. M., Eds., Elsevier, Cambridge, 1988, S47.

14. **Levin, S. A. and Harwell, M. A.**, Environmental risks and genetically engineered organisms, in *Biotechnology. Implications for Public Policy*, Panem, S., Ed., The Brookings Institution, Washington, D.C., 1985, 56.
15. **Marx, J. L.**, Assessing the risks of microbial release, *Science*, 237, 1413, 1987.
16. **Miller, R. V.**, Potential for transfer and establishment of engineered genetic sequences, in *Planned Release of Genetically Engineered Organisms (Trends in Biotechnology/Trends in Ecology and Evolution Special Publication)*, Hodgson, J. and Sugden, A. M., Eds., Elsevier, Cambridge, 1988, S23.
17. **National Academy of Sciences,** *Introduction of Recombinant DNA-Engineered Organisms into the Environment: Key Issues,* National Academy Press, Washington, D.C., 1987.
18. **Regal, P. J.**, Models of genetically engineered organisms and their ecological impact, in *Ecology of Biological Invasions of North America and Hawaii*, Mooney, H. A. and Drake, J. A., Eds., Springer-Verlag, New York, 1986, 111.
19. **Rissler, J. F.**, Research needs for biotic environmental effects of genetically-engineered microorganisms, Environmental Science and Engineering Fellows Report, American Association for the Advancement of Science/Environmental Protection Agency, Washington, D.C., 1983.
20. **Sharples, F. E.**, Regulation of products from biotechnology, *Science*, 235, 1329, 1987.
21. **Simonsen, L. and Levin, B. R.**, Evaluating the risk of releasing genetically engineered organisms, in *Planned Release of Genetically Engineered Organisms (Trends in Biotechnology/Trends in Ecology and Evolution Special Publication)*, Hodgson, J. and Sugden, A. M., Eds., Elsevier, Cambridge, 1988, S27.
22. **Stearns, S., Meyer, J., and Shykoff, J.**, COGENE 88 at Basel, in *Planned Release of Genetically Engineered Organisms (Trends in Biotechnology/Trends in Ecology and Evolution Special Publication)*, Hodgson, J. and Sugden, A. M., Eds., Elsevier, Cambridge, 1988, S2.
23. **Williamson, M.**, Potential effects of recombinant DNA organisms on ecosystems and their components, in *Planned Release of Genetically Engineered Organisms (Trends in Biotechnology/Trends in Ecology and Evolution Special Publication)*, Hodgson, J. and Sugden, A. M., Eds., Elsevier, Cambridge, 1988, S32.

Chapter 15

SAFETY TO NONTARGET INVERTEBRATES OF FUNGAL BIOCONTROL AGENTS

M. S. Goettel, T. J. Poprawski, J. D. Vandenberg, Z. Li, and D. W. Roberts

TABLE OF CONTENTS

I. INTRODUCTION

Insects, mites, nematodes, and other invertebrates as well as plants are susceptible to fungal diseases. Much research emphasis in the past was on the protection of beneficial organisms such as agricultural plants and insects from these pathogenic fungi. Recently, however, interest has shifted to the exploitation of these fungi as control agents of noxious weeds, insects, and other invertebrates. Some fungi are presently used for the control of weeds,[215] nematodes,[101] mites,[138] and insects[202] in various parts of the world. Many more are under development and some are nearing receipt of governmental registration for field use.

There are over 700 described species of entomopathogenic fungi grouped in approximately 100 genera.[182] Major reviews on their pathology are those of Madelin,[135,136] Müller-Kögler,[147] Roberts and Humber,[182] and Steinhaus.[206] Reviews on the prospects of using these fungi for insect and mite control are those of Ferron,[56,57] Hall and Papierok,[82] Lisansky and Hall,[129] Roberts,[181] Roberts and Yendol,[183] and Wraight and Roberts.[234]

This chapter reviews the relative safety to invertebrates of the presently registered as well as potential fungal control agents of insects and mites. Fungi that are specific to beneficial or other nontarget arthropods, fungal toxins, and genetically engineered fungi are not covered.

The sole previous review dealing specifically with the safety of fungi to invertebrates is a section in Müller-Kögler's book which was published over 20 years ago.[147] General reviews on the safety of microbial control agents to invertebrates are those of Bailey[10] and Flexner et al.[61] Papers by Cameron,[27] Chapman et al.,[32] Engler and Arata,[49] Laird,[117] Longworth and Kalmakoff,[131] Pimentel,[163] Pimentel et al.,[164] and Steinhaus[208] have dealt with the potential dangers of entomopathogens to the environment. Hall et al.[83] have published suggested guidelines for the registration of entomopathogenic fungi.

II. REGISTERED ENTOMOPATHOGENIC FUNGI

Four species of fungi are currently registered for use as insect control agents (Table 1); but only two are used on a wide scale. In addition, while *Paecilomyces lilacinus* is actually registered for the control of nematodes, it might also have considerable potential for insect control on rice.[188] In the People's Republic of China (PRC), approximately 7.0×10^5 ha were treated with *Beauveria bassiana* in 1979 for control of pine caterpillars (*Dendrolimus* spp.) and 1.6×10^6 ha in 1976 for control of corn borers (*Ostrinia furnacalis*).[124] In Brazil approximately 1×10^5 ha of sugar cane are treated annually with *Metarhizium anisopliae* to control spittle bugs.[178]

These products are generally safe. Effects on nontarget invertebrates are detailed in the sections to follow. Flexner et al.[61] reviewed some of the tests with the *B. bassiana* product Boverin® against nontarget insects. Mortality was moderate or high only when the beneficial insects ingested the fungus, and not when they were merely contacted by it. With the broad known host ranges of fungi such as *B. bassiana and M. anisopliae,* there is potential for infection or toxicosis of nontarget invertebrates. Continued monitoring following widespread field applications is advisable.

III. HOST RANGE AND SPECIFICITY

A "safe" fungus, with respect to invertebrates, is essentially one with a restricted host range; fungi with narrow host ranges pose the least threat to nontarget organisms (NTOs). This may not be the optimum situation, however, as far as use of such fungi for biocontrol is concerned. NTOs may serve as secondary hosts in which the inoculum is maintained and propagated, thus promoting later infections in the target host population.[55]

TABLE 1
Registered Entomopathogenic Fungi

Pathogen	Trade name	Target species	Country	Ref.
Beauveria bassiana	Boverin®	*Leptinotarsa decemli-neata, Cydia pomonella*	U.S.S.R.	58
	None	*Dendrolimus punctatus, D. tabulaeformis, Ostrinia furnacalis, Neophotettix* spp.	China	94
Hirsutella thompsonii	Mycar®	*Phyllocoptruta oleivora*	U.S.	138
Metarhizium anisopliae	Biocontrol®, Biomax®, Combio®, Metabiol®, Metapol®, Metaquino®	Spittle bugs	Brazil	178, 202
Paecilomyces lilacinus	Biocon®	Nematodes	Philippines	102
Verticillium lecanii	Vertalac®	Aphids	U.K.	80
	Mycotal®	Whiteflies		
	Thriptal®	Thrips		

Fungi have one of the widest spectra of host ranges among entomopathogens. For instance, the host list for *B. bassiana* includes over 700 species.[125] Lists of nontarget invertebrates and those found not susceptible to some common fungal entomopathogens are presented in Table 2. Any conclusions on the host ranges and apparent host specificities of isolates of entomopathogenic fungi based on such host lists must be made with caution. Many records in these lists are based on single specimens, sometimes identifications of both pathogen and host are dubious, and the host ranges have rarely been verified experimentally. Furthermore, such lists are based on fungal species, whereas laboratory studies have shown that different isolates of the same species have varying degrees of specificity.[24,52-54,156] Present evidence indicates that isolates of species with wide host ranges are most virulent to the host from which they were first isolated.[54,59,81,112,156] This difference in virulence according to the host species is usually not present when the fungus is inoculated via injection into the host hemocoel. This suggests that in such cases the resistance is at the level of the cuticle.[108,176]

Fungi are frequently even more specific under field conditions and especially during epizootics. There are several reports of fungi attacking only one host even though closely related susceptible species are present.[82,231] Such resistance is thought to occur as a result of the complex biotic and abiotic interactions which occur in the field.[55,82]

The numerous fungal species believed to be insect order- or family-specific have rarely been tested for their noninfectivity or pathogenicity to species of different orders or families. Some insects can be readily infected in the laboratory by fungi not known to attack them naturally. On the other hand, some fungal species are highly specific. For example, the only known natural host of *Entomophaga maimaiga* is the gypsy moth, *Lymantria dispar,* in Japan. In one of the few host range trials with the Entomophthorales, an isolate of *E. maimaiga* infected only a few of the lepidopteran species tested and not representatives of other insect orders. When caterpillars of the gypsy moth were inoculated by injection of protoplasts, mortalities of 87 to 100% were obtained. In contrast, injection of *E. aulicae* and *E. grylli* from various insect sources caused no mortality.[203] This demonstrates that host resistance for some host-specific fungi is not necessarily at the level of the cuticle.

Host specificity needs much further investigation, both from the point of view of the range of arthropods infected, and from that of the biotic and abiotic factors or interactions responsible for its restriction or extension. In view of the complexities of the field situation, great caution should be exercised when attempting to extrapolate laboratory results to the field situation.

TABLE 2

Host Range and Specificity of Some Fungal Pathogens to Invertebrates

Fungus/target hosts	Nontarget hosts	Challenged and not infected
Chytridiomycetes *Coelomomyces* spp./ Diptera:Culicidae(h)[37]	Ostracoda[132] Copepoda(h)[37,132] Diptera:Culicidae (*Toxorhynchites rutilus* ssp. *septentrionalis*(r)[153]	Oligochaeta:Naididae[233] Branchiopoda:Cladocera (*Daphnia carinata*[233] Ostracoda[233] Malacostraca:Amphipoda (*Orchestia* aff. *tennis*[233], Isopoda (*Notidotea lacustris*[233] Diptera:Chironomidae (*Chironomus zealandicus*[233]
Oomycetes *Lagenidium giganteum*/ Diptera:Culicidae(h)[60,139]	Branchiopoda:Cladocera (*Daphnia* sp.(u)[36] Copepoda(u)[36] Diptera:Chaoboridae (*Chaoborus astictopus*(m)[107]	Polychaeta[140] Gastropoda[139] Branchiopoda:Cladocera[140](*Daphnia pulex*[60] Ostracoda[107]:Podocopa[60] Copepoda[107]:Cyclopoida (*Cyclops* sp.[60,140] Malacostraca:Amphipoda (*Gammarus lacustris*[107], Decapoda (*Cambarus* sp.[139], *Procambarus clarkii*,[60,107] *Palaemonetes pugio*[63]) Odonata:Aecshnidae (*Aecshna* sp.[107], Calopterygidae (*Ischnura* sp.[107]), Libellulidae (*Libellula* sp.[107] Ephemeroptera:Baetidae (*Callibaetes* sp.,[107] *C. montanus*[60]) Heteroptera:Belostomatidae (*Belostoma fluminea*[60,107]), Corixidae (*Corisella* sp.[60]), Coenagrionidae (*Ishnura cervula*[107]), Notonectidae (*Notonecta unifasciata*[107] Diptera:Chaoboridae (*Chaoborus flavicans*[99]), Chironomidae[60,140] Ephydridae,[60] Syrphidae[60] Coleoptera:Hydrophilidae (*Enochrus cuspidatus*,[60] *Hydrophilus triangularis*,[107] *Tropisternus lateralis*[107]), Dytiscidae[140] (*Hygrotus* sp.,[107] *H. medialis*,[60] *Laccophilus decipiens*,[60,107] *Thermonectus basilarus*[60,107]), Coccinellidae (*Hippodamia convergens*[107]) Hymenoptera:Apidae (*Apis mellifera*[107]

Leptolegnia chapmanii[142] Diptera:Culicidae(h)[142]	None reported	Branchiopoda:Cladocera (*Daphnia* sp.[142]) Odonata:Aecshnidae, Libellulidae, Gomphidae[142] Plecoptera:Nemouridae, Peltoperlidae (*Peltoperla* sp.), Perlidae, Perlodidae[142] Diptera:Simuliidae (*Simulium* spp.) Chironomidae, Tipulidae (*Dicranota* sp., *Tipula* sp.)[142] Coleoptera:Dytiscidae, Haliplidae, Noteridae[142] Trichoptera:Hydropsychidae, Limnephilidae, Rhyacophilidae[142]
Zygomycetes *Conidiobolus coronatus* and *C. thromboides*/ Symphyla(h)[73] Acari(r)[191] Isoptera(r)[105] Homoptera:Aphididae(h)[79]	Hymenoptera:Formicidae (*Mesoponera* sp.[155])	Diptera:Syrphidae[79] Coleoptera:Coccinellidae[79] Hemiptera:Anthocoridae[79] Neuroptera:Chrysopidae (*Chrysopa* spp.[79]) Hymenoptera[79]
Entomophaga maimaiga/ Lepidoptera(r-h)[203]	None reported	Orthoptera:Acrididae (*Camnula pellucida*[203]) Coleoptera:Coccinellidae (*Epilachna varivestis*) Chrysomelidae (*Diabrotica undecimpunctata*)[203] Lepidoptera:Geometridae, Noctuidae (*Calpe canadensis, Heliothis virescens, Spodoptera eridania*), Lasiocampidae (*Malacosoma disstria*), Saturniidae (*Hemileuca maia*), Lymantriidae (*Dasychira vagans*), Pyralidae (*Plodia interpunctella*), Pieridae (*Pieris rapae*)[203]
Entomophthora muscae[162] Heteroptera(r)[47,93,211,217] Diptera(l-h)[47,93,211,217]	Diptera:Empididae (*Empis tessellata, Rhamphomyia stigmosa*(r)[122]), Muscidae (*Coenosia muscae*(r)[29]), Sarcophagidae (*Acridiophaga aculeata*(u), *Kellymyia kellyi*(h), *Pseudosarcophaga affinis*(m),[11] Syriphidae(l)[33,93,155,175,184,207,217] (*Eupeodes* sp.(r),[210] *Melanostoma scalare, Platychirus* spp.(r),[122,210] *Scopeuma stercorarium*(r),[122] *Syrphus* spp.(r)[33,217]), Tachinidae (*Dexilla vacua*(r),[210] *Tachina* sp.(r)[33]) Coleoptera:Cantharidae (*Cantharis livida*(r)[47]) Hymenoptera:Torymidae (*Torymus druparum*(r)[47]	Hymenoptera:Apidae (*Apis mellifera*[223]) Diptera:Calliphoridae (*Phormia regina, Calliphora vicina*), Syrphidae (*Allograpta obliqua, Eristalis arbustorum, E. tenax, Toxomerus geminatus, Sphaerophoria scripta, Syrphus* sp.), Tephritidae (*Rhagoletis pomonella*)[211]

TABLE 2 (continued)
Host Range and Specificity of Some Fungal Pathogens to Invertebrates

Fungus/target hosts	Nontarget hosts	Challenged and not infected
Erynia delphacis/ Homoptera(h)[93,228,90] Heteroptera:Miridae(u)[93]	Hemiptera:Miridae (*Cyrtorrhinus lividipennis*(u)[90]	None reported
Erynia pieris/ Lepidoptera(l-h)[126] Homoptera(m)[126] Diptera(l)[126]	None reported	Homoptera:Aphididae (*Aphis fabae, Brevicoryne brassicae*)[126] Coleoptera:Chrysomelidae (*Leptinotarsa decemlineata*), Coccinellidae (*Epilachna varivestis*), Tenebrionidae (*Tribolium confusum*)[126] Lepidoptera:Lymantriidae (*Lymantria dispar*), Pyralidae (*Galleria mellonella, Ostrinia nubilalis, Plodia interpunctella*)[126]
Erynia radicans/ Homoptera(r-h)[93,137,141,156,228] Heteroptera(r)[162] Diptera(r)[122,156] Coleoptera(u)[93] Hymenoptera(h)[12,93,156] Lepidoptera(h)[12,93,156,162]	Diptera:Syrphidae(r)[207,217] Coleoptera:Lampyridae[33] Hymenoptera:Aphidiidae(u),[54] Braconidae(u),[156] Chalcididae(u),[156] Ichneumonidae(u)[12,33,162,216] (*Bathyplectes tristis*(u),[41] *Cratichneumon lanius*(r),[12] *Lissonata* sp. (r)[12]) Pteromalidae (*Pteromalus* sp. (r)[12]), Formicidae(r)[162]	Homoptera:Aphididae (*Acyrthosiphon pisum, Aphis craccivora*), Cicadellidae (*Macrosteles fascifrons*[141] Coleoptera:Coccinellidae (*Coleomegilla maculata Eriopis connexa*[137] Hymenoptera:Aphidiidae (*Trioxys complanatus*[145] Ichneumonidae (*Angitia* sp.[220] Lepidoptera:Lymantriidae (*Lymantria dispar*), Pyralidae (*Galleria mellonella*), Noctuidae (*Trichoplusia ni, Spodoptera exigua, Spodoptera littoralis*)[141]
Neozygites acaridis/ Acari(u)[144,228]	Acari: Macrochelidae (*Macrochelus peregrinus*(h)[144]), Gamasidae, (*Pergamasus crassipes*(r)[144,161])	None reported
Hyphomycetes *Beauveria bassiana/* Gastropoda(u)[46] Acari(h)[46,124,128,191] Orthoptera(h)[46,76] Dermaptera(r)[93] Isoptera(u)[125] Blattaria(h)[46,190] Thysanoptera(r)[46,76]	Acari:Phytoseiidae (*Metaseiulius occidentalis*(l)[45] Araneae(r)[120,159,162,186] Diplopoda:Polydesmida (*Polydesmus* sp. (r)[74] Embioptera:Teratembiidae (*Micrutalis dorsalis*(u)[125] Dermaptera:Forficulidae (*Forficula auricilaria*(r)[217] Mantodea:Mantidae (*Mantis religiosa*(t),[42] *Tenodera capitata*[125]	Diptera:Phoridae (*Megaselia rufipes*[147]), Syrphidae,[194] Tachinidae[43] (*Compsilura concinnata*,[25,46] *Metagonistylum paratheresia*[25] Hymenoptera:Braconidae (*Apanteles congregatus*[46]), Formicidae (*Pheidole megacephala*[30], Trichogrammatidae (*Trichogramma cacoeciae, T. embryophagum, T. evanescens*,[104] *T. pallidum*[110] Neuroptera:Chrysopidae (*Chrysoperla carnea*[110]

Homoptera(h)[46,76,159,188]
Heteroptera(r)[46,76,93,122,217]
Diptera(h)[46,76,111,122]
Coleoptera(h)[2,46,53,76,91,210]
Hymenoptera(u)[76,134,210]
Siphonaptera(h)[125,150]
Lepidoptera(h)[46,76,134]

Heteroptera:Nabidae (*Nabis* spp.(r)[93,125]), Pentatomidae (*Perillus bioculatus*(l)[78,147]), Veliidae (*Mesovelia mulsanti*(r)[93])

Diptera:Bibionidae (*Bibio marci*(r)[190]), Bombyliidae (*Villa brunea*(u)[54]), Cecidomyiidae (*Aphidoletes thompsonii*(u)[100]), Chamaemyiidae (*Gremifania nigrocellulata*(u),[200] *Leucopis* spp.(u)[200]), Dolichopodidae (*Medetera* sp.(u)[230]), Syrphidae (*Eristalis latifrons*(r)[125]), Tachinidae(r)[44,93] (*Exorista sorbillans*(r),[125] *Lixophaga diatraeae*(h)[25,210]),
Tipulidae(r)[90]

Coleoptera:Carabidae[93,146,186] (*Broscus cephalotes*(l),[103] *Carabus violaceus*(u),[116] *Harpalus pubescens*(l),[103,113] *Lebia bivittata*(r),[125] *Pterostichus cupreus*(l)[113]), Chrysomelidae (*Chrysolina hypericii*(r)[207]), Cleridae (*Enoclerus sphegeus*(u),[230] Coccinellidae[44,46,159,200] (*Adalia bipunctata*(l),[46,77] *Adona notata*(l),[98] *Anatis ocellata*(r),[125] *Calvia quattuordecimguttata*(r),[122] *Chilochorus bipustulatus*(u),[200] *Coccinella* spp.,[46,77,88,93,98,110] *Coleomegilla maculata*(h),[93,137] *Cryptolaemus montrouzieri*(m),[110] *Epilachna* spp.(r),[125] *Eriopis connexa*(h),[137] *Exochomus quadripustulatus*(u),[200] *Harmonia punctata*(l),[98] *Hippodamia* spp.(u),[23,77,210] *Rodalia cardinalis*(t),[33] *Scymnus* spp.(u),[77,125] *Semiadalia undecimnotata*(l)[88]),
Colydiidae (*Dastarcus longulus*(r)[125]), Derodontidae (*Laricobius erichsonii*(u)[200]), Hydrophilidae (*Tropisternus* sp.(r)[210]), Nitidulidae (*Gischrochilus quadrasignatus*(r)[210]), Rhizophagidae (*Rhizophagus grandis*(h)[123,230]), Salpingidae (*Pytho americanus*(u)[230]), Silphidae (*Silpha* sp.(r)[39]),
Staphilinidae (*Lathrobium brunnipes*(r)[122]), Tenebrionidae (*Corticeus* sp.(u)[230])

Hymenoptera:Aphelinidae (*Encarsia formosa*(l)[20]),
Apoidea(r),[93] Apidae (*Apis mellifera*(m),[4,76,187,218,219,223,225] *Bombus pratorum*(r),[120,122,162] *B. terrestris*(h)[199]), Braconidae

TABLE 2 (continued)
Host Range and Specificity of Some Fungal Pathogens to Invertebrates

Fungus/target hosts	Nontarget hosts	Challenged and not infected
	(Chelonus annulipes(r), [210] *Coeloides rufovariegatus*(u), [230] *Cotesia glomeratus*(l), [48,190] *Microbracon* sp. (r)[125]), Formicidae[76,90,93,159,217] *(Formica* spp.(r), [125] *Lasius* spp. (r)[122,198]), Ichneumonidae[44,93,162] *(Coelichneumon rudis, Erigorgus femorator*(r)[54]), Vespida[93,162] *(Vespa* spp.(r), [33,125] *Vesperus* sp.(r), [221] *Vespula* spp.(r), [120,122,210]) Neuroptera:Chrysopidae *(Chrysoperla* spp. (u), [93,125,200](l–m)[45,78]) Lepidoptera:Arctiidae *(Hypocrita jacobaeae*(r)[120]), Bombycidae *(Bombyx mori*(h)[4,76,93,134,176,221]), Pyralidae *(Cactoblastis cactorum*(r)[149])	Gastropoda:Basommatophora *(Physa* sp. [35]) Malacostraca:Decapoda, Atyidae[214] Odonata:Anisoptera, Zygoptera[214] Heteroptera:Belostomatidae *(Diplonychus rusticus*[35]), Corixidae, [35] Nepidae, [114] Notonectidae[114] *(Anisops* sp. [35]) Diptera:Psychodidae *(Psychoda* sp., [114] *Telmatoscopus albipunctatus*[214]), Ptychopteridae *(Bittacomomorph clavipes*[114]), Simuliidae *(Simulium vittaum*[72]), Tipulidae, [213] Stratiomyidae[213] *(Stratiomyia* sp. [114]), Tabanidae *(Chrysops* sp., *Tabanus* sp. [114]), Syrphidae[213] Coleoptera:Dytiscidae[114] *(Necterosoma* sp. [35]), Hydrophilidae, Haliplidae[114] Trichoptera[214] Hymenoptera:Apidae *(Apis mellifera*[34])
Culicinomyces clavisporus/ Diptera:Culicidae(h), [214] Ceratopogonidae(h), [64,114,214] Simuliidae(l–h)[114,213]	Diptera:Culicidae *(Aedes alternans*(h)[35]), Chaoboridae *(Chaoborus* sp.(l)[114]), Chironomidae[64] *(Chironomus* sp.(h)[38,114,214]), Dixidae, [64] Syrphidae *(Eristalis aeneus*(h), [114] *E. maculatus*(h)[35]), Ephydridae *(Brachdeutra argentata*(h)[114])	

Hirsutella thompsonii/
Acari(h)[71,93,138]
Coleoptera(m)[91]

Acari:Phytosiidae (*Typhlodromalus peregrinus*(u))[138]

Diptera:Culicidae (*Aedes aegypti, Anopheles stephensi, Culex pipiens*[177])
Coleoptera:Coccinellidae (*Coccidophilus citricola, Lindorus lophanthae*[204])
Hymenoptera:Apidae (*Apis mellifera*[28])
Lepidoptera:Lymantriidae (*Lymantria dispar*[227])

Metarhizium anisopliae/
Symphyla(h)[73,217]
Orthoptera(h)[14,93,224]
Dermaptera(r)[19,158]
Isoptera(h)[84,235]
Homoptera(h)[93,188,210]
Heteroptera(u)[93,224]
Diptera(h)[33,40,111,224]
Coleoptera(h)[33,91,93,122,185,224]
Hymenoptera(r)[14,217,224]
Siphonaptera(h)[150]
Lepidoptera(h)[93,210,217,224]

Malacostraca:Amphipoda[222]
Acari:Astigmatidae (*Histogaster anops*(u), Mesostigmatidae (*Macrocheles* sp.(u))[193]
Ephemeroptera(u)[222]
Dermaptera:Forficulidae(r)[4] (*Forficula auricularia*(r),[19,33] *F. labidura*[221])
Heteroptera:Veliidae(l)[22]
Diptera:Asilidae (*Pleiomma* sp.(r)[224]), Chironomidae (*Chironomus*) sp.(u)[33]), Culicidae (*Toxorhynchites amboinensis*(h)[174], Tachinidae (*Dexia rustica*(r)[54], Tipulidae (*Tipula* sp.[224])
Coleoptera:Carabidae(r)[93,217] (*Amara obesa*(r),[185] *Colpodes japonicus*,[115] *Omus* sp.(r)[185]), Coccinellidae (*Epilachna* sp.(r)[33]), Gyrinidae(h),[179] Lampyridae (*Lamprophorus* sp.(r)[158])
Hymenoptera:Apidae (*Apis mellifera*(l-h)[26,222], Formicidae(r),[14,158] Ichneumonidae (*Amblyteles* sp.(r)[185]), Scoliidae (*Campsomermis quadrifasciata*(r)[14], *C. radula, C. tasmaniensis*(u)[100]), Sphecidae (*Sceliphron spirifex*(r)[66], Tiphiidae (*Tiphia inornata*(u)[212]), Vespidae(r)[66] (*Vespa sylvestris*(r)[159,192])
Lepidoptera:Bombycidae (*Bombyx mori*(h)[33,75,196]

Gastropoda[179]
Branchiopoda:Anostraca, Cladocera (*Daphnia* sp.[180])
Copepoda:Cyclopoida (*Cyclops* sp.[180])
Odonata:Anisoptera[179]
Diptera:Culicidae (*Aedes aegypti*[a,177], Tachinidae (*Metagonistylum minense, Paratheresia claripalpis*[62])
Coleoptera:Tenebrionidae (*Tenebrio molitor*[228])
Hymenoptera:Braconidae (*Apanteles flavipes*[62])

Nomuraea rileyi/
Heteroptera(u)[96]
Diptera(r)[217]
Lepidoptera(h)[33,93,95,192,227]
Coleoptera(h)[2,33,91]

Araneae(r)[192]
Lepidoptera:Bombycidae (*Bombyx mori*(h)[152,176])

Heteroptera:Pentatomidae (*Podisus maculiventris,*[1,95]
Diptera:Culicidae (*Aedes aegypti, Anopheles stephensi, Culex pipiens*[177], Muscidae (*Glossina morsitans*[165]), Tachinidae (*Voria ruralis*[95])

TABLE 2 (continued)
Host Range and Specificity of Some Fungal Pathogens to Invertebrates

Fungus/target hosts	Nontarget hosts	Challenged and not infected
Paecilomyces farinosus/ Homoptera(h)[93,159,217] Heteroptera(r)[93] Phasmida(r)[217] Diptera(l)[93,159,165] Coleoptera(h)[91,93] Hymenoptera(r)[93,120,217] Lepidoptera(h)[33,50,93,192,217]	Araneae(r)[122,159,192] Collembola:Entomobryidae (*Entomobrya unostrigata*(u)[217] Diptera:Cecidomyiidae (*Aphidoletes thompsonii*(u)[200], Tachinidae(r)[93] Coleoptera:Carabidae (*Broscus cephalotes*(m), [103] *Harpalus pubescens*(l), [103,113] *Pterostichus cupreus*(l)[113], Derodontidae (*Laricobius erichsonii*(u)[200] Hymenoptera:Apidae(r)[120] (*Apis mellifera*(m), [192,219] *Bombus terrestris*, [199] *Psithyrus bohemicus*(u)[199]), Formicidae (*Anoplolepsis longipes*[93]), Ichneumonidae, [226] Pteromalidae (*Dibrachys affinis*(h)[226]), Vespidae (*Vespa sylvestris*(r)[122,159,192]) Lepidoptera:Arctiidae (*Hypocrita jacobaeae*(r)[50,122]), Bombycidae (*Bombyx mori*(h)[154,160,226])	Coleoptera:Coccinellidae (*Hippodamia convergens*[95]), Chrysomelidae (*Centonia aurata*[53]) Curculionidae (*Chalcodermus aeneus*, [21] *Hypera postica*[95]), Scarabaeidae (*Melolontha melolontha*, *Oryctes rhinoceros*), [53] Tenebrionidae (*Tenebrio molitor*, *Tribolium confusum*)[2] Hymenoptera:Braconidae (*Apanteles margininventris*, [95] *Microplitis croceipes*[109,201]), Ichneumonidae (*Campoletis soronensis*[95]), Scelionidae (*Telenomus proditor*[95]) Neuroptera:Chrysopidae (*Chrysoperla carnea*[95]) Lepidoptera:Noctuidae (*Heliothis virescens*[224]), Pieridae (*Pieris rapae*[169]), Sphingidae (*Manduca quinquemaculata*, *M. sexta*)[31] Araneae:*Ciniflo ferox*[50] Lepidoptera:Lymantriidae (*Lymantria dispar*a[227]

Paecilomyces lilacinus/
Nematoda(h)[93,100]
Homoptera(h)[188]
Heteroptera(r)[192]
Coleoptera(r)[93]

Lepidoptera:Bombycidae (*Bombyx mori*(r)[93,197])

Diptera:Culicidae (*Aedes aegypti, Anopheles stephensi, Culex pipiens*[176]) Coleoptera:Tenebrionidae (*Tenebrio molitor, Tribolium confusum*[2]) Lepidoptera:Pyralidae (*Plodia interpunctella*[2])

Tolypocladium cylindrosporum/
Diptera(m)[111,118] esp.
Culicidae(h)[118]
Ceratopogonidae(h)[118]
Lepidoptera(l-h)[118]

Branchiopoda:Cladocera (*Daphnia carinata*(m)[70]) Copepoda:Harpacticidae (*Tigriopus* sp. (h)[70]) Ephemeroptera:Leptophlepiidae (*Deleatidium* sp. (m)[70]) Diptera:Chaoboridae (*Chaoborus crystallinus*(l), *C. trivittatus*(h)[118], Dixidae (*Paradixa* sp.(h)[70], Psychodidae[118]

Blattaria:Blattellidae (*Blattella germanica*[118]) Heteroptera:Lygaeidae (*Oncopeltus fasciatus*[118]) Notonectidae (*Anisops* sp.[69])

Verticillium lecanii/
Acari(r)[121,171,202]
Nematoda(r)[85]
Orthoptera(h)[87]
Thysanoptera(m)[202]
Homoptera(h)[33,81,87,93,122]
Diptera(r)[16,162]
Coleoptera(h)[18,91,95,162]
Lepidoptera(r)[93,162]

Acari:Oribatidae(r)[93] Araneae(r)[122] Collembola:Isotomidae (*Podura longicornis, Folsomia cavicola* (u)[120]) Heteroptera:Nabidae (*Nabis alternatus*(h)[87]) Hymenoptera:Aphelinidae (*Encarsia formosa*(l), [20,81] Apidae (*Apis mellifera*(l)[15] *Bombus terrestris*(h), [199] *Psithyrus bohemicus*(r)[199]), Braconidae (*Aphidius matricariae*(u), [194] *Lysiphlebis* sp.(u)[157]), Eulophidae (*Diglyphus intermedius*(l), [157] *Tetrastichus eriophyes*(u)[121]), Ichneumonidae(r), [120,162] Torymidae (*Torymus cyanimus*(r)[162])

Acari:Phytoseiidae (*Phytoseiulus persimilis*[171]) Branchiopoda:Cladocera (*Daphnia* sp. [171]) Oligochaeta:Lumbricidae[171] Diptera:Culicidae (*Aedes aegypti, Anopheles stephensi, Culex pipiens*[177]) Coleoptera:Coccinellidae (*Hippodamia quinquesignata*[87]) Lepidoptera:Pyralidae (*Ostrinia nubilalis*[87])

Note: Susceptibility: (h) = high (75 to 100%), (m) = moderate (50 to 75%), (l) = low (<50%), (u) = unknown, (r) = rare, usually known only from single records.

a   Not susceptible to certain strains.

IV. EFFECTS ON NONTARGET INVERTEBRATES

A. Bees and Other Pollinators

The honeybee, *Apis mellifera,* is used in the commercial production of honey and is the most important pollinator of agricultural crops. Naturally, approval of any agricultural pesticide must include some assessment of its potential impact on honeybees. Such safety tests are required for pesticide registration in most countries. No candidate fungal control agent has been reported to cause an epizootic among honeybees[10,147] or other important pollinators such as syrphids.[229] However, fungi have been reported as either naturally endemic or infective under laboratory conditions while others are not capable of infecting these beneficial insects (Table 2).

Several fungi have been found in *Bombus* spp., both in natural field colonies and in greenhouse test colonies.[199] In particular, *B. bassiana, Paecilomyces farinosus,* and *Verticillium* (= *Cephalosporium*) *lecanii* were isolated from *B. terrestris* queens. *V. lecanii* was frequently isolated from experimental and field colonies of several *Bombus* spp. and *Psithyrus bohemicus. Beauveria brongniartii* (= *tenella*) was isolated from a worker honeybee pupa.[168] The single observations of *B. bassiana* from adult honeybees include those of Alves[4] in Brazil and Rombach[187] in Korea. *Metarhizium anisopliae* was also isolated from adult honeybees.[26]

Several studies have been conducted attempting to infect honeybees with candidate fungal control agents. *B. bassiana* was found to infect adult bees;[5,218] Vincens[225] reported spread of the infection from inoculated to uninoculated bees. In contrast, Toumanoff[219] attributed the mortality he observed to toxins as he never found either *B. bassiana* or *P. farinosus* (= *Isaria farinosa*) colonizing the body cavity. Vandenberg[223] employed a commercial preparation of *B. bassiana* spores and obtained high infection rates among adult bees in cages using doses of ca. 1×10^4 spores per bee. In laboratory challenge tests with *M. anisopliae,* less than 10% of inoculated bees became infected.[26] *Verticillium lecanii* was reported to infect bees.[199] However, other studies with commercial *V. lecanii* preparations indicated that mortality was due to mechanical blockage of thoracic spiracles and not to infection.[9]

Adult honeybees were not susceptible to infection by several other fungi in various tests. The fungi include *Culicinomyces clavisporus,*[34] *Entomophaga maimaiga* (as *aulicae*),[223] *Hirsutella thompsonii,*[28] and *Lagenidium giganteum.*[107]

Since no natural epizootics caused by any potential fungal control agents have been reported among bees or other pollinators, there is likely a low risk to these insects if field application of fungi is pursued. However, caution is warranted if applications include direct exposure to bee colonies or foragers.

B. SILKWORMS

Fungal diseases have plagued sericulture since man first began using insects for silk production. Through modern practices such as control of rearing humidity and temperature, strain selection, and basic hygiene, fungal diseases in silkworm farms are presently either eliminated or kept below an economic threshold.[4,92,127,170] Indeed, the success in control of *B. bassiana* mycoses in Chinese sericulture resulted in a critical shortage of infected worms which are valued for their medical purposes. This shortage was overcome by the deliberate inoculation of silkworm pupae that remained after reeling the silk.[6,7]

Many species of potential fungal control agents can cause epizootics in the silkworm, *Bombyx mori.* These include *Beauveria bassiana, Nomuraea rileyi, Paecilomyces farinosus, Metarhizium anisopliae* (Table 2), and *P. fumosoroseus.*[196] For this reason, it is illegal to use microbial insecticides in certain areas of the PRC and Japan. However, *B. bassiana* epizootics in *Bombyx mori* have not occurred in many areas of the PRC where it is used on a large scale for pest control, even when applications were within 70 m of silkworm stock farms.[8,124]

The results of a Chinese study further indicate that application of *Beauveria bassiana* for pest control should not pose a threat to sericulture.[8] As part of the study, a silkworm-rearing shed was built within a pine plantation. Between 1975 and 1976, 110 ha of the plantation were treated six times with 2100 kg of a *B. bassiana* preparation. Good control of pine caterpillars (*Dendrolimus* spp.) was obtained, but incidence of *B. bassiana* in the silkworms was under 4%. This was not significantly different from mortality in nontreated control areas. Additional laboratory studies indicated that strains of *B. bassiana* isolated from pine caterpillars were 100 times less virulent to the silkworm than strains isolated from the latter. Similar differences in the susceptibility of *Bombyx mori* to various isolates of *Beauveria bassiana* were demonstrated by Riba et al.[176]

Indications are that, through careful selection of both pathogen and silkworm strains and employment of basic hygiene practices, it may be possible to use fungal control agents in the immediate vicinity of sericulture farms without endangering the silkworms. Indeed, studies in Japan suggest that a strain of *P.* (= *Spicaria*) *fumosoroseus* can be used in sericulture to control the silkworm tachinid parasite, *Blepharipa* (= *Crossocosmia*) *zebina*.[106]

C. PREDATORS AND PARASITOIDS

In order to protect invertebrate biocontrol agents of insect pests, the possible adverse effects of fungal control agents on these organisms must be considered. It has been shown in a number of cases that fungi can be detrimental, either directly or indirectly, to predators and parasitoids. Other studies show that the two may be compatible (see Chapter 5).

Mycoses have been observed in a number of predators and parasitoids[13,147] (Table 2). Little is known about the epizootiology of these mycoses and resultant effect on the control of target species; however, indications are that in nature mycoses in these insects are relatively uncommon and many species seem refractory to infection[147] (Table 2). An exception is the occurrence of *Beauveria bassiana* epizootics in hibernating coccinellids.[77,88,97,98,143] This suggests, as pointed out by Flexner et al.,[61] that application of some fungal control agents may adversely affect certain populations of predators during hibernation. *B. bassiana* also causes epizootics in the mass rearing of the *Dendroctonus* predator, *Rhizophagus grandis*;[123,230] however, such infections can probably be controlled by strict screening and facility hygiene.[92]

In one of the few field studies designed to study the effects of fungal application on NTOs, Baltensweiler and Cerutti[17] found that application of *B. brongniartii* blastospores to control the May beetle (*Melolontha melolontha*) resulted in an overall infection rate of 1.1% in the invertebrate NTOs. No infections were found in predators such as coccinellids and neuropterans, however, up to 9% of spiders were found infected. Nevertheless, the authors concluded that the application of *B. brongniartii* should not endanger the nontarget invertebrate fauna since overall mortality was low and the inoculum quickly disappeared from the environment.

Fungal control agents may affect predators or parasitoids indirectly either by depleting the host population or, in the case of parasitoids, by competing with them in the host tissue. The classical example of indirect effects by host depletion is that of Ullyett and Schonken in South Africa.[220] They found that early season natural epizootics of *Erynia radicans* among populations of the diamondback moth, *Plutella maculipennis*, resulted in a collapse of nonsusceptible predator and parasitoid populations. In Nicaragua, Falcon[51] also observed that fungal epizootics (*Entomophthora* spp., *Nomuraea rileyi*, and *Aspergillus flavus*) coincided with collapse of beneficial insect populations near the end of the wet season.

Although parasitoids are generally not infected by fungi within the host tissue, several studies have demonstrated that both parasitoids and pathogens may compete for the host tissue. Parasitoids may even predispose the host to infection. The pathogen usually overcomes early instar parasitoids whereas older parasitoids are able to complete their development.

This has been shown for fungus/parasitoid/host relationships between *Nomuraea rileyi/ Microplites croceipes/Heliothis zea,*[109] *Erynia radicans/Trioxys complanatus/Therioaphis trifolii,*[145] *Erynia neoaphidis/Aphidius sonchi/Hyperomyzus lactucae,*[145] *Erynia neoaphidis/ A. rhopalosiphi/Metopolophium dirhodum,*[167] and *Beauveria bassiana/Cotesia (Apanteles) glomerata/Pieris brassicae* (see Chapter 5).[48,67]

Predisposition to fungal infection of parasitized hosts is thought to be the result of weakening of the host cuticle which facilitates fungal entry. The subsequent development of the fungus within parasitized hosts is inhibited by an antimycotic substance secreted by the parasitoid.[48,232] An antimycotic substance is also produced by the symbiotic bacterium, *Xenorhabdus nematophilus,* in the entomopathogenic nematode, *Steinernema feltiae* (= *Neoaplectana carpocapsae).*[166] Although fungal infections of entomopathogenic nematodes have been observed,[102] it is not known if an antimycotic protects the nematode from fungal infection within its host.

Relatively few studies have dealt with the compatibility of entomopathogenic fungi with the natural or introduced species of parasitoids and predators to determine which ones can be used most effectively in an integrated manner. The effective coexistence of species of *Hirsutella, Myiophagus,* and *Entomophthora,* and of parasites and predators in the control of citrus pest arthropods has been demonstrated in Florida by Muma.[148] Ignoffo[95] showed that natural epizootics of *Nomuraea rileyi* in the noctuid, *Plathypena scabra,* did not apparently affect the efficacy of the pest's natural parasite complex. This also appeared to be true for the *N. rileyi/Microplitis croceipes/Heliothis* spp. complex in corn fields of the central Mississippi delta.[201] Introduction of the parasitoids *Microtonus aethiopoides* and *M. colesi* into eastern Ontario has resulted, in conjunction with *Erynia* sp., in long-term suppression of the weevil, *Hypera postica,*[86] even though the fungus strongly competes with *M. colesi.*[130] Dresner[46] observed that the introduction of *Beauveria bassiana* in the field for control of aphids, tomato hornworm, *Manduca quinquemaculata,* and Mexican bean beetles, *Epilachna varivestis,* did not seem to affect syrphids, chrysopids, the parasitoid, *Cotesia congregatus,* or the spined soldier bug, *Podisus maculiventris.* In Venezuela, the natural occurrence of *V. lecanii,* together with the predators *Cycloneda sanguinea, Oxyptamus gastrostactus,* and *Zelus* sp. afforded considerable control of the aphid *Toxoptera citricidus.*[189] Berisford and Tsao[22] demonstrated an inverse relationship between hyphomycetous fungi attacking the bagworm, *Thyridopteryx ephemeraeformis,* and its parasitoids; fungi were the key mortality factors during wet periods while parasitoids were the predominant ones during dry periods.

Several studies have demonstrated the compatibility of fungi and parasitoids or predators in integrated pest management programs. Studies in Brazil have shown that it may be possible to use *Metarhizium anisopliae* and three parasitoids of the sugarcane borer, *Diatraea saccharalis,* simultaneously in integrated management programs of sugarcane pests.[62] The most successful use of fungal biocontrol agents in conjunction with predators and parasitoids has been in greenhouses. *Verticillium lecanii* can be used with the predatory mite, *Phytoseiulus persimilis,* and the beneficial wasp, *Encarsia formosa.*[81,119,171] *Aschersonia aleyrodis* does not infect larvae parasitized by *E. formosa* and infects primarily the first two instars of the greenhouse whitefly, *Trialeurodes vaporariorum,* while *E. formosa* is able to discriminate between healthy and infected larvae, a situation which provides optimum utilization of the host by both parasitoid and pathogen.[65,119,172,173]

It thus appears that in many cases fungal control agents can be integrated with parasitoids and predators. The strategic release of entomogenous fungi and parasitoids appears to be important. In many cases only good synchronization will assure joint action against the host and prevent antagonism, competition, or inhibition between both types of biological control agents.

D. OTHERS

Although many species of fungi are known to infect noninsect invertebrates,[3,151,205,209] with few exceptions entomopathogenic fungi do not occur in these organisms. The most extensive nontarget challenge tests have been conducted with the mosquito pathogenic fungi *Coelomomyces opifexi*,[233] *Lagenidium giganteum*,[60,107] *Leptolegnia chapmanii*,[142] *Culicinomyces clavisporus*,[35,114,214] and *Metarhizium anisopliae*.[179,180] The results of these studies indicate that these fungi should not be harmful to most aquatic nontarget invertebrates (Table 2). Lack of natural infections in these organisms further supports this. Therefore, present evidence suggests that fungal entomopathogens with practical control potential are not a threat to invertebrates other than insects.

V. DISCUSSION AND CONCLUSIONS

A review of the present knowledge on potential fungal control agents indicates that these organisms pose a minimal risk to NTOs. Indeed, when compared with chemical insecticides, fungal biocontrol agents offer, among other advantages, a method of control that has a very narrow host range, can usually be integrated with other biocontrol agents, may provide prolonged control by establishment and recycling within the habitat, and is also biodegradable.

The best documented cases of detrimental effects of entomopathogenic fungi on nontarget invertebrates are indirect effects on the predator and parasitoid populations through host depletion. In certain instances, fungi may also adversely directly affect certain nontarget invertebrates including predators and parasitoids. It must be realized, however, that generally speaking, it is not possible to reduce the population level of a pest without also adversely affecting another component of the ecosystem. For instance, by depleting or even reducing a host population, it is inevitable that the host's predators and parasites will be adversely affected. In such situations, it is the responsibility of pest managers to integrate the use of the biocontrol agents in order to utilize all control methods in the most efficient manner. For instance, the preservation of a beneficial insect may only require the proper timing of a fungal application (see Chapter 5, Section IV).

In any case, the general consensus is that fungi do pose inherent, albeit minimal risks and therefore should be regulated in some manner. Consequently, registration is of paramount importance. Most guidelines for registration of entomopathogenic fungi require laboratory testing for infectivity to nontarget invertebrates.[83,133] However, limitations of the present knowledge of fungal specificity and how it relates to epizootiology make it impossible to extrapolate such data to the field situation. Nevertheless, limited laboratory infectivity studies with the formulated product against nontarget invertebrates may identify potential hazards that should be addressed during field trials.

With respect to safety to invertebrates, safe and optimum use of fungi can only be determined for each specific integrated management system. To ensure maximum safety to invertebrates, fungal control agents must not become just a replacement for chemical insecticides, but should be integrated with other control strategies.[51,68] This can only be accomplished if efficient formulated products are available to the researcher to test.

A major current limitation on the use of fungal biocontrol agents is the lack of formulated preparations. The private sector in turn is reluctant to develop a product which will then require extensive and costly safety testing prior to field evaluation. Therefore, it is imperative that necessary safety regulations and registration procedures do not hinder the development and consequent testing and utilization of such formulations. A better approach for ensuring long-term safety to invertebrates would be the continued monitoring during field use after preliminary laboratory and field safety testing.

ACKNOWLEDGMENTS

We thank Joyce Barmore and Kurt Haas for help in reference collection and manuscript preparation, Karen Toohey for helpful suggestions, Felix Sperling for providing translations,

Michiel Rombach for critical review, the many scientists who answered our pleas for information and especially those who provided unpublished data, and the late E. A. Steinhaus whose reprint collection, presently housed at the Boyce Thompson Institute, proved indispensable. Partial support for the preparation of this manuscript was provided by the Natural Sciences and Engineering Research Council of Canada, the USAID Bean/Cowpea Collaborative Research Support Program, and the Jesse Smith Noyes Foundation.

REFERENCES

1. **Abbas, M. S. T. and Boucias, D. G.**, Interaction between the fungas *Nomuraea rileyi*, the cabbage looper, *Trichoplusia ni*, and the predator, *Podisus maculiventris* (Say), *Proc. 17th Annu. Meet., Soc. Invertebr. Pathol.*, 1985, 33.

2. **Aizawa, K., Shimazu, T., and Shimizu, S.**, Pathogenicity of microorganisms to stored-product insects, Proc. Joint U.S.-Japan Seminar on Stored Products Insects, Manhattan, KS, 1976, 59.

3. **Alderman, D. J.**, Fungi as pathogens of non insect invertebrates, in *Fundamental and Applied Aspects of Invertebrate Pathology*, Samson, R. A., Vlak, J. M., and Peters, D., Eds., Foundation of the 4th Int. Colloq. Invertebrate Pathology, Wageningen, 1986, 354.

4. **Alves, S. B.**, personal communication, 1987.

5. **Anon.**, What effect do insect pathogens have on honeybees?, *Agric. Res.*, 14, 8, 1965.

6. **Anon.**, Mummified silkworm pupae as an alternative of larvae, *Use Control Anim.*, 5, 15, 1972 (Chinese).

7. **Anon.**, Production of muscardined pupal tablets by artificial inoculation, *Microbiology*, 3, 30, 1976 (Chinese).

8. **Anon.**, Studies on the relationship between the application of *Beauveria bassiana* against the pine caterpillar and the silkworm muscardine occurrence, in *Proceedings on the Biocontrol of Forest Pests*, Forestry Press China, Beijing, 1981, 142 (Chinese).

9. **Anon.**, Annual Report Rothamsted Experimental Station, United Kingdom, 1984, 93.

10. **Bailey, L.**, The safety of pest-insect pathogens for beneficial insects, in *Microbial Control of Insects and Mites*, Burges, H. D. and Hussey, N. W., Eds., Academic Press, New York, 1971, 491.

11. **Baird, R. B.**, Notes on a laboratory infection of Diptera caused by the fungus *Empusa muscae* Cohn, *Can. Entomol.*, 89, 432, 1957.

12. **Balazy, S.**, Entomophthoraceous fungi parasitic on Hymenoptera, *Bull. Acad. Pol. Sci. Ser. Sci. Biol.*, 29, 227, 1982.

13. **Balazy, S.**, Observations on mycosis found in some beneficial insects in forestry, *Rocz. Wyz. Szk. Roln. Poznan.*, 27, 15, 1965.

14. **Balfour-Browne, F. L.**, The green muscardine disease of insects, with special reference to an epidemic in a swarm of locusts in Eritrea, *Proc. R. Entomol. Soc. London*, (A)35, 65, 1960.

15. **Ball, B. V.**, personal communication, 1987.

16. **Ballard, E. M. and Knapp, F. W.**, Occurrence of the fungus *Verticillium lecanii* on a new host species: *Aedes triseriatus* (Diptera: Culicidae), *J. Med. Entomol.*, 21, 751, 1984.

17. **Baltensweiler, W. and Cerutti, F.**, A study of the possible side effects of using the fungus *Beauveria brogniartii* to control the May beetle on the fauna of the forest edge, *Mitt. Schweiz. Entomol. Ges.*, 59, 276, 1986 (German with English summary).

18. **Barson, G.**, Laboratory studies on the fungus *Verticillium lecanii*, a larval pathogen of the large elm bark beetle *Scolytus scolytus*, *Ann. Appl. Biol.*, 83, 207, 1976.

19. **Barss, H. P. and Stearns, H. C.**, The green muscardine fungus (*Oospora destructor*[Metschn.] Delacroix) on European earwig and other insects in Oregon, *Phytopathology*, 15, 729, 1925.

20. **Beglyarov, G. A. and Maslienko, L. V.**, The toxicity of certain pesticides to *Encarsia*, *Rev. Appl. Entomol. Ser. A*, 67, 328, 1979.

21. **Bell, J. V. and Hamalle, R. J.**, Three fungi tested for control of the cowpea Curculio, *Chalcodermus aeneus*, *J. Invertebr. Pathol.*, 15, 447, 1970.

22. **Berisford, Y. C. and Tsao, C. H.**, Parasitism, predation, and disease in the bagworm, *Thyridopteryx ephemeraeformis* (Haworth) (Lepidoptera: Psychidae), *Environ. Entomol.*, 4, 549, 1975.

23. **Billings, F. H. and Glenn, P. A.**, Results of the artificial use of the white fungus disease in Kansas, *U.S. Dep. Agric. Bull.*, 107, 1, 1911.

24. **Boucias, D. G. and Pendland, J. C.**, Host recognition and specificity of entomopathogenic fungi, in *Infection Processes of Fungi*, Roberts, D. W. and Aist, J. R., Eds., Rockefeller Foundation, New York, 1984, 185.

25. **Box, H. E. and Pontis Videla, R. E.**, Apuntes sobre el hongo entomógeno *Beauveria bassiana* (Mont.) Vuill., parásito de *Diatraea* en Venezuela, *Agron. Trop. Maracay Venez.*, 1, 233, 1951 (Spanish with English summary).

26. **Burnside, C. E.**, Fungous disease of the honeybee, *U.S. Dep. Agric. Tech. Bull.*, 149, 1, 1930.

27. **Cameron, J. W. M.**, Suitability of pathogens for biological control, in *Insect Pathology and Microbial Control*, Proc. Int. Coll. Insect Pathol. Microb. Control, Wageningen, 1967, 182.

28. **Cantwell, G. E. and Lehnert, T.**, Lack of effect of certain microbial insecticides on the honeybee, *J. Invertebr. Pathol.*, 33, 381, 1979.

29. **Carruthers, R. I., Haynes, D. L., and MacLeod, D. M.**, *Entomophthora muscae* (Entomophthorales: Entomophthoraceae) mycosis in the onion fly, *Delia antiqua* (Diptera: Anthomyiidae), *J. Invertebr. Pathol.*, 45, 81, 1985.

30. **Castineiras, A. and Calderon, A.**, Susceptibility of *Pheidole megacephala* to three microbial insecticides: Dipel, Bitoxobacillin 202 and *Beauveria bassiana* under laboratory conditions, *Ciencia Tec. Agric. Prot. Plantas Suppl.*, 61, 1982 (Spanish with English summary).

31. **Chamberlin, F. S. and Dutky, S. R.**, Tests of pathogens for the control of tobacco insects, *J. Econ. Entomol.*, 51, 560, 1958.

32. **Chapman, H. C., Davidson, E. W., Laird, M., Roberts, D. W., and Undeen, A. H.**, Safety of microbial control agents to non-target invertebrates, *Environ. Conserv.*, 6, 278, 1979.

33. **Charles, V. K.**, A preliminary check list of the entomogenous fungi of North America, *U.S. Dep. Agric. Insect Pest Survey Bull.*, 21, 207, 1941.

34. **Cooper, R., Hornitzky, M., and Medcraft, B. E.**, Non-susceptibility of *Apis mellifera* to *Culicinomyces clavosporus*, *J. Aust. Entomol. Soc.*, 23, 173, 1984.

35. **Cooper, R. W. and Sweeney, A. W.**, Effect of *Culicinomyces* against non-target organisms in the laboratory and following field application of the fungus for mosquito control, unpublished report to World Health Organization, 1981, 1.

36. **Couch, J. N.**, A new saprophytic species of *Lagenidium*, with notes on other forms, *Mycologia*, 27, 376, 1935.

37. **Couch, J. N. and Bland, C. E.**, The Genus *Coelomomyces*, Academic Press, New York, 1985, 416.

38. **Couch, J. N., Romney, S. V., and Rao, B.**, A new fungus which attacks mosquitoes and related Diptera, *Mycologia*, 66, 374, 1974.

39. **Danysz, J.**, Infestation du silphe opaque avec *Spototrichum globuliferum*, *Bull. Soc. Entomol. S.*, 181, 1894 (as cited in Reference 76).

40. **Daoust, R. A. and Roberts, D. W.**, Virulence of natural and insect-passaged strains of *Metarhizium anisopliae* to mosquito larvae, *J. Invertebr. Pathol.*, 40, 107, 1982.

41. **Dicke, F. F.**, First record of *Bathyplectes tristis* (Grav.), a parasite of the clover leaf weevil in the United States, *J. Econ. Entomol.*, 30, 375, 1937.

42. **Dieuzeide, R.**, Les champignons entomophytes du genre *Beauveria* Vuill. Contribution a l'étude de *Beauveria effusa* Vuill. parasite du doryphore, *Ann. Epiphyt.*, 11, 185, 1925 (as cited in Reference 76).

43. **Djadečko, N. P. and Sikura, A. I.**, Biologische Bekämpfung von *Tortrix viridana*, Lesn. Khoz., 10, 56, 1961 (Russian) (as cited in Reference 147).

44. **Donaubauer, E.**, Über eine Mykose der Latenzlarve von *Cephaleia abietis* L. Sydowia *Ann. Mycol. Ser. 2*, 13, 183, 1959 (as cited in Reference 147).

45. **Donegan, K.**, personal communication, 1987.

46. **Dresner, E.**, Culture and use of entomogenous fungi for the control of insect pests, *Contrib. Boyce Thompson Inst.*, 15, 319, 1949.

47. **Eilenberg, J., Bresciani, J., and Martin, J.**, *Entomophthora* species with *E. muscae* like primary spores on two new insect orders, Coleoptera and Hymenoptera *Nord. J. Bot.*, 7, 577, 1987.

48. **El-Sufty, R. and Führer, E.**, Interrelationships between *Pieris brassicae* L. (Lep., Pieridae), *Apanteles glomeratus* L. (Hym., Braconidae) and the fungus *Beauveria bassiana* (Bals.) Vuill, *Z. Angew. Entomol.*, 92, 321, 1981.

49. **Engler, R. and Arata, A. A.**, Public health and environmental safety, in *Tsetse: The Future for Biological Methods in Integrated Control*, Laird, M., Ed., International Development and Research Centre, Ottawa, 1977, 157.

50. **Evans, R. E.**, Some entomogenous fungi, *Proc. Birmingham Nat. Hist. Soc.*, 21, 33, 1967.

51. **Falcon, L. A.**, Insect pathogens: integration into a pest management system, in *Proceedings of the Summer Institute on Biological Control*, Maxwell, F. G. and Harris, F. A., Eds., University Press of Mississippi, Jackson, 1974, 618.

52. **Fargues, J.**, Sensibilité à *Beauveria bassiana* (Bals.) Vuill. (Fungi Imperfecti, Moniliales) des larves de doryphore *Leptinotarsa decemlineata* Say (Coléoptère, Chrysomelidae) soumises à des doses réduites d'insecticide chimique, Ph.D. thesis, University of Paris, Paris, 1972.

53. **Fargues, J.**, Spécificité des champignons pathogènes imparfaits (Hyphomycètes) pour les larves de Coléoptères (Scarabaeidae et Chrysomelidae) *Entomophaga*, 21, 313, 1976.

54. **Fargues, J.**, Spécificité des hyphomycètes entomopathogènes et résistance interspécifique des hyphomycètes entomopathogènes et résistance interspécifique des larves d'insectes, D.Sci. thesis, Pierre & Marie Curie University, Paris, 1981.

55. **Fargues, J. and Remaudière, G.**, Considerations on the specificity of entomopathogenic fungi, *Mycopathologia*, 62, 31, 1978.

56. **Ferron, P.**, Biological control of insect pests by entomogenous fungi, *Annu. Rev. Entomol.*, 23, 409, 1978.

57. **Ferron, P.**, Insect control, in *Comprehensive Insect Physiology, Biochemistry and Pharmacology*, Kerkut, G. A. and Gilbert, L. I., Eds., Pergamon Press, Paris, 1985, 314.

58. **Ferron, P.**, Pest control by the fungi *Beauveria* and *Metarrhizium*, in *Microbial Control of Pests and Plant Diseases 1970—1980*, Burges, H. D., Ed., Academic Press, London, 1981, 465.

59. **Ferron, P., Hurpin, B., and Robert, P.-H.**, Sur la spécificité de *Metarrhizium anisopliae* (Metsch.) Sorokin, *Entomophaga*, 17, 165, 1972.

60. **Fetter-Lasko, J. L. and Washino, R. K.**, *In situ* studies on seasonality and recycling patterns in California of *Lagenidium giganteum* Couch, an aquatic fungal pathogen of mosquitoes, *Environ. Entomol.*, 12, 635, 1983.

61. **Flexner, J. L., Lighthart, B., and Croft, B. A.**, The effects of microbial insecticides on non-target, beneficial arthropods *Agric. Ecosyst. Environ.*, 16, 203, 1986.

62. **Folegatti, M. E. G. and Alves, S. B.**, Interacção entre o fungo *Metarhizium anisopliae* (Metsch.) Sorok., 1883 e os princípais parasitóides da broca da cana-de-açúcar *Diatraea saccharalis* (Fabricius, 1794), *An. Soc. Entomol. Brasil*, 16, 351, 1987. (Portuguese with English abstract).

63. **Foss, S. S., Courtney, L. A., and Couch, J. A.**, Evaluation of a fungal agent *Lagenidium giganteum* under development as an MPCA for nontarget risks, Progress report for Office of Pesticides, EPA/600/X-86/229, Programs ORD, U.S. Environmental Protection Agency, Washington, D.C., 1986, 1.

64. **Frances, S. P., Lee, D. J., Russell, R. C., and Panter, C.**, Seasonal occurrence of the mosquito pathogenic fungus *Culicinomyces clavosporus* in a natural habitat, *J. Aust. Entomol. Soc.*, 24, 241, 1985.

65. **Fransen, J. J.**, Current status on the use of fungus *Aschersonia aleyrodis* against greenhouse whitefly, *Proc. 20th Annu. Meet. Soc. Invertebr. Pathol.*, 1987, 70.

66. **Friederichs, K.**, Über die Pleophagie des Insektenpilzes *Metarhizium anisopliae* (Metsch.) Sor. *Zentralbl. Bakteriol. Parasitenkd.*, 50, 335, 1920 (as cited in Reference 14).

67. **Führer, E. and El-Sufty, R.**, Production of fungistatic metabolites by teratocytes of *Apanteles glomeratus* L. (Hym., Braconidae), *Z. Parasitenkd.*, 59, 21, 1979 (German with English summary).

68. **Fuxa, J. R.**, Ecological considerations for the use of entomopathogens, *Annu. Rev. Entomol.*, 32, 225, 1987.

69. **Gardner, J. M.**, Mosquito larvicidal potential of the New Zealand strain of *Tolypocladium cylindrosporum*, M. Sci. diss., University of Otago, Dunedin, 1984.

70. **Gardner, J. M. and Pillai, J. S.**, *Tolypocladium cylindrosporum* (Deuteromycotina: Moniliales), a fungal pathogen of the mosquito *Aedes australis*. III. Field trials against two mosquito species, *Mycopathologia*, 97, 83, 1987.

71. **Gardner, W. A., Oetting, R. D., and Storey, G. K.**, Susceptibility of the two spotted spider mite, *Tetranychus urticae* Koch, to the fungal pathogen *Hirsutella thompsonii* Fisher, *Fla. Entomol.*, 65, 458, 1982.

72. **Gaugler, R. and Jaronski, S.**, Assessment of the mosquito-pathogenic fungus *Culicinomyces clavosporus* as a black fly (Diptera: Simuliidae) pathogen, *J. Med. Entomol.*, 20, 575, 1983.

73. **Getzin, L. W. and Shanks, C. H.**, Infection of the garden symphylan, *Scutigerella immaculata* (Newport) by *Entomophthora coronata* (Constantin) Kevorkian and *Metarrhizium anisopliae* (Metchnikoff) Sorokin, *J. Invertebr. Pathol.*, 6, 542, 1964.

74. **Giard, A.**, L' *Isaria densa* (Link) Fries' champignon parasite du hanneton commun *(Melolontha vulgaris L.)*, *Bull. Sci. Nord France Belg.*, 24, 1, 1892 (as cited in Reference 76).

75. **Glaser, R. W.**, The green muscardine disease in silkworms and its control, *Ann. Entomol. Soc. Am.*, 19, 180, 1926.

76. **Gösswald, K.**, Uber den insektentötenden Pilz, *Beauveria bassiana* (Bals.) Vuill. Bisher Bekanntes und eigene Versuche, *Arb. Biol. Reichsanst. Land. Forstwirtsch. Berlin Dahlem*, 22, 399, 1938.

77. **Grigorov, S.**, Parasites, predators and diseases of species of the family Coccinellidae *Rastenievud. Nauki*, 20, 113, 1983 (Bulgarian with English summary).

78. **Gusev, G. V., Svikle, M. J., Koval, Y. V., Zayats, Y. V., Lakhidov, A. I., and Sorokin, N. S.**, Prospects for using Colorado potato beetle entomophages in different geographical zones of the USSR, *Rev. Appl. Entomol. Ser. A.*, 65, 265, 1977.

79. **Gustafsson, M.**, On species of the genus *Entomophthora* Fres. in Sweden, *Lantbrukshogsk Ann.*, 35, 235, 1969.

80. **Hall, R. A.**, The fungus *Verticillium lecanii* as a microbial insecticide against aphids and scales, in *Microbial Control of Pests and Plant Diseases: 1970—1980*, Burges, H. D., Ed., Academic Press, London, 1981, 483.

81. **Hall, R. A.,** Control of whitefly, *Trialeurodes vaporariorum* and cotton aphid, *Aphis gossypii* in glasshouses by two isolates of the fungus, *Verticillium lecanii, Ann. Appl. Biol.,* 101, 1, 1982.

82. **Hall, R. A. and Papierok, B.,** Fungi as biological control agents of arthropods of agricultural and medical importance, *Parasitology,* 84, 205, 1982.

83. **Hall, R. A., Zimmermann, G., and Vey, A.,** Guidelines for the registration of entomogenous fungi as insecticides, *Entomophaga,* 27, 121, 1982.

84. **Hanel, H.,** A bioassay for measuring the virulence of the insect pathogenic fungus *Metarhizium anisopliae* (Metsch.) Sorok. (Fungi Imperfecti) against the termite *Nasutitermes exitiosus* (Hill) (Isoptera, Termitidae), *Z. Angew. Entomol.,* 92, 9, 1981.

85. **Hanssler, G. and Hermanns, M.,** *Verticillium lecanii* as a parasite on cysts of *Heterodera schachtii, Z. Pflanzenkr. Pflanzenschutz,* 88, 678, 1981.

86. **Harcourt, D. G., Guppy, J. C., and Binns, M. R.,** Analysis of numerical change in subeconomic populations of the alfalfa weevil, *Hypera postica* (Coleoptera: Curculionidae), in eastern Ontario, *Environ. Entomol.,* 13, 1627, 1984.

87. **Harper, A. M. and Huang, H. C.,** Evaluation of the entomophagous fungus *Verticillium lecanii* (Moniliales: Moniliaceae) as a control agent for insects, *Environ. Entomol.,* 15, 281, 1986.

88. **Hodek, I.,** *Biology of Coccinellidae,* Junk, The Hague, 1973.

89. **Hodek, I.,** Bionomics and ecology of predaceous Coccinellidae, *Annu. Rev. Entomol.,* 12, 79, 1967.

90. **Holdom, D.,** personal communication, 1987.

91. **Houle, C., Hartmann, G. C., and Wasti, S. S.,** Infectivity of eight species of entomogenous fungi to the larvae of the elm bark beetle *Scolytus multistriatus* (Marsham), *J. N.Y. Entomol. Soc.,* 95, 14, 1987.

92. **Hukuhara, T.,** Epizootiology: prevention of insect diseases, in *Epizootiology of Insect Diseases,* Fuxa, J. R. and Tanada, Y., Eds., John Wiley & Sons, New York, 1987, 497.

93. **Humber, R. A. and Soper, R. S.,** USDA-ARS collection of entomopathogenic fungal cultures, catalog of strains, U.S. Department of Agriculture, 1986, 1.

94. **Hussey, N. W. and Tinsley, T. W.,** Impressions of insect pathology in the People's Republic of China, in *Microbial Control of Pests and Plant Diseases: 1970—1980,* Burges, H. D., Ed., Academic Press, New York, 1981, 785.

95. **Ignoffo, C. M.,** The fungus *Nomuraea rileyi* as a microbial insecticide, in *Microbial Control of Pests and Plant Diseases: 1970—1980,* Burges, H. D., Ed., Academic Press, New York, 1981, 513.

96. **Ignoffo, C. M. and Garcia, C.,** Susceptibility of six species of noctuid larvae to a biotype of *Nomuraea rileyi* (Farlow) Samson from Thailand (Lepidoptera: Noctuidae), *J. Kans. Entomol. Soc.,* 60, 156, 1987.

97. **Iperti, G.,** Les coccinelles de France, *Phytoma,* 377, 14, 1980.

98. **Iperti, G.,** Les parasites des coccinelles aphidiphages dans les Alpes-Maritimes at les Basses-Alpes, *Entomophaga,* 9, 153, 1964.

99. **Jaronski, S. T. and Axtell, R. C.,** Non-susceptibility of *Chaoborus flavicans* (Chaoboridae) to the mosquito pathogen *Lagenidium giganteum* (Oomycetes), *Mosq. News,* 44, 81, 1984.

100. **Jarvis, E.,** Cane pest combat and control, *Queensl. Agric. J.,* 21, 436, 1924.

101. **Jatala, P.,** Biological control of plant-parasitic nematodes, *Annu. Rev. Phytopathol.,* 24, 453, 1986.

102. **Jatala, P.,** personal communication, 1987.

103. **Kabacik-Wasylik, D. and Kmitowa, K.,** The effect of single and mixed infections of entomopathogenic fungi on the mortality of the Carabidae (Coleoptera), *Ekol. Pol.,* 21, 645, 1973.

104. **Kapustina, O. V.,** The effect of certain pesticides in *Trichogramma, Rev. Appl. Entomol. Ser. A,* 65, 1641, 1977.

105. **Kavorkian, A. G.,** Studies on the Entomophthoraceae. I. Observations on the genus *Conidiobolus, J. Agric. Univ. P. R.,* 21, 191, 1937.

106. **Kawase, S.,** Microbiological control of insect pests, *Korean J. Plant Prot.,* 22, 67, 1983.

107. **Kerwin, J. L., Dritz, D. A., and Washino, R. K.,** Non-mammalian safety tests for *Lagenidium giganteum* (Oomycetes: Lagenidiales) *J. Econ. Entomol.,* 81, 158, 1988.

108. **Kerwin, J. L. and Washino, R. K.,** Cuticular regulation of host recognition and spore germination, in *Fundamental and Applied Aspects of Invertebrate Pathology,* Samson, R. A., Vlak, J. M., and Peters, D., Eds., Foundation of the 4th Int. Colloq. Invertebrate Pathology, Wageningen, 1986, 423.

109. **King, E. G. and Bell, J. V.,** Interactions between a braconid, *Microplitis croceipes* and a fungus, *Nomuraea rileyi,* in laboratory-reared bollworm larvae, *J. Invertebr. Pathol.,* 31, 337, 1978.

110. **Kiselek, E. V.,** The effect of biopreparations on insect enemies, *Rev. Appl. Entomol. Ser. A,* 65, 427, 1975.

111. **Kish, L. P., Terry, I., and Allen, G. E.,** Three fungi tested against the lovebug, *Plecia nearctica,* in Florida, *Fla. Entomol.,* 60, 291, 1977.

112. **Kmitowa, K. and Bajan, C.,** Pathogenicity level of various strains of *Beauveria bassiana* (Bals.) Vuill, *Pol. Ecol. Stud.,* 8, 409, 1982.

113. **Kmitowa, K. and Kabacik-Wasylik, D.,** An attempt at determining the pathogenicity of two species of entomopathogenic fungi in relation to Carabidae, *Ekol. Pol.,* 19, 727, 1971.

114. **Knight, A. L.**, Host range and temperature requirements of *Culicinomyces clavisporus*, *J. Invertebr. Pathol.*, 36, 423, 1980.

115. **Kobayasi, Y.**, The genus *Cordyceps* and its allies, *Sci. Rep. Tokyo Bunrika Daigaku, Sect. B*, 5, 53, 1941.

116. **Kozlowska, C.**, Insect killing fungi occurring on material collected for detection of biological forest injurers, *Rocz. Nauk. Roln. Lesn.*, 19, 43, 1957 (Polish with English summary).

117. **Laird, M.**, Environmental impact of insect control by microorganisms, *Ann. N. Y. Acad. Sci.*, 217, 218, 1973.

118. **Lam, N. C., Goettel, M. S., and Soares, G. G., Jr.**, Host records for the entomopathogenic hyphomycete, *Tolypocladium cylindrosporum*, *Fla. Entomol.*, 71, 86, 1988.

119. **Landa, Z.**, Protection against glasshouse whitefly *(Trialeurodes vaporariorum* Westw.) in integrated protection programmes for glasshouse cucumbers, *Sb. UVTI Zahradnictvi*, 11, 215, 1984.

120. **Leatherdale, D.**, A host catalogue of British entomogenous fungi, *Entomol. Mon. Mag.*, 94, 103, 1958.

121. **Leatherdale, D.**, A host catalogue of British entomogenous fungi: second supplement, *Entomol. Mon. Mag.*, 101, 163, 1966.

122. **Leatherdale, D.**, The arthropod hosts of entomogenous fungi in Britain, *Entomophaga*, 15, 419, 1970.

123. **Lemperière, G. R.**, Early investigations on the potential of alternate prey for the rearing of *Rhizophagus grandis*, predatory beetle of *Dendroctonus micans* (Col. Scolytidae), the great spruce bark beetle, in *Biological Control of Bark Beetles*, Proc. a seminar organized by the Commission of the European Communities and the Université Libre de Bruxelles, Brussels, 1984, 129.

124. **Li, Y. W., Lü, C. R., and Tao, H. C.**, in *Production and Application of Beauveria bassiana*, Forestry Press China, Beijing, 1981, 89 (Chinese).

125. **Li, Z.**, A list of the insect hosts of *Beauveria bassiana*, in *1st Natl. Symp. Entomogenous Fungi*, Gongzhuling, Jilin, People's Republic of China, 1987, 1.

126. **Li, Z. and Humber, R. A.**, *Erynia pieris* (Zygomycetes: Entomophthoraceae), a new pathogen of *Pieris rapae* (Lepidoptera: Pieridae): description, host range, and notes on *Erynia virescens*, *Can. J. Bot.*, 62, 653, 1984.

127. **Lin, S.**, A review on the studies of disease control in the silkworm, *Sci. Sericult.*, 7, 126, 1981 (Chinese).

128. **Lipa, J. J.**, Microbial control of mites and ticks, in *Microbial Control of Insects and Mites*, Burges, H. D. and Hussey, N. W., Eds., Academic Press, New York, 1971, 357.

129. **Lisansky, S. G. and Hall, R. A.**, Fungal control of insects, in *The Filamentous Fungi*, Vol. 4, Smith, J. E., Berry, P. R., and Kristansen, B., Eds., Edward Arnold, London, 1982, 127.

130. **Loan, C.**, Suppression of the fungi *Zoophthora* spp. by captafol: a technique to study interaction between disease and parasitism in the alfalfa weevil *Hypera postica* (Coleoptera: Curculionidae), *Proc. Entomol. Soc. Ont.*, 112, 81, 1981.

131. **Longworth, J. F. and Kalmakoff, J.**, An ecological approach to the use of insect pathogens for pest control, in *Microbial and Viral Pesticides*, Kurstak, E., Ed., Marcel Dekker, New York, 1982, 425.

132. **Lucarotti, C. J., Federici, B. A., and Chapman, H. C.**, Progress in the development of *Coelomomyces* fungi for use in integrated mosquito control programmes, in *Integrated Mosquito Control Methodologies*, Vol. 2, Laird, M. and Miles, J. W., Eds., Academic Press, New York, 1985, 251.

133. **Lundholm, B. and Stackerud, M.**, Environmental protection and biological forms of control of pest organisms, *Ecol. Bull.*, 31, 135, 1980.

134. **MacLeod, D. M.**, Investigations on the genera *Beauveria* Vuill. and *Tritirachium* Limber, *Can. J. Bot.*, 32, 818, 1954.

135. **Madelin, M. F.**, Diseases caused by hyphomycetous fungi, in *Insect Pathology: An Advanced Treatise*, Vol. 2, Steinhaus, E. A., Ed., Academic Press, New York, 1963, 233.

136. **Madelin, M. F.**, Fungal parasites of insects, *Annu. Rev. Entomol.*, 11, 423, 1966.

137. **Magalhaes, B. P., Lord, J. C., Wraight, S. P., Daoust, R. A., and Roberts, D. W.**, Pathogenicity of *Beauveria bassiana* and *Erynia radicans*, to the coccinellid predators *Coleomegilla maculata* and *Eriopis connexa*, *J. Invertebr. Pathol.*, 52, 471, 1988.

138. **McCoy, C. W.**, Pest control by the fungus *Hirsutella thompsonii*, in *Microbial Control of Pests and Plant Diseases: 1970—1980*, Burges, H. D., Ed., Academic Press, New York, 1981, 499.

139. **McCray, E. M., Jr.**, *Lagenidium giganteum* (Fungi), in *Biological Control of Mosquitoes*, Chapman, H. C., Ed., *American Mosquito Control Assoc.*, Bulletin 6, Fresno, CA, 1985, 87.

140. **McCray, E. M., Jr., Womeldorf, D. J., Husbands, R. C., and Eliason, D. A.**, Laboratory observations and field tests with *Lagenidium* against California mosquitoes, *Calif. Mosq. Control Assoc. Proc. Pap.*, 41, 123, 1973.

141. **McGuire, M. R., Maddox, J. V., and Armbrust, E. J.**, Host range studies of an *Erynia radicans* strain (Zygomycotina: Entomophthoraceae) isolated from *Empoasca fabae* (Homoptera: Cicadellidae), *J. Invertebr. Pathol.*, 50, 75, 1987.

142. **McInnis, T., Jr., Schimmel, L., Noblet, R., Guptovanij, P., and Ban, A. R.**, Host range studies with the fungus *Leptolegnia*, a parasite of mosquito larvae. (Diptera: Culicidae), *J. Med. Entomol.*, 22, 226, 1985.

143. **Mills, N. J.,** The mortality and fat content of *Adalia bipunctata* during hibernation, *Entomol. Exp. Appl.,* 30, 265, 1981.

144. **Milner, R. J.,** *Neozygites acaridis* (Petch) comb. nov., *Trans. Br. Mycol. Soc.,* 85, 641, 1985.

145. **Milner, R. J., Lutton, G. G., and Bourne, J.,** A laboratory study of the interaction between aphids, fungal pathogens and parasites, in *Proc. 4th Australian Applied Entomological Conf.,* Bailey, P. and Swincer, D., Eds., South Australian Department of Agriculture, Adelaide, 1984, 375.

146. **Moscardi, F.,** personal communication, 1987.

147. **Müller-Kögler, E.,** *Pilzkrankheiten bei Insekten,* Paul Parey, Berlin, 1965.

148. **Muma, M. H.,** Factors contributing to the natural control of citrus insects and mites in Florida, *J. Econ. Entomol.,* 48, 432, 1955.

149. **Nakao, H. K.,** Entomogenous fungi, *Proc. Hawaii. Entomol. Soc.,* 15, 7, 1953.

150. **Nel'zina, E. N., Mironov, N. P., Sorokina, L. Y., and Baktinova, N. Z.,** The pattern of parasite-host relationships of entomopathogenic fungi *Metarhizium anisopliae* (Metsch.) Sorokin and *Beauveria bassiana* (Bals.) Vuill. (Fungi Imperfecti) with fleas, *Ceratophyllus fasciatus* Bosc. (Siphonaptera), *Med. Parazitol. Parazit. Bolezni,* 47, 86, 1978 (Russian with English summary).

151. **Nentwig, W.,** Parasitic fungi as a mortality factor of spiders, *J. Arachnol.,* 13, 272, 1985.

152. **Nishimura, K.,** Diagnosis of the silkworm diseases, *Sci. Technol. Sericult.,* 23, 48, 1984 (Japanese).

153. **Nolan, R. A., Laird, M., Chapman, H. C., and Glenn, F. E., Jr.,** A mosquito parasite from a mosquito predator, *J. Invertebr. Pathol.,* 21, 172, 1973.

154. **Ono, K.,** *Paecilomyces farinosus* infection of silkworms via lanternflies, *Rep. Sericult. Stn. Guma,* 93, 1981 (Japanese).

155. **Papierok, B.,** Données écologiques et expérimentales sur les potentialités entomopathogènes de l'entomophthorale *Conidiobolus coronatus* (Costantin) Batko, *Entomophaga,* 30, 303, 1985.

156. **Papierok, B., Valadao, L., Torres, B., and Arnault, M.,** Contribution à l'étude de la spécificité parasitaire du champignon entomopathogène *Zoophthora radicans* (Zygomyctes, Entomophthorales), *Entomophaga,* 29, 109, 1984.

157. **Parrella, M. P.,** Proposed use of microbial insecticides in an integrated program for chrysanthemums, Proc. 20th Annu. Meet. Soc. Invertebr. Pathol., 1987, 60.

158. **Petch, T.,** Notes on entomogenous fungi, *Trans. Br. Mycol. Soc.,* 16, 67, 1931.

159. **Petch, T.,** A list of the entomogenous fungi of Great Britain, *Trans. Br. Mycol. Soc.,* 17, 170, 1932.

160. **Petch, T.,** *Cordyceps militaris* and *Isaria farinosa, Trans. Br. Mycol. Soc.,* 20, 216, 1936.

161. **Petch, T.,** An *Empusa* on a mite., *Proc. Linn. Soc. N.S.W.,* 65, 259, 1944.

162. **Petch, T.,** A revised list of British entomogenous fungi, *Trans. Br. Mycol. Soc.,* 31, 286, 1948.

163. **Pimentel, D.,** Environmental risks associated with biological controls, *Ecol. Bull.,* 31, 11, 1980.

164. **Pimentel, D., Glenister, C., Fast, S., and Gallahan, D.,** Environmental risks of biological pest controls, *Oikos,* 42, 283, 1984.

165. **Poinar, G. O., Jr., van der Geest, L., Helle, W., and Wassink, H.,** Pathology and nematode parasitism, in *Tsetse: The Future for Biological Methods in Integrated Control,* Laird, M., Ed., International Development Research Centre, Ottawa, 1977, 75.

166. **Poinar, G. O., Jr. and Thomas, G. M.,** unpublished, 1987.

167. **Powell, W., Wilding, N., Brobyn, P. J., and Clark, S. J.,** Interference between parasitoids (Hym.: Aphidiidae) and fungi (Entomophthorales) attacking cereal aphids, *Entomophaga,* 31, 293, 1986.

168. **Prest, D. B., Gilliam, M., Taber, S., III, and Mills, J. P.,** Fungi associated with discolored honey bee, *Apis mellifera,* larvae and pupae, *J. Invertebr. Pathol.,* 24, 253, 1974.

169. **Puttler, B., Ignoffo, C. M., and Hostetter, D. L.,** Relative susceptibility of nine caterpillar species to the fungus *Nomuraea rileyi, J. Invertebr. Pathol.,* 27, 269, 1976.

170. **Qian, Y. J., Zhuang, D. H., and Li, R. Q., et al.,** *Silkworm Disease Control,* Science Technical Press, Shanghai, 1978, 223 (Chinese).

171. **Quinlan, R.,** personal communication, 1987.

172. **Ramakers, P. M. J.,** *Aschersonia aleyrodis,* a selective biological insecticide, *Bull. SROP,* 6, 167, 1983.

173. **Ramakers, P. M. J. and Samson, R. A.,** *Aschersonia aleyrodis,* a fungal pathogen of whitefly. II. Application as a biological insecticide in glasshouses, *Z. Angew. Entomol.,* 97, 1, 1984.

174. **Ravallec, M., Robert, P.-H., and Coz, J.,** Sensibilité d'un prédateur culiciphage *Toxorhynchites amboinensis* (Doleschall) a l'hyphomycète entomopathogène *Metarhizium anisopliae* (Sorokin), *Cah. O.R.S.T.O.M. Ser. Entomol. Med. Parasitol.,* 24, 275, 1986.

175. **Remaudiere, G. and Latge, J.-P.,** Importancia de los hongos patógenos de insectos (especialmente Aphididae y Cercopidae) en Méjico y perspectivas de uso, *Bol. Serv. Plagas For.,* 11, 217, 1985.

176. **Riba, G., Katagiri, K., and Kawakami, K.,** Preliminary studies on the susceptibility of the silkworm, *Bombyx mori* (Lepidoptera: Bombycidae) to some entomogenous hyphomycetes, *Appl. Entomol. Zool.,* 17, 238, 1982.

177. **Riba, G., Keita, A., and Vincent, J. J.,** Susceptibility of mosquito larvae to different species of entomopathogenous hyphomycetes, *Cah. O.R.S.T.O.M. Ser. Entomol. Med. Parasitol.,* 22, 271, 1984.

178. **Roberts, D. W. and Wraight, S. P.,** Current status on the use of insect pathogens as biocontrol agents in agriculture: fungi, in *Fundamental and Applied Aspects of Invertebrate Pathology,* Samson, R. A., Vlak, J. M., and Peters, D., Eds., Foundation of the 4th Int. Colloq. Invertebrate Pathology, Wageningen, 1986, 510.

179. **Roberts, D. W.,** Fungal infections of mosquitoes, in *Le Contrôle des Moustiques/Mosquito Control,* Aubin, A., Bourrassa, J. P., Belloncik, S., Pellissier, M., and Lacoursière, E., Eds., University of Quebec Press, Montreal, 1974, 143.

180. **Roberts, D. W.,** Isolation and development of fungus pathogens of vectors, in Biological Regulation of Vectors, Briggs, J. D., Ed., DHEW Publ. (NIH) 77-1180, Department of Health, Education, and Welfare, Washington, D.C., 1977, 85.

181. **Roberts, D. W.,** Means for insect regulation: fungi, *Ann. N. Y. Acad. Sci.,* 217, 76, 1973.

182. **Roberts, D. W. and Humber, R. A.,** Entomogenous fungi, in *Biology of Conidial Fungi,* Vol. 2, Cole, G. T. and Kendrick, B., Eds., Academic Press, New York, 1981, 201.

183. **Roberts, D. W. and Yendol, W. G.,** Use of fungi for microbial control of insects, in *Microbial Control of Insects and Mites,* Burges, H. D. and Hussey, N. W., Eds., Academic Press, New York, 1971, 125.

184. **Rockwood, L. P.,** Entomogenous fungi of the family Entomophthoraceae in the Pacific Northwest, *J. Econ. Entomol.,* 43, 704, 1950.

185. **Rockwood, L. P.,** Entomogenous fungi of the genus *Metarrhizium* on wireworms in the Pacific northwest, *Ann. Entomol. Soc. Am.,* 43, 495, 1950.

186. **Rockwood, L. P.,** Some hyphomycetous fungi found on insects in the Pacific northwest, *J. Econ. Entomol.,* 44, 215, 1951.

187. **Rombach, M. C.,** personal communication, 1987.

188. **Rombach, M. C., Aguda, R. M., Shepard, B. M., and Roberts, D. W.,** Infection of rice brown planthopper, *Nilaparvata lugens* (Homoptera: Delphacidae), by field application of entomopathogenic hyphomycetes (Deuteromycotina), *Environ. Entomol.,* 15, 1070, 1986.

189. **Rondon, A., Arnal, E., and Godoy, F.,** Behaviour of *Verticillium lecanii* (Zimm.) Viegas, a pathogen of the aphid *Toxoptera citricidus* (Kirk.) in citrus plantations in Venezuela, *Agron. Trop.,* 30, 201, 1982.

190. **Rosypal, J.,** The sugar-beet pest *Bottynoderes punctiventris* (Germ) and its natural enemies, *Rev. Appl. Entomol.,* A19, 427, 1931.

191. **Samšiňaková, A., Kálalová, S., Daniel, M., Dusbábek, F., Honzáková, E., and Černý, V.,** Entomogenous fungi associated with the tick *Ixodes ricinus* (L.), *Folia Parasitol. (Prague)* 21, 39, 1974.

192. **Samson, R. A.,** *Paecilomyces* and some allied hyphomycetes, *Stud. Mycol.,* 6, 1, 1974.

193. **Schabel, H. G.,** Phoretic mites as carriers of entomopathogenic fungi, *J. Invertebr. Pathol.,* 39, 410, 1982.

194. **Schwangart, Fr.,** Ueber die Traubenwickler (*Conchylis ambiguella* Hübn. und *Polychrosis botrana* Schiff.) und ihre Bekämpfung, mit Berücksichtigung natürlicher Bekämpfungsfaktoren, in *Festschrift zum 60, Geburtstag Richard Hertwigs,* Jena, 1910, 464. (As cited in Reference 147).

195. **Scopes, N. E. A.,** Control of *Myzus persicae* on year-round chrysanthemums by introducing aphids parasitized by *Aphidius matricariae* into boxes of root cuttings, *Ann. Appl. Biol.,* 66, 323, 1970.

196. **Shen, Y. C.,** Occurrence and control of silkworm muscardines, *J. Seric.,* 2, 13, 1982 (Chinese).

197. **Shima, M. and Suzuki, M.,** On the metabolic products of muscardine *Spicaria rubido-purpurea* Aoki, *Bull. Seric. Exp. Stn. Japan,* 15, 587, 1963.

198. **Siemaszko, W.,** Studies on entomogenous fungi of Poland, *Arch. Biol. Soc. Sci. Warsaw,* 6, 1937, (English translation).

199. **Skou, J. P.,** Diseases in bumble-bees (*Bombus* Latr.). The occurrence, description and pathogenicity of five hyphomycetes, *R. Vet. Agric. Coll. Yearb.,* 134, 1967.

200. **Smirnoff, W. A. and Eichhorn, O.,** Diseases affecting predators of *Adelges* spp. on fir trees in Germany, Switzerland, and Turkey, *J. Invertebr. Pathol.,* 15, 6, 1970.

201. **Smith, J. W., King, E. G., and Bell, J. V.,** Parasites and pathogens among *Heliothis* species in the central Mississippi delta, *Environ. Entomol.,* 5, 224, 1976.

202. **Soper, R. S.,** Commercial mycoinsecticides, in *Invertebrate Pathology and Microbial Control,* Proc. the 3rd Int. Colloq. Invertebr. Pathol., Brighton, England, 1982, 98.

203. **Soper, R. S., Shimazu, M., Humber, R. A., Ramos, M. E., and Hajek, A. E.,** Isolation and characterization of *Entomophaga maimaiga* sp. nov., a fungal pathogen of gypsy moth, *Lymantria dispar,* from Japan, *J. Invertebr. Pathol.,* 51, 229, 1988.

204. **Sosa Gomez, D. R., Ricci, J. G., and Nasca, A. J.,** Efecto de *Hirsutella thompsonii* Fisher var. *thompsonii,* sobre larvas y adultos de *Coccidophilus citricola* Brethes y *Lindorus lophanthae* (Blaisdell) (Col.; Coccinellidae), *CIRPON Rev. Invest.,* 3, 73, 1985.

205. **Sparks, A. K.,** *Synopsis of Invertebrate Pathology,* Elsevier, Amsterdam, 1985.

206. **Steinhaus, E. A.,** *Principles of Insect Pathology,* McGraw-Hill, New York, 1949.

207. **Steinhaus, E. A.,** Report on diagnoses of diseased insects 1944—1950, *Hilgardia,* 20, 629, 1951.

208. **Steinhaus, E. A.,** Concerning the harmlessness of insect pathogens and the standardization of microbial control products, *J. Econ. Entomol.,* 50, 715, 1957.

209. **Steinhaus, E. A.,** Symposium on microbial insecticides. IV. Diseases of invertebrates other than insects, *Bacteriol. Rev., 29,* 388, 1965.

210. **Steinhaus, E. A. and Marsh, G. A.,** Report on diagnoses of diseased insects 1951—1961, *Hilgardia,* 33, 349, 1962.

211. **Steinkraus, D. C. and Kramer, J. P.,** Susceptibility of sixteen species of Diptera to the fungal pathogen *Entomophthora muscae, Mycopathologia,* 100, 55, 1987.

212. **Stevenson, J. A.,** The green muscardine fungus in Porto Rico, *J. Dep. Agric. P.R.,* 3, 19, 1918.

213. **Sweeney, A. W.,** Further observations on the host range of the mosquito fungus *Culicinomyces, Mosq. News,* 39, 140, 1979.

214. **Sweeney, A. W.,** The insect pathogenic fungus *Culicinomyces* in mosquitoes and other hosts, *Aust. J. Zool.,* 23, 59, 1975.

215. **Templeton, G. E. and Greaves, M. P.,** Biological control of weeds with fungal pathogens, *Trop. Pest Manage.,* 30, 333, 1984.

216. **Thaxter, R.,** The Entomophthoraceae of the United States, *Mem. Boston Soc. Nat. Hist.,* 4, 133, 1888.

217. **Thomas, G. M. and Poinar, G. O., Jr.,** Report of diagnoses of diseased insects 1962—1972, *Hilgardia,* 42, 261, 1973.

218. **Tokareva, N. N.,** *Puelovodstvo,* 4, 24, 1969 (as cited in Reference 117).

219. **Toumanoff, C.,** Action des champignons entomophytes sur les abeilles, *Ann. Parasitol. Hum. Comp.,* 9, 462, 1931.

220. **Ullyett, G. C. and Schonken, D. B.,** A fungus disease of *Plutella maculipennis* Curt. in South Africa, with notes on the use of entomogenous fungi, *Union S. Afr. Dep. Agric. For. Sci. Bull.,* 218, 1, 1940.

221. **Vago, C.,** Première liste de souches de germes entomopathogènes, *Entomophaga,* 4, 286, 1959.

222. **Van Leuken, W. and Roberts, D. W.,** unpublished, 1972.

223. **Vandenberg, J. D.,** Safety testing of microbial control agents versus adult worker honey bees, *Proc. 20th Annu. Meet. Soc. Invertebr. Pathol.,* 74, 1987.

224. **Veen, K. H.,** Recherches sur la maladie dû à *Metarrhizium anisopliae* chez le criquet pélerin, *Meded. Landbouwhogesch. Wageningen,* 68, 1, 1968.

225. **Vincens, F.,** Sur une muscardine à *Beauveria bassiana* produite expérimentalement sur des abeilles, *C. R. Acad. Sci.,* 177, 713, 1923.

226. **Voukassovitch, P.,** Contribution à l'étude d'un champignon entomophyte *Spicaria farinosa* (Fries) var. *veticilloides* Fron., *Ann. Epiphyt. (Paris),* 11, 73, 1925.

227. **Wasti, S. S. and Hartmann, G. C.,** Susceptibility of gypsy moth larvae to several species of entomogenous fungi, *J. N. Y. Entomol. Soc.,* 90, 125, 1982.

228. **Waterhouse, G. M. and Brady, B. L.,** Key to the species of *Entomophthora* sensu lato, *Bull. Br. Mycol. Soc.,* 16, 113, 1982.

229. **Weems, H. V., Jr.,** Natural enemies and insecticides that are detrimental to beneficial Syrphidae, *Ohio J. Sci.,* 54, 45, 1954.

230. **Whitney, H. S.,** personal communication, 1987.

231. **Wilding, N.,** Pest control by Entomophthorales, in *Microbial Control of Pests and Plant Diseases 1970—1980,* Burgess, H. D., Ed., Academic Press, New York, 1981, 539.

232. **Willers, D., Lehmann-Danziner, H., and Fuhrer, E.,** Antibacterial and antimycotic effect of a newly discovered secretion from larvae of an endoparasitic insect, *Pimpla turionellae* L. (Hym.), *Arch. Microbiol.,* 133, 225, 1982.

233. **Wong, T. L.,** The biology and life-history of *Coelomomyces opifexi* (Pillai and Smith) and its biocontrol potential against *Aedes australis* (Erichson), Ph.D. thesis, University of Otago, Dunedin, 1981, 228.

234. **Wraight, S. and Roberts, D. W.,** Insect control efforts with fungi, *J. Ind. Microbiol.,* 28(2), 77, 1987.

235. **Xie, X. Y., Dai, Z. Y., and Huang, Z. Y.,** Preliminary study on laboratory infection of termites by *Metarhizium anisopliae, Entomol. Knowl.,* 21, 223, 1984 (Chinese).

Chapter 16

SAFETY TO NONTARGET INVERTEBRATES OF NEMATODES OF ECONOMICALLY IMPORTANT PESTS

R. J. Akhurst

TABLE OF CONTENTS

I. INTRODUCTION

The range of nematodes that affect insects is large and the interactions between nematodes and insects varied.[1,2] Some species of nematode have been recognised as potentially useful for biocontrol and a few tested in the commercial sphere. Although *Romanomermis culicivorax* has been produced commercially for mosquito control, its commercial production has been discontinued.[3] The nematodes currently used for insect control are *Deladenus siricidicola,* a specific parasite used for control of the woodwasp *Sirex noctilio,* and the broad host range entomopathogens, Steinernematidae and Heterorhabditidae.

II. *DELADENUS SIRICIDICOLA*

The release of *D. siricidicola* as a classical biological control agent against *Sirex noctilio* in Australia in 1970[4] was the first successful application of nematodes for control of an insect pest. This nematode has now been distributed over hundreds of thousands of hectares.

A. LIFE HISTORY

S. noctilio is a serious pest of the introduced *Pinus radiata* (Monterey Pine, a southern Californian native) in Australia and New Zealand. The adult female is attracted to a host tree and while ovipositing, also injects into the tree spores of the fungus *Amylostereum areolatum.*[5] The siricid larvae feed on the fungus-decayed wood, tunneling through the tree.

There are three groups of natural enemies of siricid woodwasps: egg parasitoids (Hymenoptera: Ibaliidae), larval parasitoids (Hymenoptera: Ichneumonidae), and parasitic nematodes (Neotylenchidae: *Deladenus* spp.).[6] One species of these nematodes, *D. siricidicola,* is a major control agent for *S. noctilio.*[7]

The unusual life history of *D. siricidicola* was determined by Bedding.[8,9] Adult females of the entomogenous form develop in the hemocoel of infected adult *Sirex.* At host pupation, there is an explosive growth of the nematode's reproductive system and production of up to 10,000 eggs per nematode. The hatching juveniles escape from the parent into the insect hemolymph and migrate to the host ovaries or testes. In female *S. noctilio* (most strains) the juvenile nematodes are packed into the eggs, rendering the eggs nonviable. Infected *S. noctilio* oviposit normally and, in doing so, transmit the nematodes to new trees.

D. siricidicola transmitted by female *S. noctilio* feed on the insect's symbiotic fungus, *A. areolatum,* and grow into mycetophagous adults quite unlike their parasitic parents.[8] The mycetophagous form of *D. siricidicola* feeds, grows, and reproduces in the tracheids, resin canals, and between the bark and cambium, spreading throughout the tree. Since *S. noctilio* attack trees especially attractive to them, a susceptible tree is usually attacked by many *S. noctilio* females. The spreading nematode population will, therefore, encounter larvae that have emerged from eggs deposited by uninfected insects. In the microenvironment surrounding *S. noctilio* larvae, the development of the juvenile nematodes is altered. Instead of developing into the mycetophagous adults, they become either infective-stage females or males in which each spermatozoan is little more than a nucleus rather than the amoeboid spermatozoans found in the mycetophagous generation. After mating, the infective-stage females bore through the cuticle of *S. noctilio* larvae and enter the hemocoel.

B. SPECIFICITY

Bedding and Akhurst[10] examined the insect host preferences of *Deladenus* spp. associated with siricid woodwasps and their insect parasitoids. The entomogenous form of *D. siricidicola* successfully parasitizes some siricid species and *Serropalpus barbatus,* a beetle commonly associated with siricids. The mycetophagous form is far more specific. Although *D. siricidicola* will feed to a small extent on fungi of other genera, the nematodes will reproduce when feeding on only one species, *A. areolatum*. Even the symbiont of some other siricid species, *A. chailletti,* will not support reproduction by *D. siricidicola.*[10]

C. EFFECT ON NONTARGET INVERTEBRATES

D. siricidicola might be regarded either as a parasite that enhances its transfer to new hosts by multiplying in the environment of its host or as a fungal feeder that uses the insect host as a means of transport to a fresh supply of fungus.[11] Either way, it is bound to the environment of its siricid hosts by the complete specificity of the mycetophagous cycle to the fungus *A. areolatum* and its dependence on this cycle. Within that environment, it has the opportunity to infect relatively few nontarget organisms (NTOs). Although *D. siricidicola* will enter larvae of the insect parasitoids of their siricid hosts, the nematodes are unable to complete their development and the parasitoids are unharmed. The only invertebrate NTOs that might be harmed are those that infest the *Sirex*-killed trees. It is unlikely that any indigenous invertebrate is wholly dependent on *Pinus radiata* where *D. siricidicola* is used for control, because the tree is an introduced species.

III. STEINERNEMATIDAE AND HETERORHABDITIDAE

Nematodes of the families Steinernematidae and Heterorhabditidae have the potential for the control of a wide range of insect pests. Since the development of reliable and economic mass production and storage techniques,[12,13] there has been an increasing commercial interest in these nematodes.[14]

A. LIFE HISTORY

The infective-stage nematode, usually a soil or litter dweller, is a nonfeeding juvenile that carries only one species of bacterium, its specific *Xenorhabdus* associate, within its intestine.[15] In the absence of a suitable host the infective stage juveniles may survive months, even years, without feeding. The nematode is attracted to a potential insect host and it penetrates via the natural openings (mouth, anus, spiracles) and, in the case of *Heterorhabditis* spp., directly through unsclerotized cuticle.[16] The nematode invades the hemocoel where it releases its associated bacterium. The host dies as the bacterium proliferates in the cadaver, producing essential nutrients for the nematodes and antimicrobial factors that inhibit the growth of many other microorganisms.[17] The nematodes feed and reproduce within the dead insect producing a new generation of infective stage juveniles that subsequently emigrate from the cadaver.

B. SPECIFICITY

Laboratory studies indicate that these nematodes are nonspecific insect pathogens.[1] Their lack of specificity for any particular insect host is indicated by their attraction to common components of their environment (e.g., CO_2,[18] Gram-negative bacteria,[19] some ions[19]). It is, however, quite evident that, even under the ideal conditions provided in the laboratory, any nematode species is not equally infectious for all insect species[20] and stages.[21,22] Differences in susceptibility of insect species and stages may be due to a number of factors. The nematodes are restricted to invading potential hosts through natural orifices (mouth, anus, spiracles) or, for *Heterorhabditis,* unsclerotized cuticle. Consequently, their infectivity may be reduced for hosts or stages which are very small, in which the natural orifices are protected by physical or behavioral means,[23] or in which unsclerotized cuticle is not exposed to the nematodes. Nematodes cannot reach, and therefore are unable to infect, pupae encased in pupal cases,[24] and perhaps resting stages encased in compacted earthen cells. Though unable to prevent nematode entry, some insects gain some protection by their immune responses.[25] Social insects have behavioral responses that minimize the impact of the nematodes on the insect population.[26,27]

The effectiveness of the nematodes against a particular insect species may be further reduced because of the environmental niche occupied by the insect. The nematodes' susceptibility to desiccation,[28] UV radiation,[29] and high temperature[30,31] severely restricts their effectiveness on exposed surfaces. The range of temperatures over which *Steinernema* (= *Neoaplectana*) and *Heterorhabditis* species can infect and reproduce in insects[32] further limits their effectiveness to those insect species that occupy satisfactorily sheltered environments during periods where the temperature is conducive to nematode activity.

C. TARGET INSECTS

Steinernema and *Heterorhabditis* spp. have been applied successfully for the control of borer pests.[34,35] In these applications, nematodes that gain entry to the borer holes are protected from UV radiation and desiccation whereas those that remain on the surface of the plant or drift onto the soil are vulnerable.

The application of *Heterorhabditis* spp. is a commercially useful means of controlling black vine weevil, *Otiorrhynchus sulcatus,* in glasshouses and nurseries.[36] In this somewhat artificial environment, there is little danger to beneficial invertebrates.

Although the vulnerability of the nematodes on exposed surfaces appears to preclude their use in treatment of foliage feeding insects, there are some circumstances where they can be utilized. Many leaf-feeding species descend to the soil for pupation and can be controlled by the nematodes at this stage (e.g., peach borer, *Carposina nipponensis*).[37] In one situation, at least, it is possible to control a leaf-feeding species by applying nematodes directly to the foliage. One application of *Steinernema carpocapsae* (= *S. feltiae*) to chrysanthemums grown in shade houses stopped cutting damage from beet armyworm, *Spodoptera exigua,* feeding on the foliage.[38]

Apart from these pests that are currently being treated with *Steinernema* or *Heterorhabditis,* there are many others that are being evaluated. Most such insects have soil-dwelling stages and are, therefore, logical targets for control by soil-dwelling entomogenous nematodes. It is in the treatment of insects in soil that there is the greatest potential for affecting invertebrate NTOs.

D. EFFECTS ON NONTARGET INVERTEBRATES
1. Insects
a. Parasitoids

In laboratory trials, insect parasitoids were infected by *Steinernema* and *Heterorhabditis* spp. When the armyworm host of *Hyposoter exiguae* was infected with *Steinernema carpocapsae* 24 h prior to parasitoid emergence, no adult parasitoids were obtained.[24] The hymenopterous parasitoids *Glyptapanteles* (= *Apanteles*) *militaris* and *H. exiguae* and the tachinid *Compsilura concinnata* were infected as they emerged from their hosts to pupate.[24,39-41] However, *S. carpocapsae* that infected *G. militaris* apparently contacted the parasitoid fortuitously rather than by attraction.[39] Similarly, a tachinid parasitoid, *Myxexoristops* sp., of the sawfly *Cephalcia abietis* was not attractive for *Steinernema kraussei* while the parasitoid remained within the dead host.[42] Mráček and Spitzer[42] found no evidence of infection by nematodes of the tachinid or an ichneumonid parasitoid of *C. abietis* in the field. Deseö[43] found that *Leskia aurea,* a dipterous parasite of *Synanthedon myopaeformis,* was unaffected by applications of *Steinernema* spp. to control its host.

Laboratory trials have shown that adult tachinids may also be susceptible to the nematodes. Laumond et al.[44] found that adult *Metagonistylum minense* were highly susceptible when confined with *S. carpocapsae.* Adult *C. concinnata* were also susceptible to nematode infection when they emerged from pupal cases in nematode-infected soil.[41]

The nematodes may also affect insect parasitoids indirectly. When *Mythimna unipuncta* larvae parasitized by *G. militaris* were exposed to *S. carpocapsae* up to 2 d before the parasitoids were due to emerge from the host, the parasitoids were unable to complete development because of the death of the host.[40] Ishibashi et al.[45] reported only 50% emergence of the hymenopterous parasitoid *Trichomalus apanteloctenus* from *Pieris rapae* ssp. *crucivora* infected with *S. carpocapsae* before parasitoid pupation. Similar restrictions on the completion of development were shown for tachinid parasitoids.[41,42]

The nematodes pose little threat to pupal stages of parasitoids. They were unable to penetrate undamaged cocoons of the hymenopterous parasitoids *Cotesia* (= *Apanteles*) *medicaginis, H. exiguae,* and *Chelonus* sp.[30] *G. militaris* pupae were infected only if the nematodes were on or in the insect before the parasitoid completed its cocoon.[40] Pupal *T. apanteloctenus* were not infected by *S. carpocapsae*.[45]

b. Predators

In laboratory studies, dipteran predators of *C. abietis* were occasionally infected and killed by *S. kraussei*.[42] However, Mráček and Spitzer[42] were unable to detect any infection of the predators in the field.

When the hemipteran *Agriophodorus dohrni* fed on *Malacosoma neustri* ssp. *testacea* 24 h after the latter were inoculated with *S. carpocapsae,* 20% of the predators were infected.[45] This infection rate was perhaps achieved because infective stage nematodes were concentrated in and about the lepidopteran host, exposing the predator to an artificially high dosage.

c. Other Insects

Laboratory tests showed that larvae[46] and adults[47] of the honeybee are susceptible to *Steinernema carpocapsae*. Kaya et al.[48] found that spraying *S. carpocapsae* directly into hives caused some mortality in the 3 d following spraying but did not adversely affect the colony. The temperature and low humidity of the hive are apparently not conducive to nematode infection and development.[48] Although spraying nematodes directly onto workers resulted in only 4 to 10% infection, Kaya et al.[48] recommended that applications be made when workers are not foraging.

When nematodes were sprayed into apple and pear trees for the control of the boring caterpillars of the leopard moth *Zeuzera pyrina,* there was no apparent effect on flies, wasps, or ants in that environment.[43] Monthly applications of *S. carpocapsae* (500 cm^{-2}) to soil for 6 months resulted in drastic fluctuations in the levels of collembolans (also mites and other arthropods) shortly after application but there were no lasting effects on the populations; there was no definite tendency in the initial fluctuation.[49]

2. Other Arthropods

There are no records of natural infection of noninsect hosts by steinernematid or heterorhabditid nematodes. However, spiders, harvestmen, and pseudoscorpions were infected when exposed to very high concentrations (>1000 cm^{-2}) of *Steinernema* and *Heterorhabditis* on filter paper.[50,51] In contrast, soil applications of nematodes that produced high levels of mortality in weevils, *Otiorrhynchus* spp., did not kill spiders or centipedes[52] and repeated soil applications of *S. carpocapsae* had no lasting effect on myriapod populations.[45]

A mesostigmatid mite, *Eugamasus* sp., fed on infective stage *S. carpocapsae* but six applications of the nematodes to soil at monthly intervals had no lasting effect on mite populations.[45] Spray applications of nematodes for control of *Z. pyrina* did not affect *Microtrombidium pusillum* (Acarida) on tree trunks.[43]

3. Other Invertebrates

Application of high dosages of *Steinernema glaseri* adversely affected the root-knot nematode *Meloidogyne javanica* in pots.[53] Although addition of *S. carpocapsae* or *S. glaseri* to soil resulted in an initial decline in native nematode populations, the native populations recovered or surpassed their original levels within weeks.[54,55] The population densities of plant-parasitic species were depressed for a few weeks in the 1984 trials but not 1985. There was a temporary increase in the levels of predatory nematodes and native rhabditids. Perturbations in the native populations were greater when initial population densities were lower.[55]

Earthworms are not significantly affected by the nematodes.[43] Nematodes were able to reproduce on earthworm cadavers but did not kill intact worms and did not significantly increase mortality levels in damaged worms.[56]

Although *S. glaseri* killed the fever snail *Oncomelania hupensis* in laboratory and field tests,[56] Deseö[43] found no effects on other snail species following soil applications of *S. glaseri* and *Heterorhabditis heliothidis*.

IV. CONCLUSIONS

The potential host ranges of nematodes of the families Steinernematidae and Heterorhabditidae contrast sharply with the specific parasite, *D. siricidicola*. The former are, under laboratory conditions, nonspecific pathogens of insects and, on first appraisal, it would appear that they may pose problems for invertebrate NTOs. However, data gathered until now indicate that nematodes currently used for biocontrol of insects pose few problems for natural populations of invertebrate NTOs. Although laboratory studies have demonstrated the susceptibility of some nontarget invertebrates, the few field evaluations reported have failed to detect any significant impact of the nematodes on NTOs.

Almost all of the studies to date on the effects on nematodes on nontarget invertebrates have been short-term, laboratory studies. The one by Ishibashi et al.[45] is the only medium- to long-term field study reported to date. This project, involving repeated inoculation of soil with *S. carpocapsae,* detected no lasting effects on a range of nontarget invertebrates. However, under some conditions the Steinernematidae and Heterorhabditidae are capable of adversely affecting nontarget invertebrates, particularly insect parasitoids and predators. It would, therefore, be premature to conclude that all of the *Steinernema* or *Heterorhabditis* species considered for insect pest control are safe for nontarget invertebrates in all environments. Prudence dictates that field studies of these nematodes for pest control should be accompanied by long-term studies of their possible effects on the invertebrate fauna of their environment.

ACKNOWLEDGMENTS

I would like to thank Dr. K. V. Deseö, Prof. N. Ishibashi, and Dr. Z. Mráček who provided their unpublished data. My thanks also to Drs. R. J. Milner, N. Boemare, J. Curran, and R. A. Bedding for comments on the manuscript.

REFERENCES

1. **Poinar, G. O.,** *Nematodes for Biological Control of Insects,* CRC Press, Boca Raton, FL, 1979.
2. **Nickle, W. R.,** *Plant and Insect Nematodes,* Marcel Dekker, New York, 1984.
3. **Kaya, H. K.,** Constraints associated with commercialization of entomogenous nematodes, in *Fundamental and Applied Aspects of Invertebrate Pathology,* Samson, R. A., Vlak, J. M., and Peters, D., Eds., Foundation of the 4th Int. Colloq. Invertebrate Pathology, Wageningen, 1986, 661.
4. **Bedding, R. A. and Akhurst, R. J.,** Use of the nematode *Deladenus siricidicola* in the biological control of *Sirex noctilio* in Australia, *J. Aust. Entomol. Soc.,* 13, 129, 1974.

5. **Coutts, M. P.**, The mechanism of pathogenicity of *Sirex noctilio* on *Pinus radiata* I & II, *Aust. J. Biol. Sci.*, 22, 915, and 1153, 1969.

6. **Taylor, K. L.**, The introduction and establishment of insect parasitoids to control *Sirex noctilio* in Australia, *Entomophaga*, 21, 429, 1976.

7. **Neumann, F. G., Morey, J. L., and McKimm, R. J.**, The *Sirex* wasp in Victoria, Bulletin 29, Lands and Forests Division, Department of Conservation, Lands and Forests, 1987.

8. **Bedding, R. A.**, Parasitic and free-living cycles in entomogenous nematodes of the genus *Deladenus*, *Nature*, 214, 174, 1967.

9. **Bedding, R. A.**, Biology of *Deladenus siricidicola* (Neotylenchidae) an entomophagous-mycetophagous nematode parasitic in siricid woodwasps, *Nematologica*, 18, 482, 1972.

10. **Bedding, R. A. and Akhurst, R. J.**, Geographical distribution and host preferences of *Deladenus* species (Nematoda: Neotylenchidae) parasitic in siricid woodwasps and associated hymenopterous parasites, *Nematologica*, 24, 286, 1978.

11. **Bedding, R. A.**, Nematode parasites of Hymenoptera, in *Plant and Insect Nematodes*, Nickle, W. R., Ed., Marcel Dekker, New York, 1984, 755.

12. **Bedding, R. A.**, Low cost *in vitro* mass production of *Neoaplectana* and *Heterorhabditis* species (Nematoda) for field control of insect pests, *Nematologica*, 27, 110, 1981.

13. **Bedding, R. A.**, Large scale production, storage and transport of the insect-parasitic nematodes *Neoaplectana* spp. and *Heterorhabditis* spp., *Ann. Appl. Biol.*, 104, 117, 1984.

14. **Poinar, G. O.**, Entomophagous nematodes, in *Biological Plant and Health Protection*, Franz, J. M., Ed., Fortschritte der Zoologie 32, Gustav Fischer, Stuttgart, 1986, 95.

15. **Poinar, G. O. and Thomas, G. M.**, The presence of *Achromobacter nematophilus* in the infective stage of a *Neoaplectana* sp. (Steinernematidae: Nematoda), *Nematologica*, 12, 105, 1966.

16. **Bedding, R. A. and Molyneux, A. S.**, Penetration of insect cuticle by infective juveniles of *Heterorhabditis* spp. (Heterorhabditidae: Nematoda), *Nematologica*, 28, 354, 1982.

17. **Akhurst, R. J.**, Antibiotic activity of *Xenorhabdus* spp., bacteria symbiotically associated with insect pathogenic nematodes of the families Steinernematidae and Heterorhabditidae, *J. Gen. Microbiol.*, 128, 3061, 1982.

18. **Gaugler, R., LeBeck, L., Nakagaki, B., and Boush, G. M.**, Orientation of the entomogenous nematode *Neoaplectana carpocapsae* to carbon dioxide, *Environ. Entomol.*, 9, 649, 1980.

19. **Pye, A. E. and Burman, M.**, *Neoaplectana carpocapsae:* nematode accumulations on chemical and bacterial gradients, *Exp. Parasitol.*, 51, 13, 1981.

20. **Bedding, R. A., Molyneux, A. S., and Akhurst, R. J.**, *Heterorhabditis* spp., *Neoaplectana* spp. and *Steinernema kraussei;* interspecific and intraspecific differences in infectivity for insects, *Exp. Parasitol.*, 55, 249, 1983.

21. **Gaugler, R. and Molloy, D.**, Instar susceptibility of *Simulium vittatum* (Diptera: Simuliidae) to the entomogenous nematode *Neoaplectana carpocapsae*, *J. Nematol.*, 13, 1, 1981.

22. **Geden, C. J., Axtell, R. C., and Brooks, W. M.**, Susceptibility of the lesser mealworm *Alphitobius diaperinus* (Coleoptera: Tenebrionidae) to the entomogenous nematodes *Steinernema feltiae*, *S. glaseri* (Steinernematidae) and *Heterorhabditis heliothidis* (Heterorhabditidae), *J. Entomol. Sci.*, 20, 331, 1985.

23. **Akhurst, R. J.**, Controlling insects in soil with entomopathogenic nematodes, in *Fundamental and Applied Aspects of Invertebrate Pathology*, Samson, R. A., Vlak, J. M., and Peters, D., Eds., Foundation of the 4th Int. Colloq. Invertebrate Pathology, Wageningen, 1986, 265.

24. **Kaya, H. K. and Hotchkin, P. G.**, The nematode *Neoaplectana carpocapsae* Weiser and its effect on selected ichneumonid and braconid parasites, *Exp. Entomol.*, 10, 474, 1981.

25. **Welch, H. E. and Bronskill, J. F.**, Parasitism of mosquito larvae by the nematode, DD 136, *Can. J. Zool.*, 40, 1263, 1962.

26. **Kermarrec, A., Febvay, G., and Decharme, M.**, Protection of leaf-cutting ants from biohazards: Is there a future for microbiological control?, in *Fire Ants and Leaf-Cutting Ants*, Lofgren, C. S. and Vander-Meer, R. K., Eds., Westview Studies in Insect Biology, Boulder, CO, 1985, 339.

27. **Bedding, R. A.**, personal communication.

28. **Schmiege, D. C.**, The feasibility of using a neoaplectanid nematode for control of some forest insect pests, *J. Econ. Entomol.*, 56, 427, 1963.

29. **Gaugler, R. and Boush, G. M.**, Effects of ultraviolet radiation and sunlight on the entomogenous nematode, *Neoaplectana carpocapsae*, *J. Invertebr. Pathol.*, 32, 291, 1978.

30. **Kaya, H. K.**, Development of the DD-136 strain of *Neoaplectana carpocapsae* at constant temperature, *J. Nematol.*, 9, 346, 1977.

31. **Molyneux, A. S.**, Survival of infective juveniles of *Heterorhabditis* spp. and *Steinernema* spp. (Nematoda: Rhabditida) at various temperatures and their subsequent infectivity for insects, *Rev. Nematol.*, 8, 165, 1985.

32. **Molyneux, A. S.**, *Heterorhabditis* spp. and *Steinernema* (= *Neoaplectana*) spp.: temperature and aspects of behaviour and infectivity, *Exp. Parasitol.*, 62, 169, 1986.

33. **Miller, L. A. and Bedding, R. A.,** Field testing of the insect parasitic nematode, *Neoaplectana bibionis* (Nematoda: Steinernematidae) against currant borer moth, *Synanthedon tipuliformis* (Lep.: Sesiidae) in blackcurrants, *Entomophaga,* 27, 109, 1982.

34. **Lindgren, J. E. and Barnett, W. W.,** Applying parasitic nematodes to control carpenterworms in fig orchards, *Calif. Agric.,* 36, 7, 1982.

35. **Deseö, K. V. and Miller, L. A.,** Efficacy of entomogenous nematodes, *Steinernema* spp., against clearwing moths, *Synanthedon* spp., in north Italian orchards, *Nematologica,* 31, 100, 1985.

36. **Bedding, R. A. and Miller, L. A.,** Use of a nematode *Heterorhabditis heliothidis,* to control black vine weevil, *Otiorrhynchus sulcatus,* in potted plants, *Ann. Appl. Biol.,* 99, 211, 1981.

37. **Wang, J.,** Entomopathogenic nematode research in China, in *Fundamental and Applied Aspects of Invertebrate Pathology,* Samson, R. A., Vlak, J. M., and Peters, D., Eds., Foundation of the 4th Int. Colloq. Invertebrate Pathology, Wageningen, 1986, 317.

38. **Begley, J. W.,** Control of foliar feeding lepidopterous pests with entomophagous nematodes, 20th Annu. Meeting, Soc. Invertebr. Pathol., Gainesville, FL, 1987, 65.

39. **Kaya, H. K.,** Infectivity of *Neoaplectana carpocapsae and Heterorhabditis heliothidis* to pupae of the parasite *Apanteles militaris, J. Nematol.,* 10, 241, 1978.

40. **Kaya, H. K.,** Interaction between *Neoaplectana carpocapsae* (Nematoda: Steinernematidae) and *Apanteles militaris* (Hymenoptera: Braconidae), a parasitoid of the armyworm, *Pseudaletia unipuncta, J. Invertebr. Pathol.,* 31, 358, 1978.

41. **Kaya, H. K.,** Effect of the entomogenous nematode *Neoaplectana carpocapsae* on the tachinid parasite *Compsilura concinnata* (Diptera: Tachinidae), *J. Nematol.,* 16, 9, 1984.

42. **Mráček, Z. and Spitzer, K.,** Interaction of the predators and parasitoids of the sawfly, *Cephalcia abietis* (Pamphiliidae: Hymenoptera) with its nematode *Steinernema kraussei, J. Invertebr. Pathol.,* 42, 397, 1983.

43. **Deseö, K. V.,** personal communication.

44. **Laumond, C., Mauleon, H., and Kermarrec, A.,** Données nouvelles sur le spectre d'hôtes et le parasitisme du nématode entomophage *Neoaplectana carpocapsae, Entomophaga,* 24, 13, 1979.

45. **Ishibashi, N., Fah-zu Young, Nakashima, M., Abiru, C., and Haraguchi, N.,** Effects of application of DD-136 on silkworm, *Bombyx mori,* a predatory insect, *Agriosphodorus dohrni,* a parasitoid *Trichomalus apanteloctenus,* soil mites and other non-target soil arthropods, with brief notes on feeding behaviour and predatory pressure of soil mites, tardigrades, and predatory nematodes on DD-136 nematodes, in Recent Advances in Biological Control of Insect Pests by Entomogenous Nematodes in Japan, Report to Ministry of Education, Culture and Science, Japan, 1987, 158.

46. **Cantwell, G. E., Lehnert, T., and Fowler, J.,** Are biological insecticides harmful to the honey bee?, *Am. Bee J.,* 112, 255, 1972.

47. **Hackett, K. J. and Poinar, G. O.,** The ability of *Neoaplectana carpocapsae* Weiser (Steinernematidae: Rhabditoidea) to infect adult honeybees (*Apis mellifera,* Apidae: Hymenoptera), *Am. Bee J.,* 113, 100, 1973.

48. **Kaya, H. K., Marston, J. M., Lindegren, J. E., and Peng, Y. S.,** Low susceptibility of the honey bee, *Apis mellifera* L. (Hymenoptera: Apidae) to the entomogenous nematode, *Neoaplectana carpocapsae* Weiser, *Environ. Entomol.,* 11, 920, 1982.

49. **Ishibashi, N.,** personal communication.

50. **Poinar, G. O. and Thomas, G. M.,** Laboratory infection of spiders and harvestmen (Arachnida: Araneae and Opiliones) with *Neoaplectana* and *Heterorhabditis* nematodes (Rhabditoidea), *J. Arachnol.,* 13, 297, 1985.

51. **Poinar, G. O. and Thomas, G. M.,** Laboratory infection of *Garypus californicus* (Pseudoscorpionida, Garypidae) with neoaplectanid and heterorhabditid nematodes (Rhabditoidea), *J. Arachnol.,* 13, 400, 1985.

52. **Deseö, K. V., Kovacs, A., Lercari, G., and Constanzi, M.,** Possibilita di applicazione di nematodi entomoparassiti contro insetti dannosi nella floricoltura, *Inf. Fitopatol.,* 11, 37, 1985.

53. **Bird, A. F. and Bird, J.,** Observations on the use of insect parasitic nematodes as a means of biological control of root-knot nematodes *Int. J. Parasitol.,* 16, 511, 1986.

54. **Ishibashi, N. and Kondo, E.,** *Steinernema feltiae* (DD-136) and *S. glaseri:* persistence in soil and bark compost and their influence on native nematodes, *J. Nematol.,* 18, 310, 1986.

55. **Ishibashi, N. and Kondo, E.,** Dynamics of the entomogenous nematode *Steinernema feltiae* applied to soil with and without nematocide treatment, *J. Nematol.,* 19, 404, 1987.

56. **Capinera, J. L., Blue, S. L., and Wheeler, G. S.,** Survival of earthworms exposed to *Neoaplectana carpocapsae* nematodes, *J. Invertebr. Pathol.,* 39, 419, 1982.

57. **Li, P., Deng, C., Zhang, S., and Yang, H.,** Laboratory studies on the infectivity of the nematode *Steinernema glaseri* to *Oncomelania hupensis,* a snail intermediate of blood fluke, *Schistosoma japonicum, Chin. J. Biol. Control,* 2, 1986.

V. Conclusions

Chapter 17

CONCLUSIONS

M. Laird

Like synthetic chemical insecticides, microbial ones pose safety issues with respect to toxicity, allergenicity, and irritancy. Their unique capacities for (in some cases) replicating following field applications, and for multiplying within a suitable host (Chapter 8) raise issues of infectivity to nontarget organisms (NTOs), too.

In their addendum to Chapter 8, Siegel and Shadduck comment that one school of thought has defined "infection" as the mere presence of microorganisms within living tissues (a circumstance not necessarily occasioning harm, while sometimes being responsible for "false positives"); while another school "inextricably links infection to disease by defining the former as occurring when living agents 'enter an animal body and set up a disturbance' either through multiplication in tissue, production of toxins, or both. From the viewpoint of safety testing, the second definition of infection, which links the presence of a microorganism to tissue damage, is the more useful one." These co-authors submit "that infection in mammalian safety studies be considered established when there is evidence of multiplication of the microbial control agent in mammalian tissue, coupled with tissue damage" (Chapter 8). They also point out that the whole question is one which, to date, has been sidestepped in safety guidelines issued by regulatory agencies.

That said, it is evident from Chapters 1 through 4, supported by information elsewhere in the book such as that concerning the history of registration and reregistration of *Bacillus popilliae* spore powder in the U.S. (Chapter 13), that in much of the Northern Hemisphere protocols for the safety testing and registration of microbial control agents and that of formulated microbial insecticides based upon them have been established and are under constant review. A lack of relevant information for much of the Southern Hemisphere is equally evident from Section I. This is not to imply that the subject is being ignored in major food- and fiber-producing Southern Hemisphere countries.

In Australia, for example, the Commonwealth Technical Committee on Agricultural Chemicals is the body responsible for all safety and registration issues in the agricultural chemical field. Federal and state authorities there are currently examining the stance to be adopted over microbial insecticides against the background of, e.g., the regulations discussed in Chapters 1 and 2, and recommendations drawn up by the International Organization for Biological Control (IOBC) (personal communication, Professor D. E. Pinnock). Meanwhile, relevant Australian safety and registration matters are being considered on a case-by-case basis only. Much the same can be said for New Zealand, and doubtless for various countries of the Neotropical and Ethiopian Regions.

In so far as the numerous recently independent states of the tropical South Pacific are concerned, the position is clearly similar to that of greatly more populous countries of the "developing world" north of the Equator. Without an adequate body of local expertise, they find themselves without other recourse than to base their decisions "on the fact of the registration of the preparations in advanced countries" (personal communication, Professor Keio Aizawa). Examples are Malaysia, the Philippines, Taiwan, and Thailand.

When evaluating this problem from their individual standpoints, countries of the Southern Hemisphere and the Third World at large will do well to ponder Quinlan's remarks (Chapter 2) that "In practice, most of the countries of the EEC, with respect to most of the organisms under development for use as microbial pesticides, will accept safety data produced to meet the guidelines and specifications of the EPA"; and that "Because of this, commercial companies will always tend to develop their data package in response to the EPA rather than any other registration authority.'' At the same time, the U.S., members of the European Economic Community, the U.S.S.R., and Japan (all among the vanguard with respect to appropriate legislation), as well as the multitude of states looking to them for leadership, must keep it constantly in mind that "Perfection is the child of Time".[1] For, although beyond its infancy, our subject remains in its first youth. A very great deal is still to be learned before all questions of the safety of microbial insecticides come close to resolution. Indeed, all such questions, notably the central one of "How safe *is* safe?", may never be finally resolved. This is because as genetically engineered entomopathogens (Chapter 14) enter more and more purposefully into the microbial aspects of integrated control of arthropod pests and vectors, novel problems posing fresh challenges must continue to present themselves.

Turning to matters of environmental health considered in Section II, it is thought provoking that the fungal entomopathogen, *Beauveria bassiana,* was the subject of the seminal demonstration by Agostino Bassi,[2] just over a century and a half ago, that infectious disease can be caused in an animal by a microorganism.

Therefore, long before the development of the discipline of invertebrate pathology, a candidate microbial control agent (then unrecognized as such) was seen to be harmful to one of humankind's most ancient and lucrative beneficial arthropods. Goettel et al. (Chapter 15) expand on the vulnerability of *Bombyx mori* to *Beauveria bassiana* among other fungal entomopathogens. They draw attention to the consequent illegality of applying microbial insecticides in certain areas of the People's Republic of China (PRC) and Japan. Nevertheless, careful selection of appropriate strains of both *Bombyx mori* and its fungal pathogen, combined with basic hygiene practices, suggest the feasibility of employing *Beauveria bassiana* in the immediate vicinity of sericulture farms without endangering silkworms. Moreover, recent Japanese work suggests that a strain of *Paecilomyces fumosoroseus* may prove serviceable to sericulture through its capacity to control the tachinid parasitoid of silkworms, *Blepharipa zebina* (Chapter 15).

Silkworms and our other prime beneficial insect, *Apis mellifera* (the honeybee), received the lion's share of attention as prospective NTOs during the early days of microbial control. However, a review[3] published as recently as 1973 stressed that "Most of the data concerning this topic are still to be collected." Since then, though, a great deal more research has been undertaken. This is illustrated by the fact that of the 206 references cited in Vinson's overview of the potential impact of microbial insecticides on beneficial arthropods in the terrestrial environment (Chapter 5), 65% postdate the 1973 review. There has been particularly significant progress as regards the impact of microbial insecticides on parasitoids, the evidence suggesting a general incompatibility of hymenopteran ones with microbial insecticides. This largely results from the vulnerability of the parasitoids to the premature death of their hosts, occasioned by entomopathogens. The problem demands critical risk-benefit evaluation as well as appropriate fine tuning of the total package of control measures being employed (Chapter 15). In some specific instances, carefully timed integrated control practices can correct such situations. Thus, compatability between the hymenopteran, *Encarsia formosa,* and the fungus, *Aschersonia aleyrodis,* against the greenhouse whitefly, *Trialeurodes vaporarium,* was achieved by delaying release of the parasitoid until 4 or more d after fungal application (Chapter 5).

Incompatibility between hymenopteran and dipteran parasitoids of economic insect pests and nematodes is discussed in Chapter 16. Results to date suggest that predators are generally resistant to contact with diseased hosts and to most pathogens. Certain entomopathogenic protozoans appear to offer the chief exception to this statement, while generalist nematode parasites (or parasitoids, as some regard them) may be hazardous to predators of economic insect pests in moist situations (Chapter 16).

Perhaps the most encouraging conclusion of Chapter 5 is that most entomopathogens, when used carefully, seem safe to *Apis mellifera*, which, however, is the only pollinator thoroughly investigated in this context. A priority area for further research thus concerns the impact of microbial insecticides on solitary bees and a wide range of other pollinators. From the latter, though, there are good grounds for exempting mosquitoes; even though the nectar-feeding habits of most male and many female Culicidae remain unknown, and the heads of circumboreal mosquitoes in the act of biting not infrequently bear the pollinia of terrestrial orchids, *Habenaria* spp.![4]

Dejoux and Elouard (Chapter 6) review the impact of microbials on the freshwater environment, paying particular attention to the United Nations Development Programme/ World Bank/World Health Organization Onchocerciasis Control Programme (WHO/OCP) in West Africa. Reassuringly, they found no evidence there of *Bacillus thuringiensis* ssp. *israelensis,* as used against larvae of the vectors of onchocerciasis, *Simulium damnosum* s.l., harming more than one other element of stream ecosystems. In the many streams concerned, sprayed weekly throughout the dry season, the only organisms other than Simuliidae proving decidedly susceptible to this microbial agent as applied in the Teknar® formulation were molluscs of the family Ancyclidae. Indeed, Dejoux and Elouard were able to report that these WHO/OCP treatments left important fish-food insect larvae (Diptera/ Chironomidae) and predators on immature simuliids (Trichoptera/Hydropsychidae) substantially unaffected.

While still on the subject of WHO/OCP, it should be mentioned that none of this book's chapters pay any attention to the mermithid nematodes of Culicidae and Simuliidae. The best-known of these parasitic helminths in so far as mosquitoes and blackflies are concerned, are *Romanomermis culicivorax* and *Isomermis lairdi,* respectively. Despite a promising entry into field trials in the 1960s and 1970s, the former, according to the latest available paper,[5] "would be of limited use in antimalaria campaigns in southern Iran." While the latter, as tested against *Simulium damnosum* s.l. in West Africa,[6] is "difficile à utiliser, d'une manière efficace, pour la lutte contre les espèces vectrices de l'onchocercose en Afrique de l'ouest." In both cases, inability to mass produce *in vitro* and to deliver cheaply enough to remote field areas for repetitive, inoculative application, constitute the stumbling block. Yet, the assertion of Petersen[7] that "Few biological agents are as selective as mermithid nematodes" has been amply born cut by experimental studies[8,9] in West Africa as elsewhere. Available evidence indicates that once the production/supply problems instanced have been solved, it is unlikely in the extreme that mermithids will pose health or environmental hazards.

In summarizing the impact of microbials on the estuarine and marine environments, Couch and Foss (Chapter 8) properly insist that care must always be exercised not to lump together all subspecies and modes of action of, e.g., *Bacillus thuringiensis*. The importance of this is very evident from the review of safety of microbial insecticides by Burges.[10] The overviews of Section II agree upon a general satisfaction with the environmental safety of *B. thuringiensis,* also that of baculoviruses generally. The last-mentioned issue is addressed authoritatively by Gröner (Chapter 10), who roundly declares that "In contrast to experience with most chemical insecticides, directly adverse effects of baculoviruses on beneficial insects have never been reported from field tests."

With regard to *B. thuringiensis* and its β-exotoxin, Melin and Cozzi (Chapter 11) acknowledge that "populations of *Apis mellifera* are unlikely to be affected in the field because of the low application rate, dilution, and insect feeding habits", but recommend the inclusion in future field trials of long-term evaluations of any possible delayed adverse effect of low doses on honeybees. Siegel and Shadduck (Chapter 8) go further, submitting that "Exotoxins are a separate issue and should be evaluated as chemicals, using the standard chemical safety protocols."

All co-authors herein are in good agreement that available and pending microbial insecticides, even those with a broad spectrum of hosts such as certain entomopathogenic fungi (Chapters 5, 6, and 15) are demonstrably safer in terms of environmental impact than are any of the conventional chemical pesticides. With respect to the fungi (as indeed with all microbial control agents), care must obviously be taken not to extrapolate from data on some particular taxon to a closely related entity; e.g., from a marine/estuarine species of *Lagenidium* to a freshwater one, or *vice versa*. As well expressed by Goettel et al. (Chapter 15), "With respect to safety to invertebrates, safe and optimum use of fungi can only be determined for each specific integrated management system." These authors proceed to stress that "A major current limitation on the use of fungal biocontrol agents is the lack of formulated preparations;" and urge that "it is imperative that necessary safety regulations and registration procedures do not hinder the development and consequent testing and utilization of such formulations."

Furthermore, the environmental safety record of economic entomology's first commercialized microbial insecticide, *Bacillus popilliae*, remains exemplary (Chapter 13). So indeed does that of its innocuousness as regards human health. Obenchain and Ellis (Chapter 13) reported that clinical, medical, and serological examination of eight employees of Fairfax Biological Laboratories, whose combined full- or part-time employment exposure to this bacterium amounted to 85 years, revealed "no evidence of illness associated with occupational exposure to *B. popilliae* and that continued exposure does not induce antibody production in the human."

The outlook for the soon-to-be-commercialized *B. sphaericus,* a mosquito pathogen well suited for use against culicine pests and vectors tolerant of polluted-water larval habitats, and having the potential of recycling therein,[11] seems similarly assured (Chapters 6 and 12). As regards safety of entomopathogens in general to humans and other vertebrates, Saik et al. (Chapter 9) cite Ignoffo's words, "Absolute safety cannot be guaranteed in all living systems for all time," followed by his: "Toxicity or pathogenicity can generally be demonstrated if no limitation is imposed on dosage or type of vertebrate system." In this connection, though, Siegel and Shadduck (Chapter 8) produce evidence on the frequent impossibility of achieving an LD_{50} when using microbial agents experimentally. For on their administration by standard procedures to nonhost species, "The quantities of material needed to produce death are so large as to suffocate the test animal or block its gastrointestinal tract." Siegel and Shadduck warn that using the results of individual maximum challenge testing such as demonstration of death following injection of large quantities of a candidate microbial agent into a vulnerable site such as the brain, may lead to premature rejection of what may really be a perfectly safe microbial insecticide. Importantly, too, they emphasize that their exhaustive "laboratory data support the conclusion that entomopathogens are not infectious to humans, despite their persisting for various lengths of time in mammalian tissue," and that "immunodeficient humans may not be at greater risk of infection from these agents than immune intact humans." Nevertheless, in their overview of impact of microbial insecticides on freshwater ecosystems, and with specific respect to *B. sphaericus,* Dejoux and Elouard (Chapter 6) correctly insist that despite the encouraging lack of demonstrated adverse effects of this microorganism upon NTOs, "pending the production of normalized formulations more studies of toxicity are still required."

The various specialized contributions of Section IV provide additional reassurances concerning points touched on above. However, a warning note with reference to NTOs other than beneficial ones is struck by Melin and Cozzi (Chapter 11), who draw attention to the paucity of environmental impact information on *B. thuringiensis* and its subspecies as regards NTOs often neglected because they are neither "useful" nor "harmful" from the human standpoint. Future environmental impact assessments might well be extended to, e.g., endangered populations within specific target area ecosystems. For example, some rare butterfly may be potentially at risk in its last refuge to applications of a lepidopteran-active entomopathogen against caterpillars of forestry importance. Other topics to which more attention must be directed, clearly include the possible adverse effects of steinernematid and heterorhabditid nematodes upon the invertebrate fauna of their environment, as recommended by Akhurst (Chapter 16); despite the fact that "the few field evaluations reported have failed to detect any significant impact of the nematodes on NTOs."

There is wide agreement on the low probability of environmental harm being caused by genetically engineered microbial insecticides. Fuxa (Chapter 14) declares that in this context "Zero risk will be unachievable for the foreseeable future." A future possibility that must never be lost sight of in view of the increasing interest in applying relevant technology to making highly specific biocontrol microorganisms lethal to a broader range of pests and hence more commercially attractive, is that the resultant products might also threaten a greater diversity of NTOs.

REFERENCES

1. **Hall, J.,** *Works* (of Bishop Joseph Hall), London, 1625, 670.
2. **Bassi, A.,** *Del mal del Segno, Calicinaccio o Moscardino, malattia che afflugge i bachi da Seta. Parta Prima,* Dalla Tipografia Orcesti, Lodi, 1835.
3. **Laird, M.,** Environmental impact of insect control by microorganisms, *Ann. N. Y. Acad. Sci.,* 217, 218, 1973.
4. **Twinn, C. R., Hocking, B., McDuffie, W. C., and Cross, H. F.,** A preliminary account of the biting flies at Churchill, Manitoba, *Can. J. Res. Sect., D,* 26, 334, 1948.
5. **Zaim, M., Ladonni, H., Ershadi, M. R. Y., Manouchehri, A. V., Sahabi, Z., Nazari, M., and Shahmohammadi, H.,** Field application of *Romanomermis culicivorax* (Mermithidae: Nematoda) to control anopheline larvae in southern Iran, *J. Am. Mosq. Control Assoc.,* 4, 351, 1988.
6. **Mondet, B.,** *Études sur* Isomermis lairdi *(Nematoda, Mermithidae) Parasite de* Simulium damnosum s. l. *(Diptera, Simuliidae) en* Afrique de l'Ouest, Travaux et Documents de l'O.R.S.T.O.M., Paris, 1981.
7. **Petersen, J. J.,** Nematode parasites, in *Biological Control of Mosquitoes,* Chapman, H. C., Eds., American Mosquito Control Association, Bull. No. 6, Fresno, CA, 1985, 110.
8. **Colbo, M. H., Laird, M., and Petersen, J.,** *Simulium damnosum* penetrations by *R. culicivorax,* in *Annu. Rep. Reseach Unit Vector Pathology,* Memorial University, Newfoundland, 1978, 23.
9. **Laird, M., and Petersen, J.,** *R. culicivorax* and non-target organisms, in *Annu. Rep. Research Unit Vector Pathology,* Memorial University, Newfoundland, 1978, 28.
10. **Burges, H. D.,** Safety, safety testing and quality control of microbial pesticides, in *Microbial Control of Pests and Plant Diseases 1970—1980,* Burges, H. D., Ed., Academic Press, New York, 1981, 737.
11. **Singer, S.,** *Bacillus sphaericus* (Bacteria), in *Biological Control of Mosquitoes,* Chapman, H. C., Ed., American Mosquito Control Association, Bull. No. 6, Fresno, CA,, 1985, 110.
12. **Burton, M., Ed.,** *Systematic Dictionary of Mammals of the World,* 2nd ed., Museum Press Ltd., London, 1965, 235.

Index

INDEX

A

ABC-6146, safety studies of, 21
Abiotic factors, 67—68
Acarina
  B. thuringiensis lepidopteran-active strain effects on, 157
  β-exotoxin toxicity to, 160—161
Acquired Immunodeficiency Syndrome (AIDS), 104
Agricultural chemicals, see also Chemical pesticides
  registration by manufacturer and importer, 32—33
  regulation of in Japan, 32—33
Agriculture, arthropod pest threat to, 116
Allergenicity testing, 14, 117
All-Union Institute of Hygienic and Toxic Aspects of Pesticides and Polymers (USSR), 24
All-Union Scientific Research Institute of Biological Methods of Plant Protection (USSR), 24
Aphthorvirus, 124
Apis mellifera, see Honeybee
Aquatic organisms, see also Nontarget organisms (NTOs), aquatic
  benthic, β-exotoxin toxicity to, 158
  invertebrate, baculovirus effect on, 143
  safety of baculoviruses to, 141
Arachnida, susceptibility of to β-exotoxin, 162
Araneida, β-exotoxin toxicity to, 161
Arthropods
  effects of *Steinernema* and *Heterorhabditis* spp. on, 237
  exposure of to microbial agents, 86
  potential impact of microbial insecticides on, 44—56
Arthus, 116—117
Ascovirus, 52
Aspergillus, 49, 126
Autographa californica-NPV, 143—144

B

Bacillus moritai preparation
  allergy and anaphylaxis tests of with guinea pigs, 35
  control of in Japan, 32—36
  human feeding tests of, 35—36
  safety testing with birds, 35
  safety testing with cattle, 35
  safety testing with fish, 35
  safety testing with pigs, 35
  safety test with laying hens for, 35
  tests of with mice, 33—34
  tests of with rabbits, 34—35
Bacillus popilliae
  data requirements of EPA reregistration program for, 197—198
  discovery and application of, 190—191
  Dutky, 45

host-range and pathogen specificity/virulence of, 191—192
infectivity studies of, 191—192
in vitro growth and sporulation of, 192—193
isolation of, 116
powder form of, 191
product and residue analysis data on, 193—196
registration of in U.S., 102
registration requirements for use of on pasture grass, 193
safety considerations in use of, 190—198
safety data on formulations of, 194—195
safety of to vertebrates, 111
tier I toxicological data on, 196—197
vegetative stage formulations of, 192, 197
Bacillus sphaericus
  future outlook for, 246
  impact of on NTOs in lenitic habitats, 175—184
  infectivity of to mammals, 104—106
  Neide 1904
    impact of on freshwater biota, 73—75
    toxicity studies of, 73
  safety of to nontarget aquatic organisms, 170—184
  safety testing of, 120
  toxicity of to mammals, 110
Bacillus thuringiensis formulations, 8, 161
  Berliner, 45, 67
    field trials for, 68—72
    impact of on freshwater biota, 67—72
    laboratory tests for, 67—68
    toxicity of in bioassay, 68
  β-exotoxin of, 119—120, 157, 245
    factors affecting toxicity of, 158
    mode of action of, 157
    toxicity of to nontarget invertebrates, 158—162
  δ-endotoxin production by, 118—119
  EEC regulation of, 15
  effect of on silkworm, 36—37
  effects of on Ecnomidae density, 74
  environmental impact of, 38
  environmental pollution due to, 21
  infectivity of to mammals, 104—106
  interaction of with beneficial insects, 54
  israelensis
    efficacy of against chironomid larvae, 182
    field studies of, 171
    impact of on NTO invertebrates in lotic habitats, 170—175
    impact of on NTOs in lenitic habitats, 175—184
    laboratory studies of, 171
    long-term lotic ecosystem effects of, 174—175
    safety of to nontarget aquatic organisms, 170—184
    short-term effects of, 172—174
  Lepidopteran-active strains of
    effects of on freshwater organisms, 150—152
    effects of on marine organisms, 152
    effects of on soil organisms, 152